SPECTRAL ANALYSIS
of
RELATIVISTIC
OPERATORS

SPECTRAL ANALYSIS

of

RELATIVISTIC OPERATORS

A A Balinsky and **W D Evans**

Cardiff University, UK

$$\mathbb{B} := \Lambda_+ \left(\mathbb{D}_0 + V \right) \Lambda_+$$

$$\mathbb{H}_0 := \sqrt{-\Delta + 1}$$

$$\mathbb{D}_0 := \alpha \cdot (-i\nabla) + \beta$$

Imperial College Press

Published by

Imperial College Press
57 Shelton Street
Covent Garden
London WC2H 9HE

Distributed by

World Scientific Publishing Co. Pte. Ltd.
5 Toh Tuck Link, Singapore 596224
USA office: 27 Warren Street, Suite 401-402, Hackensack, NJ 07601
UK office: 57 Shelton Street, Covent Garden, London WC2H 9HE

British Library Cataloguing-in-Publication Data
A catalogue record for this book is available from the British Library.

SPECTRAL ANALYSIS OF RELATIVISTIC OPERATORS

ISBN-13 978-1-84816-218-1
ISBN-10 1-84816-218-9

Printed in Singapore.

To our Parents

Preface

The symbiotic relationship between mathematics and physics has seldom been more apparent than in the development of quantum mechanics and the spectral theory of self-adjoint operators in Hilbert spaces since the early years of the last century. The lofty position that quantum mechanics has attained in science has corresponded with the creation of powerful and elegant theories in functional analysis and operator theory, each feeding voraciously on the problems and discoveries of the other. At the core of this activity, the spectral analysis of the Schrödinger operator has been intensively studied by many and the achievements have been impressive. These range from the detailed description of the spectral properties of atoms subject to electrostatic and magnetic forces, to the study initiated by Dyson and Lenard, and then by Lieb and Thirring, concerning the stability of matter governed by systems of particles under the influence of internal Coulomb forces and external fields.

Attempts to incorporate relativistic effects when appropriate in the theory have encountered many difficulties. Dirac's equation describes the electron and positron as a pair and this yields an operator which is unbounded above and below. Such operators are harder to deal with than those which are semi-bounded, as is typically the case with the Schrödinger operator. Also it results in the Dirac operator not being a suitable model to describe relativistic systems of many particles because their spectrum occupies the whole of the real line and bound states are not defined. In an effort to bypass these problems with the Dirac operator, various alternatives and approximations have been suggested and studied for the kinetic energy term in the total energy Hamiltonian, which preserve some essential features. The so-called quasi-relativistic operator $\sqrt{-\Delta + 1}$ (in appropriate units) shares with the free Dirac operator $\mathbb{D}_0$ the property that its square is the Schrödinger operator, and has the advantage that $\sqrt{-\Delta + 1} - \gamma/|\mathbf{x}|$, where $\gamma/|\mathbf{x}|$ represents the Coulomb potential due to the electron-nuclear interaction, is bounded below for a range of constants γ. This operator was studied by Herbst in [Herbst (1977)] and Weder in [Weder (1974, 1975)], and their work is included within the discussion in this book. The other main operator studied in depth in this book is that introduced by Brown and Ravenhall in [Brown and Ravenhall (1951)], studied by Hardekopf and Sucher in [Hardekopf and Sucher (1985)], which attempts to split the Dirac operator into positive and negative spectral parts. The basic Brown–Ravenhall operator is of the form $\Lambda_+ \left(\mathbb{D}_0 - \gamma/|\mathbf{x}| \right) \Lambda_+$, where Λ_+ is the projection onto the positive spectral subspace of the free Dirac operator, and sensationally, it is bounded below, and indeed positive, for all known elements. It is the restriction of the quasi-relativistic operator to a subspace of the underlying L^2 space and there is justification in regarding it as a better physical model than the quasi-relativistic operator.

The book is primarily designed for the mathematician with an interest in the spectral analysis of the operators of mathematical physics, but we hope that other scientists will find topics of interest here, and we have written the book with that in mind. The topics covered naturally reflect our own interests and areas of expertise, and are mainly those with which we have been closely associated during the last fifteen years. A knowledge of basic functional analysis and operator theory is assumed, but the first chapter gives a brief survey of the necessary background material to help the reader who is not familiar with, or needs reminding of, the material and techniques in the following chapters. Much of Chapter 2 is taken up by precise descriptions and the establishment of basic properties of the Dirac, quasi-relativistic and Brown–Ravenhall operators with Coulomb potentials. This involves the definition of self-adjoint realisations in an appropriate Hilbert space, these being either defined uniquely in the case of essential self-adjointness, or otherwise as a Friedrichs extension, or some other physically relevant self-adjoint operator, associated with a lower semi-bounded quadratic form. Of particular concern is the determination of optimal conditions on the Coulomb potential for which the different types of self-adjoint realisations are valid. The nature of the spectrum of these operators in turn is addressed in Chapter 3, in particular the location of the essential spectrum, and the existence of eigenvalues, which are either isolated from the essential spectrum or embedded in it. The analysis of embedded eigenvalues is based on a simple abstract virial theorem, modelled on a celebrated result of J. Weidmann for Schrödinger operators, and this is then applied to each of the three types of operator in turn. The stability of matter is a problem that has attracted a great deal of attention, and in this context the Pauli operator in particular presents some interesting challenges. Chapter 4 deals with some of these. There is a brief outline of some of the highlights of what has been achieved over the last three decades, but the focus is mainly on important auxiliary issues and techniques, which are of intrinsic interest. In particular the existence or otherwise of magnetic fields that give rise to zero modes of the Pauli operator is examined in detail. Zero modes have some profound physical and mathematical consequences, and because of their importance they merit substantial coverage. Topics covered include the following: a discussion of some known examples with a description of techniques developed for their construction based on quaternions; a detailed analysis of a class of magnetic potentials that give rise to zero modes; growth rates and asymptotic limits of the magnetic potentials; and the relevance of zero modes to some spectral, Dirac–Sobolev and Dirac–Hardy inequalities. Also techniques that have proved to be effective in establishing stability of matter results, including Lieb–Thirring inequalities, are discussed.

Chapters are divided into sections and most sections into subsections. Theorems, corollaries, lemmas, remarks and equations are numbered consecutively within a section. Thus equation (3.2.15) is the fifteenth equation in Section 3.2, which is the second section in Chapter 3. Section 3.2.4 refers to the fourth subsection within Section 3.2.

We are grateful to Tomio Umeda for his valuable comments on an earlier draft.

A. A. Balinsky and W. D. Evans
Cardiff University
May 2010

Basic Notation

- $\mathbb{C}$ complex plane; $\mathbb{C}^n$: n-dimensional complex space;
- $\mathbb{C}_+ = \{z \in \mathbb{C} : \operatorname{Im} z > 0\}$; $\mathbb{C}_- = \{z \in \mathbb{C} : \operatorname{Im} z < 0\}$;
- $\mathbb{R}$ real line; $\mathbb{R}^n$: n-dimensional Euclidean space;
- $\mathbb{R}_+ = (0, \infty)$, $\mathbb{R}_- = (-\infty, 0)$;
- $\mathbb{S}^n$: n-dimensional sphere;
- $\omega_n = \frac{\pi^n/2}{\Gamma(1+\frac{1}{2}n)}$, the volume of the unit ball in $\mathbb{R}^n$;
- $\mathbb{N}$: positive integers; $\mathbb{N}_0 = \mathbb{N} \cup \{0\}$; $\mathbb{Z}$: all integers;
- $f(\mathbf{x}) \asymp g(\mathbf{x}) : c_1 \leq f(\mathbf{x})/g(\mathbf{x}) \leq c_2$ for some positive constants c_1, c_2.

Contents

Chapter 1

Preliminaries

In this chapter we collect concepts and results, which will be essential tools for describing and proving what follows in subsequent chapters. The discussion is brief and few proofs are given; rather we give references to appropriate sources in the literature. Unless mentioned otherwise, our Hilbert spaces are infinite dimensional and over the complex field $\mathbb{C}$.

1.1 Linear operators

We shall assume familiarity with the notions of bounded, closable, closed, symmetric and self-adjoint operators: for a comprehensive treatment of what is needed, see [Edmunds and Evans (1987)]. The following topics and facts are included in any basic coverage, we collect them here for ease of reference as they will be important later.

The first topic concerns von Neumann's theory of extensions of a symmetric operator and the supporting background material. The *numerical range* of a linear operator T with domain $\mathcal{D}(T)$ and range $\mathcal{R}(T)$ in a Hilbert space H is the set

$$\Theta(T) := \{(Tu, u) : u \in \mathcal{D}(T), \|u\| = 1\},$$

where $(\cdot, \cdot), \|\cdot\|$ are, respectively, the inner product and norm of H. It is a convex subset of $\mathbb{C}$ and, hence, so is its closure $\overline{\Theta(T)}$. The complement $\tilde{\Delta}(T) := \mathbb{C} \setminus \overline{\Theta(T)}$, has either one or two connected components. If T is bounded, $\Theta(T)$ is a bounded set and, hence, $\tilde{\Delta}(T)$ is connected; if $\overline{\Theta(T)}$ is an infinite strip, $\tilde{\Delta}(T)$ has two connected components, $\tilde{\Delta}_1(T)$ and $\tilde{\Delta}_2(T)$, say, both being half-planes. Let T be closed, denote by I the identity on H and let $\lambda \in \tilde{\Delta}(T)$. Then $T - \lambda I$ has closed range $\mathcal{R}(T - \lambda I)$ with trivial null space (or kernel) $\mathcal{N}(T - \lambda I)$ and its range has constant co-dimension in each connected component of $\tilde{\Delta}(T)$: in the standard terminology, $T - \lambda I$ is therefore a *semi-Fredholm* operator with zero *nullity*, $\operatorname{nul}(T - \lambda I) := \dim \mathcal{N}(T - \lambda I)$, and constant *deficiency*, $\operatorname{def}(T - \lambda I) := \operatorname{codim} \mathcal{R}(T - \lambda I)$, in each of $\tilde{\Delta}_1(T)$ and $\tilde{\Delta}_2(T)$. If T is a closed symmetric operator, $\Theta(T)$ is a closed subinterval of the real line and so $\tilde{\Delta}(T)$ includes the upper and lower half-planes $\mathbb{C}_\pm$. In this case the constant values of $\text{def}(T - \lambda I)$ for $\lambda \in \mathbb{C}_{\pm}$ are called the *deficiency indices* (m_+, m_-) of T :

$$m_{\mp}(T) := \text{def}(T - \lambda I), \quad \lambda \in \mathbb{C}_{\pm}$$

and also

$$\text{def}(T - \lambda I) = \dim[\mathcal{R}(T - \lambda I)^{\perp}] = \text{nul}(T^* - \overline{\lambda}I),$$

where $\perp$ denotes the *orthogonal complement*, and T^* the *adjoint* of T. The closed subspaces $\mathcal{N}_{\pm} := \mathcal{N}(T^* \mp iI)$ are called the *deficiency subspaces* of T, their dimension being the deficiency indices $m_{\pm}(T)$. An important result of von Neumann is that if T is a closed symmetric operator, the domain $\mathcal{D}(T^*)$ of its adjoint T^* has the direct sum decomposition

$$\mathcal{D}(T^*) = \mathcal{D}(T) \dotplus \mathcal{N}_+ \dotplus \mathcal{N}_-.$$

This gives

$$\dim\left\{\mathcal{D}(T^*)/\mathcal{D}(T)\right\} = m_+(T) + m_-(T), \qquad (1.1.1)$$

where $\mathcal{D}(T^*)/\mathcal{D}(T)$ denotes the quotient space.

A closed symmetric operator T is self-adjoint if and only if $\mathcal{R}(T - \lambda I) = \mathcal{R}(T - \overline{\lambda}I) = H$ for some, and consequently all, $\lambda \notin \mathbb{R}$. Therefore T is self-adjoint if and only if $m_+(T) = m_-(T) = 0$. Another consequence of von Neumann's theory of extensions of symmetric operators is that T has a self-adjoint extension if and only if $m_+(T) = m_-(T)$. If T is not closed, but only assumed to be closable, it is said to be *essentially self-adjoint* if its closure $\overline{T}$ is self-adjoint. Equivalently, $\overline{T} = T^*$ is the unique self-adjoint extension of T.

The following Kato–Rellich theorem, concerning the stability of self-adjointness or essential self-adjointness under perturbations will have an important role to play. Recall that a linear operator P in H is said to be *relatively bounded with respect to* T, or T-*bounded*, if $\mathcal{D}(T) \subseteq \mathcal{D}(P)$ and there exist non-negative constants a, b, such that

$$\|Pu\| \leq a\|u\| + b\|Tu\|, \quad \text{for all} \quad u \in \mathcal{D}(T). \qquad (1.1.2)$$

The infimum of the constants b satisfying (1.1.2) for some $a \geq 0$ is called the T-bound of P. If P is T-bounded, satisfies (1.1.2) and is closable, then $\overline{P}$ is $\overline{T}$-bounded and

$$\|\overline{P}u\| \leq a\|u\| + b\|\overline{T}u\|, \quad \text{for all} \quad u \in \mathcal{D}(\overline{T}).$$

P is said to be T-*compact* if $\mathcal{D}(T) \subseteq \mathcal{D}(P)$ and for any sequence $\{u_n\}$ in $\mathcal{D}(T)$ which is such that $\|Tu_n\| + \|u_n\|$ is bounded, $\{Pu_n\}$ contains a convergent subsequence. If we endow $\mathcal{D}(T)$ with its graph norm, namely,

$$\|u\|_T := (\|Tu\|^2 + \|u\|^2)^{1/2},$$

then P is T-compact if its restriction to $\mathcal{D}(T)$ is a compact map from $\mathcal{D}(T)$ into H. If T is self-adjoint, this is equivalent to saying that $P(T + i)^{-1}$ is compact on H.

Theorem 1.1.1. *Let T be a symmetric operator in H and P a symmetric operator, which is T-bounded with T-bound < 1. Then*

(1) if T is self-adjoint, so is $T + P$;

(2) if T is essentially self-adjoint, so is $T + P$ and $\overline{T + P} = \overline{T} + \overline{P}$;

(3) if P is T-compact and T is self-adjoint, then $T + P$ is self-adjoint.

The theorem does not hold in general if P is assumed to have T-bound 1. However, the following is proved by Wüst in [Wüst (1971)]; see also [Kato (1976)], Theorem V.4.6.

Theorem 1.1.2. *Let T be essentially self-adjoint and P a symmetric operator with $\mathcal{D}(T) \subseteq \mathcal{D}(P)$ and $\|Pu\| \leq a\|u\| + \|Tu\|$ for some $a \geq 0$ and all $u \in \mathcal{D}(T)$. Then $T + P$ is essentially self-adjoint.*

1.2 Quadratic forms

In quantum mechanics, the differential operators encountered, like the Schrödinger and Dirac operators, are required by the theory to be self-adjoint in the underlying Hilbert space H. This is often not an easy property to establish, and it is customary to start by first restricting the operator on $C_0^\infty(\mathbb{R}^3)$, for instance, so defining a symmetric operator. If this operator is essentially self-adjoint, then there is no ambiguity about the self-adjoint operator to be taken, since the closure is the unique self-adjoint extension. However, it is often the case that there are many self-adjoint extensions and then physical considerations come into play in selecting the appropriate self-adjoint extension. The physically relevant extension is the Friedrichs extension. This is defined in terms of a symmetric quadratic (or sesquilinear) form associated with the initial symmetric operator; see [Edmunds and Evans (1987)], Chapter IV. More generally, one can start with a symmetric quadratic form t (usually referred to merely as a form), which is densely defined, bounded below and closed in H. Closed means that if $t[u] := t[u, u] \geq \delta\|u\|^2$ for all u in the domain $\mathcal{D}(t)$ of t, then $\mathcal{D}(t)$ endowed with the norm $\|u\|_t := (t - \delta + 1)^{1/2}[u]$ is complete: the form is *closable* if the completion of $\mathcal{D}(t)$ with respect to the aforementioned norm can be identified with a subspace of H, i.e., this completion of $\mathcal{D}(t)$ is continuously embedded in H. We set H_t to be the normed space $(\mathcal{D}(t); \|\cdot\|_t)$: it is in fact an inner-product space as the norm is obviously generated by the inner product $(t - \delta + 1)[\cdot, \cdot]$.

A form p is said to be *relatively bounded with respect to* a form t, or simply *t-bounded*, if $\mathcal{D}(p) \supseteq \mathcal{D}(t)$ and

$$|p[u]| \leq a\|u\|^2 + b|t[u]|, \quad u \in \mathcal{D}(t), \tag{1.2.1}$$

where a, b are non-negative constants. The infimum of the constants b satisfying (1.2.1) for some $a \geq 0$ is called the *t-bound* of p. The analogue of Theorem 1.1.1 for forms is

Theorem 1.2.1. *Let t be a densely defined, symmetric quadratic form, which is bounded below, and p a symmetric form, which is t-bounded with t-bound < 1. Then*

(1) t is closable;

(2) t + p is bounded below;

(3) t + p is closable with the closures of t and t + p having the same domain;

(4) t + p is closed if and only if t is closed.

There is a physically distinguished self-adjoint operator associated with t, given by Kato's first representation theorem:

Theorem 1.2.2. *Let t be a closed, densely defined quadratic form in H, which is bounded below. Then there exists a self-adjoint operator T in H such that*

(1) $u \in \mathcal{D}(T)$ if and only if there exists $f \in H$ such that, for all $v \in \mathcal{D}(t)$,

$$t[u, v] = (f, v)$$

in which case $f = Tu$;

(2) $\mathcal{D}(T)$ is dense in H_t;

(3) if $u \in \mathcal{D}(t), f \in H$ and

$$t[u, v] = (f, v)$$

for all v in a dense subspace of H_t, then $f = Tu$.

The space H_t is referred to as the form domain of T, and is usually denoted by $\mathbb{Q}(T)$.

If T_0 is a given symmetric operator in H which is bounded below, the form

$$t_0[u, v] = (T_0 u, v), \quad u, v \in \mathcal{D}(T_0),$$

is closable, densely defined and bounded below and its closure satisfies the first representation theorem. In this case the operator T in the theorem is the *Friedrichs extension* of T_0. It is in fact the restriction of T_0^* to $\mathcal{D}(T_0^*) \cap \mathcal{D}(t)$ and has the same lower bound as T_0.

Suppose now that $t = t_1 + t_2$, with domain $\mathcal{D}(t_1) \cap \mathcal{D}(t_2)$, where t_1, t_2 are densely defined, closed forms, which are bounded below, and that t is also densely defined, closed and bounded below. Then, self-adjoint operators T, T_1, T_2 are associated with the forms t, t_1, t_2. In this case T is called the *form sum* of T_1 and T_2, written $T = T_1 \dotplus T_2$. If $\mathcal{D}(T_1) \cap \mathcal{D}(T_2)$ is dense in H, the Friedrichs extension of $T_1 + T_2$ is also defined, but in general this differs from the form sum; see [Kato (1976)], Example VI-2.19.

A non-negative self-adjoint operator T has a unique square root $T^{1/2}$ which is also non-negative and self-adjoint. Furthermore, $\mathcal{D}(T)$ is a core of $T^{1/2}$, i.e., $\mathcal{D}(T)$ is dense in the Hilbert space $H_+(T^{1/2})$ defined by $\mathcal{D}(T^{1/2})$ with the graph norm

$$\|u\|_{H_+(T^{1/2})} := \left(\|T^{1/2}u\|^2 + \|u\|^2 \right)^{1/2} = \|(T + 1)^{1/2}u\|.$$

These observations lead to the *second representation theorem*:

Theorem 1.2.3. *Let* t *be a closed, densely defined, non-negative symmetric form, and let* T *be the associated self-adjoint operator. Then* $\mathcal{D}(t) = \mathcal{D}(T^{1/2})$ *and*

$$t[u, v] = (T^{1/2}u, T^{1/2}v), \quad u, v \in \mathcal{D}(T^{1/2}).$$

Thus $\mathbb{Q}(T) = H_+(T^{1/2})$.

The operator $(T+1)^{1/2}$ is an isometric isomorphism of $H_+(T^{1/2})$ onto H and its adjoint is $(T+1)^{1/2}$, which is an isometry of H onto $H_-(T^{1/2})$, the completion of H with respect to the norm

$$\|u\|_{H_-(T^{1/2})} = \|((T+1)^{-1/2}u\|^2.$$

We have the triplet of spaces

$$H_+(T^{1/2}) \hookrightarrow H \hookrightarrow H_-(T^{1/2}), \tag{1.2.2}$$

with embeddings defined by the identification maps, which are continuous and have dense ranges.

Let P be a non-negative, self-adjoint operator, which is a *form-bounded* perturbation of the non-negative self-adjoint operator T in the sense that $\mathbb{Q}(P) \supseteq \mathbb{Q}(T) = H_+(T^{1/2})$, i.e., $\mathcal{D}(P^{1/2}) \supseteq \mathcal{D}(T^{1/2})$ and

$$\|P^{1/2}u\| \leq K\|u\|_{H_+(T^{1/2})}$$

for some non-negative constant K. Then P is said to be *form compact, relative to* T, or *form* T-*compact*, if it is compact from $H_+(T^{1/2})$ to $H_-(T^{1/2})$. Equivalently, this means that $(T+1)^{-1/2}P(T+1)^{-1/2} : H \to H$ is compact. Note that, on setting $A = P^{1/2}(T+1)^{-1/2}$, we have $(T+1)^{-1/2}P(T+1)^{-1/2} = A^*A$, which is compact on H if and only if A is compact.

Theorem 1.2.3 yields the *polar decomposition* of a general densely defined, closed operator S acting between two Hilbert spaces H_+ and H_-. In this case

$$s[u, v] := (Su, Sv)_{H_-}, \quad u, v \in \mathcal{D}(S) \subseteq H_+$$

is densely defined, non-negative and closed, since

$$(s+1)[u] = \|Su\|_{H_-}^2 + \|u\|_{H_+}^2 = \|u\|_{H_+(S)}^2$$

and $H_+(S)$ is complete. It follows that the non-negative, self-adjoint operator associated with s in Theorem 1.2.3 is $T = S^*S$ and

$$s[u, v] = (T^{1/2}u, T^{1/2}v)_{H_+}, \quad u, v \in \mathcal{D}(T^{1/2}) = \mathcal{D}(S).$$

The operator $T^{1/2} = (S^*S)^{1/2}$ is called the *absolute value* of S and written $|S|$. The *form domain* of a general self-adjoint operator S is defined to be $\mathbb{Q}(|S|)$. The operators $S_\pm := (1/2)(|S| \pm S)$ are called the *positive* and *negative parts* of S. The map

$$|S|u \mapsto Su : \mathcal{R}(|S|) \text{ onto } \mathcal{R}(S)$$

is an isometry, and extends by continuity to an isometry U from $\overline{\mathcal{R}(|S|)} = \mathcal{N}(|S|)^\perp$ onto $\overline{\mathcal{R}(S)} = \mathcal{N}(S)^\perp$. On setting $Uu = 0$ for $u \in \mathcal{R}(|S|)^\perp = \mathcal{N}(|S|)$, U is a *partial isometry* with *initial set* $\overline{\mathcal{R}(|S|)}$ and *final set* $\overline{\mathcal{R}(S)}$. Its adjoint U^* is a partial isometry with initial set $\overline{\mathcal{R}(S)}$ and final set $\overline{\mathcal{R}(|S|)}$. The formula

$$S = U|S|, \quad \mathcal{D}(S) = \mathcal{D}(|S|), \tag{1.2.3}$$

is called the *polar decomposition* of S. See [Edmunds and Evans (1987)], Section IV.3 for further details.

An elegant way of establishing (1.2.3) and consequent properties, is based on the notion of a *supercharge* in *supersymmetric quantum mechanics*; see [Thaller (1992)] Chapter 5. Abstractly, a supercharge Q is a self-adjoint operator in a Hilbert space $\mathcal{H}$ which anti-commutes with a unitary involution τ defined on $\mathcal{H}$, i.e., τ is a bounded self-adjoint operator on $\mathcal{H}$ which is such that $\tau\tau^* = \tau^*\tau = \tau^2 = I$ and $Q\tau + \tau Q = 0$. The operators $P_\pm = 1/2(1 \pm \tau)$ are orthogonal projections onto subspaces $\mathcal{H}_+$ and $\mathcal{H}$ has the orthogonal sum decomposition $\mathcal{H} = \mathcal{H}_+ \oplus \mathcal{H}_-$. On expressing $u = u_+ + u_-$ as the column vector $(u_+, u_-)^t$, where the superscript denotes the transpose, the standard representation for τ is

$$\tau = \begin{pmatrix} 1 & 0 \\ 0 & -1 \end{pmatrix}$$

with respect to which, a supercharge Q is an off-diagonal matrix operator. It is readily shown that there is a 1-1 correspondence between densely defined closed operators S acting between $\mathcal{H}_+$ and $\mathcal{H}_-$, and supercharges of the form

$$Q = \begin{pmatrix} 0 & S^* \\ S & 0 \end{pmatrix}$$

on $\mathcal{D}(Q) = \mathcal{D}(S) \oplus \mathcal{D}(S^*)$ in $\mathcal{H} = \mathcal{H}_+ \oplus \mathcal{H}_-$. The operator Q is self-adjoint and

$$\mathcal{D}(Q^2) = \{f \in \mathcal{D}(Q) : Qf \in \mathcal{D}(Q)\} = \mathcal{D}(S^*S) \oplus \mathcal{D}(SS^*).$$

Since

$$\mathcal{N}(Q) = \mathcal{N}(S) \oplus \mathcal{N}(S^*),$$

and

$$\mathcal{N}(Q) = \mathcal{N}(Q^2) = \mathbb{R}(Q)^\perp,$$

it follows that

$$\mathcal{N}(S) = \mathcal{N}(S^*S) = \mathcal{R}(S^*)^\perp = \mathcal{R}(S^*S)^\perp \tag{1.2.4}$$

and

$$\mathcal{N}(S^*) = \mathcal{N}(SS^*) = \mathcal{R}(S)^\perp = \mathcal{R}(SS^*)^\perp. \tag{1.2.5}$$

Also, by (1.2.3) and $S^* = |S|U^*$, we have on $\mathcal{N}(Q)^\perp$ that

$$\begin{pmatrix} S^*S & 0 \\ 0 & SS^* \end{pmatrix} = Q^2 = \begin{pmatrix} U^*SS^*U & 0 \\ 0 & US^*SU^* \end{pmatrix}.$$

From this we can infer that S^*S on $(\mathcal{N}(S))^\perp$ is unitarily equivalent to SS^* on $(\mathbb{N}(S^*))^\perp$ and so

$$\sigma(SS^*) \setminus \{0\} = \sigma(SS^*) \setminus \{0\}, \qquad (1.2.6)$$

where σ denotes the spectrum. In particular, if $\lambda \neq 0$, is an eigenvalue of S^*S with eigenvector f, then λ is an eigenvalue of SS^* with eigenvector Sf, since

$$SS^*(Sf) = S(S^*Sf) = \lambda Sf.$$

Conversely, if $SS^*g = \lambda g$, $\lambda \neq 0$, then $g \in \mathcal{D}(S^*)$ and $S^*SS^*g = \lambda S^*g$. Hence, $f = S^*g \neq 0$ and satisfies $S^*Sf = \lambda f$. Thus g is of the form $g = Sf$, where f is an eigenvector of S^*S corresponding to λ. See [Thaller (1992)], Section 5.2 for a more comprehensive and detailed discussion.

1.3 Spectra of self-adjoint operators

To fix notation, we recall the basic definitions. The *resolvent set* $\rho(T)$ of a closed operator T in H is the open subset of $\mathbb{C}$ defined by

$$\rho(T) := \{\lambda \in \mathbb{C} : (T - \lambda I)^{-1} \text{ is bounded on } H\};$$

$(T - \lambda I)^{-1}$ is called the *resolvent* of T. The *spectrum* $\sigma(T)$ is the complement $\mathbb{C} \setminus \rho(T)$. If T is self-adjoint, the half-planes $\mathbb{C}_\pm$ lie in $\rho(T)$ and

$$\|(T - \lambda I)^{-1}\| \leq |\mathrm{Im}\lambda|^{-1}, \quad \lambda \notin \mathbb{R}.$$

Hence, the spectrum of a self-adjoint operator is a closed subset of the real line. Its *discrete spectrum* $\sigma_D(T)$ consists of the isolated eigenvalues of finite multiplicity. The complement $\sigma(T) \setminus \sigma_D(T)$ is the *essential spectrum* $\sigma_e(T)$ of T. The essential spectrum is a closed subset of $\mathbb{R}$ and can contain eigenvalues, which are either not isolated or are of infinite multiplicity. The set of all eigenvalues of T is called the *point spectrum* $\sigma_p(T)$. If $\{E(\lambda)\}$ is the (right continuous) spectral family of projections associated with T, then $P(\lambda) := E(\lambda) - E(\lambda - 0) \neq 0$ if and only if λ is an eigenvalue, in which case $P(\lambda)$ is the projection of H onto the eigenspace corresponding to λ. Let H_p denote the closed subspace of H spanned by all the $P(\lambda)H$ and $H_c := H_p^\perp$. It can be shown that $u \in H_c$ if and only if $(E(\lambda)u, u) = \|E(\lambda)u\|^2$ is a continuous function of λ. The spectrum of the restriction of T to H_p is $\sigma_p(T)$ and the spectrum of the restriction of T to H_c is called the continuous spectrum of T and written $\sigma_c(T)$.

It is helpful to subdivide $\sigma_c(T)$ into two parts, determined by the properties of the Borel measure $m_u(S) := (E(S)u, u)$ for $u \in H$, where, for instance, $E([a, b]) = E(b) - E(b - 0)$. The *absolutely continuous (singular)* subspace $H_{ac}(H_s)$ of T is the set of $u \in H$ for which the measure m_u is absolutely continuous (singular). The *absolutely continuous spectrum* $\sigma_{ac}(T)$ (*singular continuous spectrum* $\sigma_{sc}(T)$) of T are the spectra of the restrictions of T to H_{ac}, H_s respectively. It can be shown that H has the orthogonal decomposition $H = H_{ac} \oplus H_{sc} \oplus H_p$ and $H_c = H_{ac} \oplus H_{sc}$. See [Reed and Simon (1978)] for further details.

1.4   Compact operators

A particularly important role for compact self-adjoint operators in spectral analysis is as a tool for locating essential spectra. This is a consequence of a celebrated result of H. Weyl, namely, that if a symmetric operator P is T-compact, where T is self-adjoint, then $\sigma_e(T + P) = \sigma_e(T)$. From this it follows that if T, S are self-adjoint operators and $(T - \lambda I)^{-1} - (S - \lambda I)^{-1}$ is compact for some (and hence all) $\lambda \in \rho(T) \cap \rho(S)$, then $\sigma_e(T) = \sigma_e(S)$. The following two improvements are useful in applications. The first concerns an operator sum and the second a form sum. In the first we use the notion of T^n-compactness, with $T \geq 0$ assumed if $n \notin \mathbb{N}$: an operator P is said to be T^n-*compact* if $\mathcal{D}(P) \supset \mathcal{D}(T)$ and for any sequence $\{u_m\}$ which is such that $\|T^n u_m\|^2 + \|u_m\|^2$ is bounded, then $\{P u_m\}$ contains a convergent subsequence. In other words, P is compact as a map from $\mathcal{D}(T^n)$ with the graph norm $(\|T^n u\|^2 + \|u\|^2)^{1/2}$ into H or, equivalently, $P(T^n + i)^{-1}$ is compact on H. In the second P is form T^n-*compact*, which means that $T \geq 0$ and $(T + 1)^{-n/2}|P|(T + 1)^{-n/2}$ is compact in H.

Theorem 1.4.1. *Let T be a self-adjoint operator, P a symmetric operator defined on $\mathcal{D}(T)$, which is T^n-compact for some $n \in \mathbb{N}$, and suppose that $T + P$ is self-adjoint. Then*

$$\sigma_e(T + P) = \sigma_e(T).$$

Gustafson and Weidmann proved in [Gustafson and Weidmann (1969)] that Theorem 1.4.1 for $n \in \mathbb{N}$ is no more general than Schechter's original result in [Schechter (1966)] for the case $n = 2$.

Theorem 1.4.2. *Let T be a positive self-adjoint operator and P a self-adjoint operator with $\mathbb{Q}(P) \equiv \mathbb{Q}(|P|) \supset \mathbb{Q}(T)$. Suppose the form associated with the sum $T + P$ is bounded below and closed on $\mathbb{Q}(T)$ and let S be the form sum $T \dotplus P$. Suppose that at least one of the following is satisfied:*

(1) P is T^n-compact for some $n \in \mathbb{N}$;
(2) $\mathbb{Q}(|P|) \supset \mathbb{Q}(T)$ and $|P|$ is form T^n-compact for some $n \in \mathbb{N}$.

Then $\sigma_e(S) = \sigma_e(T)$.

See [Reed and Simon (1978)] p. 116, Corollary 4.

For semi-bounded, self-adjoint operators, the *max-min principle* quoted in the next theorem is an invaluable analytical and computational tool. In Chapter 4, we shall give a result that establishes a modified version for operators with a spectral gap, and thus, in particular, the Dirac operator. The eigenvalues of the self-adjoint operator T below are counted according to their multiplicity and arranged in increasing order.

Theorem 1.4.3. *Let T be a lower semi-bounded, self-adjoint operator in H and, for each $n \in \mathbb{N}$, define*

$$\mu_n(T) := \sup_{M_{n-1}} \inf_{\substack{\psi \in \mathcal{D}(T) \cap M_{n-1}^\perp, \\ \|\psi\| = 1}} (T\psi, \psi), \tag{1.4.1}$$

where the supremum is taken over all linear subspaces M_{n-1} of H of dimension at most $n - 1$. Then, for each $n \in \mathbb{N}$, the following hold.

(i) $\mu_n(T) < \lambda_e(T) := \inf\{\lambda \in \sigma_e(T)\}$ if and only if T has at least n eigenvalues less than $\lambda_e(T)$. In this case, $\mu_n(T)$ is the nth eigenvalue of T and the infimum in (1.4.1) is attained when M_{n-1} is the linear span of $e_1, e_2, \cdots, e_n$, where e_j is the eigenvector of T corresponding to the jth eigenvalue.

(ii) $\mu_n(T) = \lambda_e(T)$ if and only if T has at most $n - 1$ eigenvalues less than $\lambda_e(T)$, and in this case $\mu_m(T) = \mu_n(T)$ for all $m > n$.

In (1.4.1), (Tu, u) and $\mathcal{D}(T)$ may be replaced by the form and form domain of T, respectively.

For non-negative, self-adjoint operators A, B on a Hilbert space H, we write $A \leq B$ if $\mathbb{Q}(B) \subseteq \mathbb{Q}(A)$ and $a[u] \leq b[u]$ for all $u \in \mathbb{Q}(B)$, where a, b are the forms of A, B, respectively. It follows from the form version of the max-min principle that the eigenvalues $\lambda_n(A), \lambda_n(B), n \in \mathbb{N}$, satisfy $\lambda_n(A) \leq \lambda_n(B)$ for all n and hence $N(\lambda, B) \leq N(\lambda, A)$, where

$$N(\lambda, T) := \sharp\{n : \lambda_n(T) \leq \lambda\};$$

see [Edmunds and Evans (1987)], Lemma X1.2.3.

1.5 Fourier and Mellin transforms

The *Fourier transform* is given by

$$(\mathbb{F}u)(\mathbf{p}) := \frac{1}{(2\pi)^{n/2}} \int_{\mathbb{R}^n} e^{-i\mathbf{p}\cdot\mathbf{x}} u(\mathbf{x}) d\mathbf{x}, \tag{1.5.1}$$

where $\mathbf{p} \cdot \mathbf{x} = \sum_{j=1}^n p_j x_j$. To analyse and describe some of its important properties we shall use the following standard terminology throughout: for a multi-index $\alpha = (\alpha_1, \alpha_2, \cdots, \alpha_n) \in \mathbb{N}_0^n$,

$$D^\alpha := \prod_{j=1}^n \left(\frac{1}{i}\frac{\partial}{\partial x_j}\right)^{\alpha_j}, \tag{1.5.2}$$

$$\mathbf{x}^j := \prod_{j=1}^n x_j^{\alpha_j}, \tag{1.5.3}$$

and $|\boldsymbol{\alpha}| = \alpha_1 + \alpha_2 + \cdots + \alpha_n$.

We shall denote by $\mathbf{S}(\mathbb{R}^n)$ the Schwartz space of rapidly decreasing functions, i.e., $u \in \mathbf{S}(\mathbb{R}^n)$ if $u \in C^\infty(\mathbb{R}^n)$ and for every $\ell \in \mathbb{N}_0$ and $\boldsymbol{\alpha} = (\alpha_1, \alpha_2, \cdots, \alpha_n) \in \mathbb{N}_0^n$, there exists a positive constant $K(\boldsymbol{\alpha}, \ell)$ such that

$$|\mathbf{x}|^\ell |D^{\boldsymbol{\alpha}} u(\mathbf{x})| \leq K(\boldsymbol{\alpha}, \ell).$$

The Fourier transform has the following well-known properties (see [Weidmann (1980)], Chapter 10):

(i) $\mathbb{F}$ is a linear bijection of $\mathbf{S}(\mathbb{R}^n)$ onto itself and its inverse is given by

$$\left(\mathbb{F}^{-1} v\right)(\mathbf{x}) = \frac{1}{(2\pi)^{n/2}} \int_{\mathbb{R}^n} e^{i\mathbf{x} \cdot \mathbf{P}} v(\mathbf{p}) d\mathbf{p}. \tag{1.5.4}$$

Moreover, $(\mathbb{F}^{-1} v)(\mathbf{x}) = (\mathbb{F}v)(-\mathbf{x})$ and $\mathbb{F}^4 = I$, the identity on $\mathbf{S}(\mathbb{R}^n)$.

(ii) For $v \in \mathbf{S}(\mathbb{R}^n)$ and $\boldsymbol{\alpha} \in \mathbb{N}^n$

$$\{\mathbb{F} D^{\boldsymbol{\alpha}} \mathbb{F}^{-1}\} v(\mathbf{p}) = \mathbf{p}^{\boldsymbol{\alpha}} v(\mathbf{p}). \tag{1.5.5}$$

(iii) $\mathbb{F}$ has a unique extension, which is a unitary operator on $L^2(\mathbb{R}^n)$. We shall continue to call this unitary operator the Fourier transform and to write it as $\mathbb{F}$. We therefore have

$$\mathbb{F}^{-1} = \mathbb{F}^*. \tag{1.5.6}$$

In particular, this gives the *Parseval formula*

$$(u, v)_{L^2(\mathbb{R}^n)} = (\mathbb{F}u, \mathbb{F}v)_{L^2(\mathbb{R}^n)}. \tag{1.5.7}$$

(iv) $\mathbb{F}$ and $\mathbb{F}^{-1}$ are continuous injections of $L^1(\mathbb{R}^n)$ into the space of bounded continuous functions on $\mathbb{R}^n$ with the supremum norm:

$$\sup_{\mathbf{p} \in \mathbb{R}^n} |(\mathbb{F}f)(\mathbf{p})| \leq \frac{1}{(2\pi)^{n/2}} \|f\|_{L^1(\mathbb{R}^n)}, \quad \sup_{\mathbf{x} \in \mathbb{R}^n} |(\mathbb{F}^{-1} f)(\mathbf{x})| \leq \frac{1}{(2\pi)^{n/2}} \|f\|_{L^1(\mathbb{R}^n)}.$$

(v) For $f, g \in \mathbf{S}(\mathbb{R}^n)$, define the *convolution* $f * g$ by

$$(f * g)(\mathbf{x}) := \frac{1}{(2\pi)^{n/2}} \int_{\mathbb{R}^n} f(\mathbf{y}) g(\mathbf{x} - \mathbf{y}) d\mathbf{y}$$

$$= \frac{1}{(2\pi)^{n/2}} \int_{\mathbb{R}^n} f(\mathbf{x} - \mathbf{y}) g(\mathbf{y}) d\mathbf{y} = (g * f)(x). \tag{1.5.8}$$

Then

$$\mathbb{F}(f * g) = \widehat{f}\widehat{g}, \tag{1.5.9}$$

where $\widehat{f} = \mathbb{F}f$. If $f, g \in L^1(\mathbb{R}^n)$, $f * g$ is well-defined and belongs to $L^1(\mathbb{R}^n)$, and (1.5.9) continues to hold. Also, if $f, g \in L^2(\mathbb{R}^n)$, $\widehat{f}\widehat{g} \in L^2(\mathbb{R}^n)$ if and only if $f * g \in L^2(\mathbb{R}^n)$ and in this case

$$f * g = \mathbb{F}^{-1}(\widehat{f}\widehat{g}). \tag{1.5.10}$$

We also have

$$\mathbb{F}(fg) = \hat{f} * \hat{g}. \tag{1.5.11}$$

The Fourier transform can also be defined on distributions; see [Stein and Weiss (1971)]. It is sufficient for our needs to consider linear functionals on $\mathbf{S}(\mathbb{R}^n)$ of the form

$$u(\phi) = \int_{\mathbb{R}^n} u(\mathbf{x})\phi(\mathbf{x})d\mathbf{x}, \quad \phi \in \mathbf{S}(\mathbb{R}^n),$$

where $u(\cdot)/(1 + |\cdot|^2)^k \in L^1(\mathbb{R}^n)$ for some positive integer k. Such a u lies in the space $\mathbf{S}'(\mathbb{R}^n)$ of *tempered* distribution. The Fourier transform $\hat{u}$ is an element of $\mathbf{S}'(\mathbb{R}^n)$ defined by

$$\hat{u}(\phi) = u(\hat{\phi}), \quad \phi \in \mathbf{S}(\mathbb{R}^n).$$

The convolution $u * \phi$ for $u \in \mathbf{S}'(\mathbb{R}^n), \phi \in \mathbf{S}(\mathbb{R}^n)$ is defined by

$$(u * \phi)\psi = u(\tilde{\phi} * \psi), \quad \psi \in \mathbf{S}(\mathbb{R}^n),$$

where $\tilde{\phi}(\mathbf{x}) = \phi(-\mathbf{x})$ and an analogue of (1.5.10) holds, namely,

$$\mathbb{F}(u * \phi) = \hat{u}\hat{\phi}, \quad u \in \mathbf{S}'(\mathbb{R}^n), \quad \phi \in \mathbf{S}(\mathbb{R}^n) \tag{1.5.12}$$

in the sense that

$$[\mathbb{F}(u * \phi)]\psi = (\hat{u}\hat{\phi})\psi, \quad \psi \in \mathbf{S}(\mathbb{R}^n).$$

The following important examples will be needed later.

Example 1 Let $f(\mathbf{x}) = e^{-\frac{1}{2}|\mathbf{x}|^2}$. Then $\hat{f} = f$ and for $a > 0$,

$$[\mathbb{F}(e^{-\pi a|\cdot|^2})](\mathbf{p}) = \frac{1}{(2\pi a)^{n/2}}e^{-|\mathbf{p}|^2/4\pi a}. \tag{1.5.13}$$

Proof. By Fubini's Theorem, for all $\mathbf{p} \in \mathbb{R}^n$,

$$\hat{f}(\mathbf{p}) = \frac{1}{(2\pi)^{n/2}} \int_{\mathbb{R}^n} \exp\{-i\mathbf{p} \cdot \mathbf{x} - (1/2)|\mathbf{x}|^2\}d\mathbf{x}$$

$$= \prod_{j=1}^{n} \frac{1}{(2\pi)^{n/2}} \int_{\mathbb{R}} \exp\{-ip_j x_j - (1/2)x_j^2\}dx_j$$

$$= \prod_{j=1}^{n} \frac{1}{(2\pi)^{n/2}} e^{-\frac{1}{2}p_j^2} \int_{\mathbb{R}} \exp\{-\frac{1}{2}(x_j + ip_j)^2\}dx_j.$$

It follows that $\hat{f} = f$ since, by Cauchy's Theorem from complex analysis,

$$\int_{\mathbb{R}} \exp\{-\frac{1}{2}(x_j + ip_j)^2\}dx_j = \int_{\mathbb{R}} \exp\{-\frac{1}{2}x_j^2\}dx_j = \sqrt{2\pi}.$$

The identity (1.5.13) follows by a change of variables. $\qquad\square$

Example 2 For $0 < \alpha < n$ and $c_\alpha = (2)^{\alpha/2}\Gamma(\alpha/2)$ where Γ is the Gamma function,

$$\mathbb{F}(|\cdot|^{\alpha-n}) = \frac{c_\alpha}{c_{n-\alpha}}|\cdot|^{-\alpha}, \tag{1.5.14}$$

in the sense that for all $\phi \in \mathbf{S}(\mathbb{R}^n)$,

$$\frac{c_\alpha}{c_{n-\alpha}}\mathbb{F}^{-1}[|\cdot|^{-\alpha}\hat{\phi}](\mathbf{x}) = \frac{1}{(2\pi)^{n/2}}\int_{\mathbb{R}^n}|\mathbf{x}-\mathbf{y}|^{\alpha-n}\phi(\mathbf{y})dy. \tag{1.5.15}$$

In particular, with $n = 3$ and $\alpha = 2$,

$$\mathbb{F}(|\cdot|^{-1})(\mathbf{p}) = \sqrt{2/\pi}|\mathbf{p}|^{-2}, \tag{1.5.16}$$

in the sense that, for all $\phi \in \mathbf{S}$,

$$\mathbb{F}^{-1}[|\cdot|^{-2}\hat{\phi}](\mathbf{x}) = \frac{1}{4\pi}\int_{\mathbb{R}^n}|\mathbf{x}-\mathbf{y}|^{-1}\phi(\mathbf{y})dy. \tag{1.5.17}$$

Proof. Let $\phi \in \mathbf{S}(\mathbb{R}^n)$. We shall give the proof from [Lieb and Loss (1997)], Theorem 5.9, based on the identity

$$\int_0^\infty e^{-\pi kt}t^{\alpha/2-1}dt = (\pi k)^{-\alpha/2}\int_0^\infty e^{-t}t^{\alpha/2-1}dt$$

$$= (2\pi)^{-\alpha/2}c_\alpha k^{-\alpha/2}$$

which follows by change of variable. Also, on using (1.5.10) and (1.5.13), we have

$$\int_{\mathbb{R}^n}e^{i\mathbf{x}\cdot\mathbf{P}}[e^{-\pi|\mathbf{p}|^2 t}\hat{\phi}(\mathbf{p})]dp = (2\pi)^{n/2}\mathbb{F}^{-1}[e^{-\pi|\cdot|^2 t}\hat{\phi}](\mathbf{x})$$

$$= (2\pi)^{n/2}\left(\mathbb{F}^{-1}[e^{-\pi|\cdot|^2 t}] * \phi\right)(\mathbf{x})$$

$$= t^{-n/2}\left[e^{-|\cdot|^2/4\pi t} * \phi\right](\mathbf{x}).$$

From these last two identities and Fubini's Theorem we derive

$$(2\pi)^{(n-\alpha)/2}c_\alpha\mathbb{F}^{-1}[|\cdot|^{-\alpha}\hat{\phi}](\mathbf{x}) = \int_{\mathbb{R}^n}e^{i\mathbf{x}\cdot\mathbf{P}}\left(\int_0^\infty e^{-\pi|\mathbf{p}|^2 t}t^{\alpha/2-1}dt\right)\hat{\phi}(\mathbf{p})dp$$

$$= \frac{1}{(2\pi)^{n/2}}\int_0^\infty t^{(\alpha-n)/2-1}\left[\int_{\mathbb{R}^n}e^{-|\mathbf{x}-\mathbf{y}|^2/4\pi t}\phi(\mathbf{y})dy\right]dt$$

$$= \frac{1}{(2\pi)^{n/2}}\int_{\mathbb{R}^n}\left(\int_0^\infty t^{(\alpha-n)/2-1}e^{-|\mathbf{x}-\mathbf{y}|^2/4\pi t}dt\right)\phi(\mathbf{y})dy$$

$$= \frac{1}{(2\pi)^{n/2}}\int_{\mathbb{R}^n}\left(\int_0^\infty t^{(n-\alpha)/2-1}e^{-|\mathbf{x}-\mathbf{y}|^2 t/4\pi}dt\right)\phi(\mathbf{y})dy$$

$$= (2\pi)^{-\alpha/2}c_{n-\alpha}\int_{\mathbb{R}^n}|\mathbf{x}-\mathbf{y}|^{\alpha-n}\phi(\mathbf{y})dy.$$

Note that (1.5.15) is consistent with (1.5.12). For on setting $u(\mathbf{x}) = |\mathbf{x}|^{\alpha-n}$, $\hat{u}(\mathbf{x}) = (c_\alpha/c_{n-\alpha})|\mathbf{x}|^{-\alpha}$, we have that $u, \hat{u} \in \mathbf{S}'(\mathbb{R}^n)$ and for all $\phi, \psi \in \mathbf{S}(\mathbb{R}^n)$,

$$[\mathbb{F}(u * \phi)]\psi = u(\hat{\phi} * \hat{\psi})$$

$$= \int_{\mathbb{R}^n}v(\mathbf{x})\hat{\psi}(\mathbf{x})dx$$

$$= \hat{v}(\psi)$$

where

$$v(\mathbf{x}) = \frac{1}{(2\pi)^{n/2}} \int_{\mathbb{R}^n} u(\mathbf{y})\phi(\mathbf{x} - \mathbf{y})d\mathbf{y} = \frac{1}{(2\pi)^{n/2}} \int_{\mathbb{R}^n} |\mathbf{x} - \mathbf{y}|^{\alpha-n}\phi(\mathbf{y})d\mathbf{y}.$$

Hence (1.5.15) implies that $\hat{u}\hat{\phi} = \hat{v}$. $\qquad\qquad\square$

The *Mellin transform* $\mathcal{M}$ is defined on $L^2(\mathbb{R}_+)$ with Lebesgue measure dx by

$$\psi^\sharp(s) := \mathcal{M}\psi(s) = \frac{1}{\sqrt{2\pi}} \int_0^\infty t^{-1/2-is}\psi(t)dt, \quad s \in \mathbb{R}. \qquad (1.5.18)$$

It is a unitary map from $L^2(\mathbb{R}_+)$ onto $L^2(\mathbb{R})$ and has inverse

$$\mathcal{M}^{-1}\psi^\sharp(t) = \frac{1}{\sqrt{2\pi}} \int_{-\infty}^\infty s^{-1/2+it}\psi^\sharp(s)ds, \quad t \in \mathbb{R}_+. \qquad (1.5.19)$$

Moreover, with respect to the *convolution*

$$(\psi \star \phi)(s) := \frac{1}{\sqrt{2\pi}} \int_0^\infty \psi(t)\phi(s/t)dt/t,$$

on $L^2(\mathbb{R}_+)$, the Mellin transform satisfies

$$[\mathcal{M}(\psi \star \phi)](s) = \psi^\sharp(s)\phi^\sharp(s). \qquad (1.5.20)$$

We shall need the Mellin transform of the function $\tilde{Q}_l(x) := Q_l\left(\frac{1}{2}[x + \frac{1}{x}]\right)$, $l \in \mathbb{N}_0$, where Q_l is the Legendre function of the second kind, namely

$$Q_l(z) := \frac{1}{2} \int_{-1}^1 \frac{P_l(t)}{z - t}dt,$$

and P_l is the Legendre polynomial. The following formula is established in [Yaouanc *et al.* (1997)], Section VI; see also [Magnus *et al.* (1966)], p. 72, 8.17.

$$[\mathcal{M}(\tilde{Q}_l)](s) = \sqrt{\frac{\pi}{2}}V_l(s + i/2), \qquad (1.5.21)$$

where

$$V_l(z) = \frac{1}{2}\frac{\Gamma([l + 1 - iz]/2)\Gamma([l + 1 + iz]/2)}{\Gamma([l + 2 - iz]/2)\Gamma([l + 2 + iz]/2)}. \qquad (1.5.22)$$

It has the expansion

$$V_l(z) = \frac{2}{\pi}\sum_{j=0}^\infty \frac{\Gamma(n + 1/2)\Gamma(n + l + 1)}{\Gamma(n + 1)\Gamma(n + l + 3/2)}\frac{2n + l + 1}{(2n + l + 1)^2 + z^2}, \qquad (1.5.23)$$

see [Oberhettinger (1974)], p. 5.

It is instructive to verify (1.5.21) and (1.5.22) by the method in [Yaouanc *et al.* (1997)]. By the definition, (1.5.21) is satisfied with

$$V_l(z) = \frac{1}{\pi} \int_0^\infty x^{-1-iz}\tilde{Q}_l(x)dx.$$

From Rodrigues' formula

$$P_l(t) = \frac{1}{2^l l!} (-1)^l \frac{d^l}{dx^l} (1 - x^2)^l,$$

it follows by integration by parts that

$$\tilde{Q}_l(x) = \int_{-1}^{1} \frac{(1 - t^2)^l}{[(x + x^{-1}) - 2t]^{l+1}} dt.$$

The substitution $y = [(x + x^{-1}) - 2t]/(1 - t^2)$ yields

$$\tilde{Q}_l(x) = \int_{\max(x, 1/x)} y^{-l-1} (y - x)^{-1/2} (y - 1/x)^{-1/2} dy$$

$$= \int_0^\infty y^{-1} F(yx) F(y/x) dy, \tag{1.5.24}$$

where $F(t) = \theta(t - 1) t^{-l/2} (t - 1)^{-1/2}$ and θ is the Heaviside function: $\theta(t) = 1$ for $t > 0$ and is otherwise 0. Thus, with $G(x) := F(1/x)$,

$$V_l(z) = \frac{1}{2\pi} \int_0^\infty t^{-1-iz/2} \left[\int_0^\infty p^{-1} F(p) F(p/t) dp \right] dt$$

$$= [\mathcal{M}(F \star G)] ([z - i]/2)$$

$$= [\mathcal{M}(F)] ([z - i]/2) [\mathcal{M}(G)] ([z - i]/2)$$

$$= \frac{1}{2\pi} \tilde{F}(z/2) \tilde{F}(-z/2),$$

where

$$\tilde{F}(z) = \int_0^\infty x^{-1-iz} F(x) dx$$

$$= \int_1^\infty x^{-1-iz} x^{-l/2} (x - 1)^{-1/2} dx$$

$$= \int_0^1 t^{l/2+iz-1/2} (1 - t)^{-1/2} dt$$

$$= \sqrt{\pi} \frac{\Gamma ([l + 1 + 2iz]/2)}{\Gamma ([l + 2 + 2iz]/2)}.$$

The formula (1.5.22) follows.

The following pairs of Mellin transforms will be needed in Chapter 2, Section 2.2.3, and are easily verified by simple integration:

$$\psi^\sharp(s) = \frac{\mu^{-is}}{(s - ia)}, \quad \psi(p) = \begin{cases} i\sqrt{2\pi} \mu^a p^{-a-1/2} \theta(p - \mu) & \text{if } \mathrm{Re}[a] > 0 \\ -i\sqrt{2\pi} \mu^a p^{a-1/2} \theta(\mu - p) & \text{if } \mathrm{Re}[a] < 0. \end{cases} \tag{1.5.25}$$

1.6 Sobolev spaces

Let Ω be a non-empty, open subset of $\mathbb{R}^n$ with closure $\overline{\Omega}$ and boundary $\partial\Omega$. The norm in the Lebesgue space $L^p(\Omega)$, $1 \leq p \leq \infty$, is written

$$\|u\|_{p,\Omega} := \begin{cases} \left(\int_\Omega |u(\mathbf{x})|^p d\mathbf{x}\right)^{1/p}, & \text{if } 1 \leq p < \infty, \\ \text{ess sup}_{\mathbf{x} \in \mathbb{R}^3} |u(\mathbf{x})|, & \text{if } p = \infty. \end{cases}$$

If there is no ambiguity we omit the subscript Ω and write $\|u\|_p$. In fact, in subsequent chapters, the underlying space is $L^2(\mathbb{R}^3)$ in which case we simply write $\|u\|$ for the norm. If the Lebesgue measure $d\mathbf{x}$ is replaced by a measure of the form $w(\mathbf{x})d\mathbf{x}$, we get the weighted space $L^p(\Omega; w(\mathbf{x})d\mathbf{x})$ with norm (when $1 \leq p < \infty$)

$$\|u\|_{L^p(\Omega; w(\mathbf{x})d\mathbf{x})} := \left(\int_\Omega |u(\mathbf{x})|^p w(\mathbf{x})d\mathbf{x}\right)^{1/p}.$$

We preserve this notation even for $L^p(\Omega)$ if there is the risk of confusion with different norms.

For points $\mathbf{x} = (x_1, x_2, \cdots, x_n) \in \mathbb{R}^n$ and n-tuples $\alpha = (\alpha_1, \alpha_2, \cdots, \alpha_n) \in \mathbb{N}_0^n$, we write

$$|\mathbf{x}| = \left(\sum_{j=1}^n x_j^2\right)^{1/2}, \; |\alpha| = \sum_{j=1}^n \alpha_j, \; D^\alpha = \prod_{j=1}^n D_j^{\alpha_j},$$

where $D_j = (1/i)\partial/\partial x_j$. For $k \in \mathbb{N}$ and $p \in [1, \infty]$, the *Sobolev space* $W^{k,p}(\Omega)$ is defined as

$$W^{k,p}(\Omega) := \{u : \Omega \to \mathbb{C}, u, D^\alpha u \in L^p(\Omega) \text{ for } |\alpha| \leq k\},$$

where the derivatives $D^\alpha u$ are taken to be in the weak or distributional sense. It is endowed with the norm

$$\|u\|_{k,p,\Omega} := \begin{cases} \left(\sum_{0 \leq |\alpha| \leq k} \|D^\alpha u\|_{p,\Omega}^p\right)^{1/p}, & \text{if } 1 \leq p < \infty, \\ \sum_{0 \leq |\alpha| \leq k} \|D^\alpha u\|_{\infty,\Omega}, & \text{if } p = \infty, \end{cases} \tag{1.6.1}$$

where $\|u\|_{p,\Omega}$ denotes the standard $L^p(\Omega)$ norm of u, namely $(\int_\Omega |u(\mathbf{x})|^p d\mathbf{x})^{1/p}$. $W^{k,p}(\Omega)$ is a separable Banach space if $p \in [1, \infty)$ and is reflexive if $p \in (1, \infty)$. When $p = 2$, $W^{k,2}(\Omega)$ is a Hilbert space with inner product

$$(u, v)_{k,2,\Omega} = \int_\Omega \sum_{|\alpha| \leq k} D^\alpha u (\overline{D^\alpha v}) d\mathbf{x}.$$

When $\Omega = \mathbb{R}^n$, we shall omit Ω in the notation and write $\| \cdot \|_{k,p}$ and $(\cdot, \cdot)_{k,2}$.

Meyers and Serrin proved in [Meyers and Serrin (1964)] that for $p \in [1, \infty)$, $W^{k,p}(\Omega)$ coincides with the completion $H^{k,p}(\Omega)$ of the linear space $C^\infty(\Omega) \cap W^{k,p}(\Omega)$ with respect to $\| \cdot \|_{k,p,\Omega}$, i.e., $C^\infty(\Omega) \cap W^{k,p}(\Omega)$ is dense in $W^{k,p}(\Omega)$. As is now standard, we shall use the notation $H^k(\Omega)$ for $H^{k,2}(\Omega) \equiv W^{k,2}(\Omega)$.

We shall occasionally need the *Bessel potential spaces* $H^{s,p}(\mathbb{R}^n), s \in \mathbb{R}_+, p \in (1, \infty)$. These are defined in terms of the Bessel potential g_s, whose Fourier transform is given by

$$\hat{g}_s(\mathbf{p}) := (1 + |\mathbf{p}|^2)^{-s/2};$$

see [Stein (1970)] for the properties of g_s. We have that

$$H^{s,p}(\mathbb{R}^n) := \{u : u = g_s * f \text{ for some } f \in L^p(\mathbb{R}^n)\} \qquad (1.6.2)$$

with norm

$$\|u\|_{s,p} = \|f\|_p. \qquad (1.6.3)$$

For all $s \geq 0$ and $p \in (1, \infty)$, the Schwartz space $\mathbf{S}(\mathbb{R}^n)$ is dense in $H^{s,p}(\mathbb{R}^n)$. When $s = k \in \mathbb{N}$, $H^{s,p}(\mathbb{R}^n)$ coincides with the Sobolev space $W^{k,p}(\mathbb{R}^n) \equiv H^{k,p}(\mathbb{R}^n)$. Since $\hat{f}(\mathbf{p}) = (1 + |\mathbf{p}|^2)^{s/2} \hat{u}(\mathbf{p})$ if $u = g_s * f$, by (1.5.9), it follows from the Parseval formula (1.5.7) when $p = 2$ that

$$\|u\|_{s,2} = \left(\int_{\mathbb{R}^n} (1 + |\mathbf{p}|^2)^s |\hat{u}(\mathbf{p})|^2 d\mathbf{p} \right)^{1/2}. \qquad (1.6.4)$$

In particular, as $\mathbb{F}(D_j u)(\mathbf{p}) = p_j \hat{u}(\mathbf{p})$, by (1.5.5),

$$\|u\|_{1,2}^2 = \int_{\mathbb{R}^n} \left(|u(\mathbf{x})|^2 + |\nabla u(\mathbf{x})|^2 \right) d\mathbf{x} \qquad (1.6.5)$$

$$= \int_{\mathbb{R}^n} (1 + |\mathbf{p}|^2) |\hat{u}(\mathbf{p})|^2 d\mathbf{p}, \qquad (1.6.6)$$

where ∇ is the gradient and $|\nabla u|^2 = \sum_{j=1}^{n} |D_j u|^2$.

Important properties of the Sobolev and Bessel potential spaces are now listed; see [Edmunds and Evans (1987)], Chapter V, for proofs and further details. We shall assume that $p \in [1, \infty)$ unless otherwise stated.

(i) For $k \in \mathbb{R}$, $C_0^\infty(\mathbb{R}^n)$ is dense in $H^{k,p}(\mathbb{R}^n)$, and, for all $s \geq 0$, and $p \in (1, \infty)$, the Schwartz space $\mathbf{S}(\mathbb{R}^n)$ is dense in $H^{s,p}(\mathbb{R}^n)$.

(ii) The closure of $C_0^\infty(\Omega)$ in $H^{k,p}(\Omega)$ is denoted by $H_0^{k,p}(\Omega)$. Thus, by (i), $H^{k,p}(\mathbb{R}^n) = H_0^{k,p}(\mathbb{R}^n)$. If Ω is bounded, $H_0^{k,p}(\Omega) \neq H^{k,p}(\Omega)$. Also if Ω is bounded,

$$\|u\|_{p,\Omega} \leq \left(\frac{|\Omega|}{\omega_n} \right)^{1/n} \|\nabla u\|_{p,\Omega}, \quad \text{for all} \quad u \in H_0^{1,p}(\Omega), \qquad (1.6.7)$$

where

$$\|\nabla u\|_{p,\Omega} = \||\nabla u|\|_{p,\Omega},$$

and $\omega_n = \pi^{n/2}/\Gamma(1 + n/2)$ is the volume of the unit ball in $\mathbb{R}^n$. Hence

$$\|u\|_{1,p,\Omega} \lesssim \left(\sum_{|\alpha|=1} \|D^\alpha u\|_{p,\Omega}^p \right)^{1/p}.$$

It follows that the norm $\| \cdot \|_{1,p,\Omega}$ on $H_0^{1,p}(\Omega)$ is equivalent to the norm

$$\|u\|_{H_0^{1,p}(\Omega)} := \left(\sum_{|\alpha|=1} \|D^\alpha u\|_{p,\Omega}^p \right)^{1/p}. \tag{1.6.8}$$

(iii) Let $1 \leq p < n$ and set $p^* = np/(n-p)$, the so-called *Sobolev conjugate* of p. Then the identification map is a continuous injection of $H_0^{1,p}(\Omega)$ into $L^{p^*}(\Omega)$: this continuous embedding is indicated by $H_0^{1,p}(\Omega) \hookrightarrow L^{p^*}(\Omega)$. This constitutes the case $1 \leq p < n$ of the *Sobolev embedding theorem* for $H_0^1(\Omega)$.

(iv) For $1 \leq p < n/k$, $H_0^{k,p}(\Omega)$ is continuously embedded in $L^s(\Omega)$, where $s = np/(n-kp)$.

(v) If $q \in [1, np/(n-kp))$ and Ω is bounded, $H_0^{k,p}(\Omega)$ is compactly embedded in $L^q(\Omega)$; this is the *Rellich–Kondrachov property*.

(vi) If Ω is bounded, $H_0^{1,n}(\Omega)$ is continuously embedded in the Orlicz space $L^\phi(\Omega)$, where $\phi(t) = \exp\left(t^{n/(n-1)} - 1\right), t \geq 0$, and, in particular, $H_0^{1,n}(\Omega) \hookrightarrow L^q(\Omega)$ for all $q \in [n, \infty)$. Note that $L^\phi(\Omega)$ is the linear span of the set of Lebesgue measurable functions u on Ω which are such that $\int_\Omega \phi(|u(\mathbf{x})|)d\mathbf{x} < \infty$, endowed with the norm

$$\|u\|_{L^\phi(\Omega)} := \inf\{\lambda : \int_\Omega \phi\left(\frac{|u(\mathbf{x})|}{\lambda}\right) d\mathbf{x} \leq 1\}.$$

It is a Banach space containing $L^\infty(\Omega)$ which, in general, is neither reflexive nor separable.

(vii) If Ω is bounded and $p > n$, then $H_0^{1,p}(\Omega)$ is continuously embedded in the space $C^{0,\gamma}(\overline{\Omega})$ of functions u which are Hölder continuous on $\overline{\Omega}$ with exponent $\gamma = 1 - n/p$ and norm

$$\|u\|_{C^{0,\gamma}(\overline{\Omega})} := \|u\|_{L^\infty(\Omega)} + \sup_{\mathbf{x},\mathbf{y} \in \overline{\Omega}, \mathbf{x} \neq \mathbf{y}} \frac{|u(\mathbf{x}) - u(\mathbf{y})|}{|\mathbf{x} - \mathbf{y}|^\gamma}.$$

The embedding of $H_0^{1,p}(\Omega)$ into $C^{0,\lambda}(\overline{\Omega})$ is compact for any $\lambda \in (0, \gamma)$.

(viii) If the boundary of Ω is sufficiently smooth (see [Edmunds and Evans (1987)], Section V.4) the above results in (iii)–(vii) continue to hold for $H^{k,p}(\Omega)$.

1.7 Inequalities

Three inequalities make regular appearances throughout the book, namely, the well-known inequalities of Sobolev, Hardy and Kato. We shall need only the L^2 versions of these inequalities, but mention some L^p versions for completeness.

- (Sobolev) For $1 \leq p < n$, there exists a constant $C_{n,p}$, depending only on n and p, such that for all $f \in C_0^\infty(\mathbb{R}^n)$,

$$\|f\|_{p^*} \leq C_{n,p}\|\nabla f\|_p = C_{n,p}\left(\int_{\mathbb{R}^n} [\sum_{j=1}^n |D_j u|^2]^{p/2} dx\right)^{1/p}. \tag{1.7.1}$$

The best possible value of the constant $C_{n,p}$ in (1.7.1) for $1 < p < n$ is

$$\pi^{-1/2} n^{-1/p}\left(\frac{p-1}{n-p}\right)^{(p-1)/p}\left\{\frac{\Gamma(1+n/2)\Gamma(n)}{\Gamma(n/p)\Gamma(1+n-n/p)}\right\}$$

and equality is attained for functions f of the form

$$f(\mathbf{x}) = [a + b|\mathbf{x}|^{p/(p-1)}]^{1-n/p},$$

where a and b are positive constants: these functions are obviously not in $C_0^\infty(\mathbb{R}^n)$ but lie in $L^p(\mathbb{R}^n)$ and the completion of $C_0^\infty(\mathbb{R}^n)$ with respect to the norm $\|\nabla \cdot \|_p$. When $p = 1$, $C_{n,1} = 1/n\omega_n^{1/n}$ is the optimal constant, and equality is never attained unless f vanishes identically. However, in this case the inequality has an extended version

$$\|f\|_{1^*} \leq (n\omega_n^{1/n})^{-1}\|Df\|$$

on the set of functions f of bounded variation on $\mathbb{R}^n$, with Df the distributional gradient (a vector-valued Radon measure) and $\|Df\|$ the total variation of f in $\mathbb{R}^n$. For this inequality, the characteristic functions of arbitrary balls are extremals.

For $1 \leq p < n$, (1.7.1) was established by Sobolev [Sobolev (1938)], the case $p = 1$ being later proved by Gagliardo [Gagliardo (1958)] and Nirenberg [Nirenberg (1959)]. The optimal constant for $p = 1$ was determined independently by Federer and Fleming in [Federer and Fleming (1960)] and Maz'ya in [Maz'ya (1960)], but for $p > 1$, the best constant was only found 10 years later, independently by Aubin [Aubin (1976)] and Talenti [Talenti (1976)].

- (Sobolev inequality for $\sqrt{-\Delta}$) For all $f \in \mathbf{S}(\mathbb{R}^n)$, $n \geq 2$ and $q = 2n/(n-1)$,

$$\|f\|_q^2 \leq C_n \int_{\mathbb{R}^n} |\mathbf{p}||\hat{f}(\mathbf{p})|^2 d\mathbf{p}, \tag{1.7.2}$$

where the optimal value of the constant C_n is

$$C_n = \left\{\left(\frac{n-1}{2}\right) 2^{1/n} \pi^{(n+1)/2n}\right\}^{-1} \Gamma\left(\frac{n+1}{2}\right)^{1/n}.$$

There is equality if and only if f is a constant multiple of a function of the form $[(\mu^2 + (\mathbf{x} - \mathbf{a})^2]^{-(n-1)/2}$ with $\mu > 0$ and $\mathbf{a} \in \mathbb{R}^n$ arbitrary. This is proved in [Lieb and Loss (1997)], Theorem 8.4.

- (Hardy) For all $f \in C_0^\infty(\mathbb{R}^n), n \geq 3$,

$$\int_{\mathbb{R}^n} \frac{|f(\mathbf{x})|^2}{|\mathbf{x}|^2}dx \leq \left(\frac{2}{(n-2)}\right)^2 \int_{\mathbb{R}^n} |\nabla f(\mathbf{x})|^2 dx. \tag{1.7.3}$$

The constant is sharp and equality is only valid for $f = 0$. The inequality is determined by the radial part of ∇ and in fact

$$\int_{\mathbb{R}^n} \frac{|f(\mathbf{x})|^2}{|\mathbf{x}|^2}dx \leq \left(\frac{2}{(n-2)}\right)^2 \int_{\mathbb{R}^n} |\frac{\partial}{\partial r}f(\mathbf{x})|^2 dx. \tag{1.7.4}$$

By completion, (1.7.3) and (1.7.4) hold for all functions f which are such that ∇f or $(\partial/\partial r)f$ (in the weak, or distributional sense) lie in $L^2(\mathbb{R}^n)$. When $n = 2$, (1.7.3) is of course trivial, while when $n = 1$, we have that, for all f that are locally absolutely continuous on $(0, \infty)$, $f' \in L^2(0, \infty)$ and such that $\lim_{r\to 0+} f(r) = 0$,

$$\int_0^\infty \frac{|f(r)|^2}{r^2}dr \leq 4 \int_0^\infty |f'(r)|^2 dr. \tag{1.7.5}$$

The L^p version of the Hardy inequality is, for $1 \leq p < n$,

$$\int_{\mathbb{R}^n} \frac{|f(\mathbf{x})|^p}{|\mathbf{x}|^p}dx \leq \left(\frac{p}{(n-p)}\right)^p \int_{\mathbb{R}^n} |\nabla f(\mathbf{x})|^p dx. \tag{1.7.6}$$

The constant is optimal and there are no non-trivial cases of equality.

- (Kato) For all $f \in \mathbf{S}(\mathbb{R}^n), n \geq 2$,

$$\int_{\mathbb{R}^n} \frac{|f(\mathbf{x})|^2}{|\mathbf{x}|}dx \leq c_n^2 \int_{\mathbb{R}^n} |\mathbf{p}||\hat{f}(\mathbf{p})|^2 dp. \tag{1.7.7}$$

The best possible constant c_n for general values of $n \geq 2$ will be included in Theorem 1.7.1 below. In particular

$$c_3 = \sqrt{\pi/2}, \quad c_2 = \Gamma(1/4)/\sqrt{2}\Gamma(3/4).$$

There are no non-trivial cases of equality.

The above Hardy and Kato inequalities are special cases of the following inequality obtained by Herbst in [Herbst (1977)], Theorem 2.5. With $\mathbf{p} = -i\nabla$ denoting the momentum operator, Herbst determines the norm of the operator $C_\alpha := |\mathbf{x}|^{-\alpha}|\mathbf{p}|^{-\alpha}$ as a map from $L^q(\mathbb{R}^n)$ into itself. The precise result is:

Theorem 1.7.1. *Let $\alpha > 0$ and $n\alpha^{-1} > q > 1$. Then C_α can be extended to a bounded operator from $L^q(\mathbb{R}^n)$ into itself, with norm*

$$\|C_\alpha : L^q(\mathbb{R}^n) \to L^q(\mathbb{R}^n)\| = \gamma(n, \alpha) := \frac{\Gamma(\frac{1}{2}[nq^{-1} - \alpha])\Gamma(\frac{1}{2}n(q')^{-1})}{2^\alpha\Gamma(\frac{1}{2}[n(q')^{-1} + \alpha])\Gamma(\frac{1}{2}nq^{-1})}, \tag{1.7.8}$$

where $q' = q/(q-1)$. If $q \geq n\alpha^{-1}$ or $q = 1$, C_α is unbounded.

In the case $q = 2$, $\mathbf{p}$ has adjoint $(1/i)\mathrm{div}$ and absolute value $|\mathbf{p}| = (\mathbf{p}^*\mathbf{p})^{1/2} = \sqrt{-\Delta}$. Also $|\mathbf{p}|$ is self-adjoint, injective and has dense range in $L^2(\mathbb{R}^n)$. Hence

$$\|C_\alpha\| = \sup_{\phi \in \mathcal{R}(|\mathbf{p}|^\alpha)} \frac{\|C_\alpha\phi\|_2}{\|\phi\|_2} = \sup_{\psi \in \mathcal{D}(|\mathbf{p}|^\alpha)} \frac{\||\mathbf{x}|^{-\alpha}\psi\|_2}{\||\mathbf{p}|^\alpha\psi\|_2},$$

where $\mathcal{R}, \mathcal{D}$ denote the range and domain respectively of the exhibited operator. Since $\mathbb{F}(|\mathbf{p}|^\alpha\psi)(\boldsymbol{\xi}) = |\boldsymbol{\xi}|^\alpha\hat{\psi}(\boldsymbol{\xi})$, it follows by Parseval's formula that (1.7.8) becomes

$$\int_{\mathbb{R}^n} \frac{1}{|\mathbf{x}|^{2\alpha}}|\psi(\mathbf{x})|^2 d\mathbf{x} \le \gamma^2(n, \alpha) \int_{\mathbb{R}^n} |\boldsymbol{\xi}|^{2\alpha}|\hat{\psi}(\boldsymbol{\xi})|^2 d\boldsymbol{\xi},$$

which is Hardy's inequality in the case $\alpha = 1$ and Kato's inequality when $\alpha = 1/2$. In Section 2.2.1 we give an alternative proof to that of Herbst for the Kato inequality when $n = 2, 3$, using spherical harmonics.

Another inequality which features prominently throughout the book is the following generalisation of Hilbert's double series theorem due to Hardy, Littlewood and Pólya, see [Hardy *et al.* (1959)], Chapter IX, Section 319.

Theorem 1.7.2. *Let $K(\cdot, \cdot)$ be a non-negative function on $\mathbb{R}_+ \times \mathbb{R}_+$ that is homogeneous of degree -1, i.e., for any $\lambda \in \mathbb{R}_+, K(\lambda x, \lambda y) = \lambda^{-1}K(x, y)$. Suppose also that for $p > 1$, we have*

$$\int_0^\infty K(x, 1)x^{-1/p}dx = \int_0^\infty K(1, y)y^{-1/p'}dy = k, \tag{1.7.9}$$

where $p' = p/(p - 1)$. Then

$$\int_0^\infty \int_0^\infty K(x, y)|f(x)g(y)|dxdy \le k \left(\int_0^\infty |f(x)|^p dx\right)^{1/p} \left(\int_0^\infty |g(y)|^{p'} dy\right)^{1/p'}, \tag{1.7.10}$$

$$\int_0^\infty dy \left(\int_0^\infty K(x, y)|f(x)|dx\right)^p \le k^p \int_0^\infty |f(x)|^p dx, \tag{1.7.11}$$

$$\int_0^\infty dx \left(\int_0^\infty K(x, y)|g(y)|dy\right)^{p'} \le k^{p'} \int_0^\infty |g(y)|^{p'} dy. \tag{1.7.12}$$

If $K(\cdot, \cdot)$ is positive, then there is inequality in (1.7.11) unless $f = 0$, in (1.7.12) unless $g = 0$, and in (1.7.10) unless either $f = 0$ or $g = 0$.

1.8   CLR and related inequalities

The CLR refers to Cwikel–Lieb–Rosenbljum who proved the inequality independently and by very different methods in [Cwikel (1977)], [Lieb (1976)] and [Rozenbljum (1972)]. The names are listed alphabetically, but in reverse chronological order of discovery. In its original form, the inequality concerns the self-adjoint operator $-\Delta - V$ defined as a form sum, whose spectrum is discrete below 0 and with V the operator of multiplication by a function V with a positive part $V_+ := (1/2)(|V|+V) \in L^{n/2}(\mathbb{R}^n)$. It asserts that, for $n \geq 3$, the number $N(-\Delta - V)$ of negative eigenvalues of $-\Delta - V$ satisfies

$$N(-\Delta - V) \leq c(n) \int_{\mathbb{R}^n} V_+(\mathbf{x})^{n/2} d\beta x, \qquad (1.8.1)$$

for some constant $c(n)$ depending only on n. Other proofs have also been given, notably those of Li and Yau [Li and Yau (1983)] and Conlon [Conlon (1984)]. A treatment, which is particularly suitable for our needs, is that given by Rozenbljum and Solomyak in [Rozenbljum and Solomyak (1998)]. This is motivated by Lieb's proof and gives an abstract version of the inequality, which can be applied to other operators that feature in the book.

In order to state the main results in [Rozenbljum and Solomyak (1998)] it is necessary to recall some basic facts about the operator semigroup $e^{-tT}, 0 \leq t < \infty$, associated with a non-negative, self-adjoint operator T acting in a Hilbert space H.

(1) $Q(t) := e^{-tT}, 0 \leq t < \infty$, is *strongly continuous*, i.e.

$$\lim_{t \to s} \|Q(t)f - Q(s)f\| = 0, \quad \text{for all } f \in H,$$

and *contractive*, i.e.

$$\|Q(t)f\| \leq \|f\|, \quad \text{for all } f \in H.$$

(2) T is the *infinitesimal generator* of $Q(t)$, i.e.

$$Tf = \lim_{t \to 0+} \frac{1}{t}\{f - Q(t)f\}, \quad \text{for all } f \in \mathcal{D}(T).$$

(3) For $f \in \mathcal{D}(T), Q(t)f \in \mathcal{D}(T)$ and

$$\frac{\partial}{\partial t}[Q(t)f] := \lim_{\varepsilon \to 0}\left\{\frac{1}{\varepsilon}[Q(t+\varepsilon) - Q(t)]f\right\} = -TQ(t)f = -Q(t)Tf, \quad (t > 0),$$

$$Q(0)f := \lim_{t \to 0+} Q(t)f = f.$$

The result in [Rozenbljum and Solomyak (1998)] deals with the set $\mathcal{P}$ of positive, self-adjoint operators in a space $L^2(\Omega)$ say, which are positivity preserving and have the $(2,\infty)$ mapping property. An operator B is *positivity preserving* (or its associated semi-group $\mathcal{Q}_B(t) := e^{-tB}, 0 \leq t < \infty$, is *positivity preserving*) if $\mathcal{Q}_B(t)u \geq 0$ for all non-negative functions $u \in L^2(\Omega)$. It has the $(2,\infty)$ *mapping property* if $\mathcal{Q}_B(t) : L^2(\Omega) \to L^\infty(\Omega)$ is bounded: for Markov semigroups, the $(2,\infty)$ property is usually called *ultracontractivity*. Semigroups which are positivity preserving and have the $(2,\infty)$ property are known to be integral operators. The kernel $Q_B(t;x,y)$ of $\mathcal{Q}_B(t)$ satisfies the symmetry condition $Q_B(t;x,y) = \overline{Q_B(t;y,x)}$ and it follows from the fact that $\mathcal{Q}_B(t)$ is a semigroup that, for a.e. $x \in \Omega$,

$$Q_B(t;x,x) = \int_\Omega Q_B(t_1;x,y)Q_B(t_2;x,y)dy, \quad t_1, t_2 > 0, \ t = t_1 + t_2$$

is well defined as an element of $L^\infty(\Omega)$, which does not depend on the particular choice of t_1 and t_2. Moreover,

$$M_B(t) := \|e^{-tB} : L^2 \to L^\infty\| = \text{esssup}_{x\in\Omega} \int |Q_B(t/2; x, y)|^2 dy < \infty.$$

$$= \text{esssup}_{x\in\Omega} \int Q_B(t/2; x, y)Q_B(t/2; y, x)dy$$

$$= \text{esssup}_{x\in\Omega} Q_B(t; x, x).$$

For $B \in \mathcal{P}$, the kernel $Q_B(t; x, y)$ is non-negative, a.e. on $\mathbb{R}_+ \times \Omega \times \Omega$. If $B \in \mathcal{P}$, and $B \geq c$ for some $c \geq 0$, then, for any $r \geq -c$, $B_r := B + r \in \mathcal{P}$, $Q_{B_r}(t) = e^{-rt}Q_B(t)$ and $M_{B_r} = e^{-rt}M_B(t)$.

The following is Theorem 2.1 in [Rozenbljum and Solomyak (1998)] and is given in terms of a non-negative, convex function G on $[0, \infty)$ which grows polynomially at infinity and is such that $G(z) = 0$ near $z = 0$. Let

$$g(r) := \int_0^\infty z^{-1}G(z)e^{-z/r}dz.$$

In the theorem, multiplication by V is assumed to be such that self-adjoint operators $B - V$ and $A - V$ can be defined as form sums; see Section 1.2.

Theorem 1.8.1. *Let $B \in \mathcal{P}$ be such that $M_B \in L^1(a, \infty)$, $a > 0$ and $M_B(t) = O(t^{-\alpha})$ at zero for some $\alpha > 0$. Then, in the above notation,*

$$N(B - V) \leq \frac{1}{g(1)} \int_0^\infty \frac{dt}{t} \int_\Omega M_B(t)G[tV_+(x)]dx. \tag{1.8.2}$$

The inequality continues to hold for $N(A - V)$ when A is a non-negative, self-adjoint operator, which is such that e^{-tA} is dominated by a positivity preserving semigroup e^{-tB} in the sense that

$$|e^{-tA}\psi| \leq e^{-tB}|\psi| \quad \text{a.e. on } \Omega. \tag{1.8.3}$$

We denote the class of such operators A by $\mathcal{PD}(B)$.

If $A \in \mathcal{PD}(B)$ then $A + r \in \mathcal{PD}(B + r)$. Also, a particular result from [Bratelli et al. (1980)] is that if e^{-tB} is positivity preserving and $A \in \mathcal{PD}(B)$, then for $0 < \alpha < 1$, e^{-tB^α} is positivity preserving and $A^\alpha \in \mathcal{PD}(B^\alpha)$. Any $A \in \mathcal{PD}(B)$ is an integral operator and (1.8.3) is equivalent to

$$|Q_A(t; x, y)| \leq Q_B(t; x, y) \quad \text{a.e. on } \mathbb{R}_+ \times \Omega \times \Omega. \tag{1.8.4}$$

We refer to [Rozenbljum and Solomyak (1998)] for details and full references. Of particular interest to us are the following examples discussed in [Rozenbljum and Solomyak (1998)], which are special cases of Theorem 1.8.1. In all cases $\Omega = \mathbb{R}^n$.

1.8.1 The Schrödinger operator

The semigroup associated with the Laplace operator $B := -\Delta$ is the heat operator, which is an integral operator with kernel

$$Q_B(t; \mathbf{x}, \mathbf{y}) = \frac{1}{(4\pi t)^{n/2}} \exp\left\{\frac{-|\mathbf{x} - \mathbf{y}|^2}{4t}\right\}. \tag{1.8.5}$$

Thus $Q_B(t)$ is positivity preserving. Furthermore, an application of the Cauchy–Schwarz inequality readily yields

$$\|Q_B(t)f\|_{L^\infty(\mathbb{R}^n)} \le 2^{-n}\|f\|_{L^2(\mathbb{R}^n)}$$

and so $Q_B(t)$ has the $(2, \infty)$ property. Consequently $B = -\Delta \in \mathcal{P}$ and $M_B(t) = \frac{1}{(4\pi t)^{n/2}}$, so that $M_B \in L^1(a, \infty), a > 0$, if and only if $n \ge 3$. Hence, to apply Theorem 1.8.1 we need $n \ge 3$.

Following Lieb in [Lieb (1976)], the choice $G(z) = (z - a)_+$ in (1.8.1), where $(\cdot)_+$ denotes the positive part and a is a positive constant to be chosen, yields

$$N(-\Delta - V) \le C(G) \int_{\mathbb{R}^n} V_+^{n/2}(\mathbf{x})d\mathbf{x}, \tag{1.8.6}$$

where $C(G) = (2\pi)^{-n/2}g(1)^{-1} \int_a^\infty (t - a)t^{-n/2-1}dt$. The optimal value of $C(G)$ as a function of a is .1156, attained when $a = 1/4$. This coincides with Lieb's constant in [Lieb (1976)], which is the best value achieved to date.

1.8.2 The magnetic Schrödinger operator

This is of the form $-\Delta_{\mathbf{A}} - V$, where $\Delta_{\mathbf{A}}$ is the *magnetic Laplacian*

$$\Delta_{\mathbf{A}} = (\boldsymbol{\nabla} + i\mathbf{A})^2 = \sum_{j=1}^{n} (\partial_j + iA_j)^2,$$

and $\mathbf{A} = \{A_j : j = 1, \cdots, n\}$ is the magnetic potential. A number of proofs exist of the result that $A \in \mathcal{PD}(-\Delta)$; see the discussion and a list of references after Theorem 2.3 in [Avron *et al.* (1978)]. It then follows from Theorem 1.8.1 that

$$N(-\Delta_{\mathbf{A}} - V) \le C(G) \int_{\mathbb{R}^n} V_+^{n/2}(\mathbf{x})d\mathbf{x}, \tag{1.8.7}$$

with the same constant as in (1.8.6). It is not true in general that $N(-\Delta_{\mathbf{A}} - V) \le N(-\Delta - V)$; see [Avron *et al.* (1978)], Example 2 following Theorem 2.14.

1.8.3 The quasi-relativistic Schrödinger operator

The pseudo-differential operator $B = \sqrt{-\Delta}$ is non-negative and self-adjoint, with domain $H^1(\mathbb{R}^n)$. The kernel of the associated semigroup $Q_B(t)$ is the Poisson kernel

$$Q_B(t; \mathbf{x}, \mathbf{y}) = \int_{\mathbb{R}^n} \exp[-2\pi|\mathbf{k}|t + 2\pi i\mathbf{k} \cdot (\mathbf{x} - \mathbf{y})]d\mathbf{k}$$

$$= \Gamma\left(\frac{n+1}{2}\right)\pi^{-(n+1)/2}\frac{t}{[t^2 + |\mathbf{x} - \mathbf{y}|^2]^{(n+1)/2}};$$

see [Stein and Weiss (1971)], Theorem 1.14, and [Lieb and Loss (1997)], Section 7.11. Thus,

$$\|Q_B(t)f\|_{L^\infty(\mathbb{R}^n)} \le c_n \sup_{\mathbf{x}\in\mathbb{R}^n} \left(\int_{\mathbb{R}^n} \frac{t^2}{[t^2+|\mathbf{x}-\mathbf{y}|^2]^{(n+1)}}d\mathbf{y}\right)^{1/2} \|f\|_{L^2(\mathbb{R}^n)}$$
$$< c_n(t)\|f\|_{L^2(\mathbb{R}^n)}$$

for a positive constant $c_n(t)$ which is finite for all $t > 0$. Thus $B = \sqrt{-\Delta}$ has the $(2,\infty)$ property and lies in $\mathcal{P}$, with $M_B(t) = c_n t^{-n}$. Theorem 1.8.1 therefore applies for all $n \ge 2$ and leads to *Daubechies' inequality*

$$N(\sqrt{-\Delta} - V) \le C_n(G)\int_{\mathbb{R}^n} V_+(\mathbf{x})^n d\mathbf{x}, \qquad (1.8.8)$$

where in the above notation and the same choice of G, $C_n(G) = g(1)^{-1}\int_0^\infty t^{-n-1}G(t)dt$.

The pseudo-differential operator $\mathbb{H}_0 = \sqrt{-\Delta + 1}$, is self-adjoint with domain $H^1(\mathbb{R}^n)$ and $\mathbb{H}_0 \ge 1$. It will feature prominently throughout the book. From the result in [Bratelli *et al.* (1980)] noted above, it follows that with $B := \mathbb{H}_0 - 1, e^{-tB}$ is positivity preserving. It is an integral operator and on the diagonal, the kernel is given by

$$Q_B(t; \mathbf{x}, \mathbf{x}) = (2\pi)^{n/2}\int_{\mathbb{R}^n} e^{-t[(|\xi|^2+1)^{1/2}-1]}d\xi, \qquad (1.8.9)$$

from which we can easily infer that $e^{-t[\mathbb{H}_0-1]}$ has the $(2,\infty)$ property and

$$M_B(t) \le C(t^{-n/2} + t^{-n}).$$

For $n \ge 3$, Theorem 1.8.1 yields Daubechies' inequality

$$N(\mathbb{H}_0 - 1 - V) \le C_n(G)\int_{\mathbb{R}^n} V_+(\mathbf{x})^n d\mathbf{x} + C_{n/2}(G)\int_{\mathbb{R}^n} V_+(\mathbf{x})^{n/2}d\mathbf{x}, \qquad (1.8.10)$$

where, in the above notation and the same choice of G, $C_a(G) = g(1)^{-1}\int_0^\infty t^{-a-1}G(t)dt$.

1.8.4 *The magnetic quasi-relativistic Schrödinger operator*

The operator $\sqrt{-\Delta_{\mathbf{A}}}$ lies in $\mathcal{PD}(-\Delta_{\mathbf{A}})$ by the result from [Bratelli *et al.* (1980)] quoted above and also satisfies (1.8.8):

$$N(\sqrt{-\Delta_{\mathbf{A}}} - V) \le C_n(G)\int_{\mathbb{R}^n} V_+(\mathbf{x})^n d\mathbf{x}. \qquad (1.8.11)$$

Similarly,

$$N([(-\Delta_{\mathbf{A}} + 1)^{1/2} - 1] - V) \le C_n(G)\int_{\mathbb{R}^n} V_+(\mathbf{x})^n d\mathbf{x} + C_{n/2}(G)\int_{\mathbb{R}^n} V_+(\mathbf{x})^{n/2}d\mathbf{x}.$$
$$(1.8.12)$$

1.9 Lieb–Thirring inequalities

These inequalities have a pivotal role in problems such as the stability of many-body systems composed of fermions. In many cases, the Hamiltonian describing the system is bounded below in terms of the sum of the negative eigenvalues of a one-body Hamiltonian. We shall see examples of this in Section 4.5 of Chapter 4. Since the seminal papers [Lieb and Thirring (1975)] and [Lieb and Thirring (1976)], Lieb–Thirring inequalities have come to mean estimates for moments of the negative eigenvalues of operators of Schrödinger type in terms of external electric and magnetic fields. The original result proved by Lieb and Thirring in [Lieb and Thirring (1975)], is that if $\gamma > \max(0, 1 - n/2)$, then there exists a universal constant $L_{\gamma,n}$ such that

$$\sum_{n \in \mathbb{N}} |\lambda_j(S)|^\gamma \leq L_{\gamma,n} \int_{\mathbb{R}^n} V_+(\mathbf{x})^{\gamma+n/2} dx, \tag{1.9.1}$$

where $\lambda_1(S) \leq \lambda_2(S) \leq \cdots$ are the negative eigenvalues of the Schrödinger operator $S := -\Delta - V$ counting multiplicities: the assumptions on V imply that the negative spectrum of S is discrete and its essential spectrum fills the positive half-line. Note that the case $n \geq 3, \gamma = 0$ is the CLR inequality associated with $-\Delta - V$.

If $V \in L^{\gamma+n/2}(\mathbb{R}^n)$, one has the Weyl-type asymptotic formula

$$\lim_{\alpha \to \infty} \alpha^{-(\gamma+n/2)} \sum_{n \in \mathbb{N}} |\lambda_j(S_\alpha)|^\gamma = (2\pi)^{-n} \lim_{\alpha \to \infty} \alpha^{-(\gamma+n/2)} \int \int_{\mathbb{R}^n \times \mathbb{R}^n} (|\boldsymbol{\xi}|^2 - \alpha V)_+^\gamma dx d\boldsymbol{\xi}$$

$$= L_{\gamma,n}^{cl} \int_{\mathbb{R}^n} V_+(\mathbf{x})^{\gamma+n/2} dx, \tag{1.9.2}$$

where $S_\alpha : -\Delta - \alpha V$ and the so-called classical constant $L_{\gamma,n}^{cl}$ is given by

$$L_{\gamma,n}^{cl} = (2\pi)^{-n} \int_{\mathbb{R}^n} (|\boldsymbol{\xi}| - 1)_+^\gamma d\boldsymbol{\xi} = \frac{\Gamma(\gamma+1)}{2^n \pi^{n/2} \Gamma(\gamma + n/2 + 1)}.$$

Therefore

$$L_{\gamma,n}^{cl} \leq L_{\gamma,n}.$$

The precise value of $L_{\gamma,n}$ is unknown in general, but a great deal is known in special cases. We refer to [Hundertmark *et al.* (2000)] for an up to date account.

The inequality (1.9.1) remains valid if a magnetic potential $\mathbf{A} = (A_1, \cdots, A_n), A_j \in L_{\text{loc}}^2(\mathbb{R}^n)$ for $j = 1, \cdots, n$, is introduced. That is, with

$$S(\mathbf{A}) := (i\nabla + \mathbf{A})^2 - V = \sum_{j=1}^n (i\partial_j + A_j)^2 - V$$

we have

$$\sum_{n \in \mathbb{N}} |\lambda_j(S(\mathbf{A}))|^\gamma \leq L_{\gamma,n} \int_{\mathbb{R}^n} V_+(\mathbf{x})^{\gamma+n/2} dx, \tag{1.9.3}$$

Chapter 2

Operators

2.1 The Dirac operator

We denote by $C_0^\infty(\mathbb{R}^3, \mathbb{C}^4)$ the set of $\mathbb{C}^4$-valued functions whose components are in $C_0^\infty(\mathbb{R}^3)$. The free Dirac operator, describing the motion of a relativistic electron or positron without external forces, is the unique self-adjoint extension of the symmetric operator defined on $C_0^\infty(\mathbb{R}^3, \mathbb{C}^4)$ by

$$\mathbb{D}_0'(m, c) := -c\boldsymbol{\alpha} \cdot (i\hbar \boldsymbol{\nabla}) + mc^2 \beta$$

$$= \frac{c\hbar}{i} \sum_{j=1}^{3} \alpha_j \partial_j + mc^2 \beta, \quad \partial_j := \frac{\partial_j}{\partial x_j}, \tag{2.1.1}$$

where $\boldsymbol{\nabla} = (\partial_1, \partial_2, \partial_3)$ is the gradient, $2\pi\hbar$ is Planck's constant, c the velocity of light, m the electron mass and the $\alpha_1, \alpha_2, \alpha_3, \beta$ are Hermitian 4×4 matrices which, on setting $\beta = \alpha_4$, and denoting by I_4 the 4×4 identity matrix, satisfy

$$\alpha_j \alpha_k + \alpha_k \alpha_j = 2\delta_{jk} I_4, \quad j, k = 1, 2, 3, 4, \quad \alpha_4 = \beta. \tag{2.1.2}$$

The operator (2.1.1) acts on $\mathbb{C}^4$-valued functions, which are called *spinors*, the derivatives ∂_j acting on each component of the spinor. We simplify notation by replacing $\mathbf{x}$ by $(\hbar/mc)\mathbf{x}$ and consider

$$\mathbb{D}_0' := \boldsymbol{\alpha} \cdot (-i\boldsymbol{\nabla}) + \beta. \tag{2.1.3}$$

The α_js and β are the *Dirac matrices*, which in the standard representation are given by

$$\alpha_j = \begin{pmatrix} 0_2 & \sigma_j \\ \sigma_j & 0_2 \end{pmatrix}, \quad j = 1, 2, 3, \quad \beta = \begin{pmatrix} I_2 & 0_2 \\ 0_2 & -I_2 \end{pmatrix}, \tag{2.1.4}$$

where the σ_j are the *Pauli matrices*

$$\sigma_1 = \begin{pmatrix} 0 & 1 \\ 1 & 0 \end{pmatrix}, \quad \sigma_2 = \begin{pmatrix} 0 & -i \\ i & 0 \end{pmatrix}, \quad \sigma_3 = \begin{pmatrix} 1 & 0 \\ 0 & -1 \end{pmatrix}. \tag{2.1.5}$$

In $\mathbb{R}^2$ we have

$$\mathbb{D}_0' = -i\sigma_1 \partial_1 - i\sigma_2 \partial_2 + \sigma_3. \tag{2.1.6}$$

To analyse the free Dirac operator we use the Fourier transform

$$(\mathbb{F}u)(\mathbf{p}) := \frac{1}{(2\pi)^{n/2}} \int_{\mathbb{R}^n} e^{-i\mathbf{p}\cdot\mathbf{x}} u(\mathbf{x}) d\mathbf{x}, \tag{2.1.7}$$

on $\mathbb{C}^4$-valued functions u. It follows from property (ii) of the Fourier transform in Section 1.5, that for $v \in C_0^\infty(\mathbb{R}^3, \mathbb{C}^4)$ and $\hat{v} := \mathbb{F}v$,

$$\{\mathbb{F}D_0'\mathbb{F}^{-1}\}\hat{v}(\mathbf{p}) = (\mathbb{M}_0\hat{v})(\mathbf{p}) \tag{2.1.8}$$

where $\mathbb{M}_0$ is the matrix multiplication operator

$$(\mathbb{M}_0\hat{v})(\mathbf{p}) := \{\boldsymbol{\alpha}.\mathbf{p} + \beta\}\hat{v}(\mathbf{p}). \tag{2.1.9}$$

The matrix $\mathbb{M}_0(\mathbf{p}) = \boldsymbol{\alpha}.\mathbf{p} + \beta$ has two double eigenvalues $\pm e(p)$, where $e(p) = \sqrt{p^2 + 1}$, $p = |\mathbf{p}|$, and is diagonalised by the unitary matrix

$$U(\mathbf{p}) := u_+(p)I_4 + u_-(p)\frac{(\boldsymbol{\alpha} \cdot \mathbf{p})}{p}\beta,$$

where

$$u_\pm(\mathbf{p}) = \frac{1}{\sqrt{2}}\sqrt{1 \pm 1/e(p)}.$$

We have

$$U(\mathbf{p})\mathbb{M}_0(\mathbf{p})U^{-1}(\mathbf{p}) = e(p)\beta,$$

with

$$U^{-1}(\mathbf{p}) = u_+(p)I_4 - u_-(p)\frac{(\boldsymbol{\alpha} \cdot \mathbf{p})}{p}\beta.$$

We denote the standard inner product and norm on $\mathbb{C}^4$ by $\langle \cdot, \cdot \rangle$ and $|\cdot|$ respectively, namely, for $\mathbf{a} = (a_1, \cdots, a_4), \mathbf{b} = (b_1, \cdots, b_4) \in \mathbb{C}^4$,

$$\langle \mathbf{a}, \mathbf{b} \rangle = \sum_{j=1}^4 a_j \bar{b}_j, \quad |\mathbf{a}| = \langle \mathbf{a}, \mathbf{a} \rangle^{1/2}.$$

Since $\mathbb{M}_0(\mathbf{p})^2 = (p^2 + 1)I_4$, it follows that

$$\begin{aligned}
|\mathbb{M}_0\hat{v}(\mathbf{p})|^2 &= \langle \mathbb{M}_0(\mathbf{p})\hat{v}(\mathbf{p}), \mathbb{M}_0(\mathbf{p})\hat{v}(\mathbf{p}) \rangle \\
&= \langle ((|\mathbf{p}|^2 + 1)\hat{v}(\mathbf{p}), \hat{v}(\mathbf{p}) \rangle \\
&= (|\mathbf{p}|^2 + 1)|\hat{v}(\mathbf{p})|^2.
\end{aligned}$$

The space of $\mathbb{C}^4$-valued functions u whose components lie in $L^2(\mathbb{R}^3)$ is denoted by $L^2(\mathbb{R}^3, \mathbb{C}^4)$ and

$$\|u\|_{L^2(\mathbb{R}^3, \mathbb{C}^4)} := \||u|\|_{L^2(\mathbb{R}^3)}.$$

Hence, the domain of the matrix multiplication operator $\mathbb{M}_0$ is given by

$$\begin{aligned}
\mathcal{D}(\mathbb{M}_0) &= \{\hat{v} : \hat{v}, \mathbb{M}_0\hat{v} \in L^2(\mathbb{R}^3, \mathbb{C}^4)\} \\
&= \{\hat{v} : \int_{\mathbb{R}^3} (p^2 + 1)|\hat{v}(\mathbf{p})|^2 d\mathbf{p} < \infty\}.
\end{aligned}$$

As $\mathbb{M}_0$ is unitarily equivalent to the operator of multiplication by $e(p)\beta$, it is self-adjoint and its spectrum and essential spectrum coincide with the set of values attained by $\pm e(p)$ as p ranges over $\mathbb{R}_+$, namely $\mathbb{R}\setminus(-1,1)$. Furthermore its spectrum is purely absolutely continuous. The free Dirac operator $\mathbb{D}_0$ is defined by

$$\mathbb{D}_0 = \mathbb{F}^{-1}\mathbb{M}_0\mathbb{F}. \tag{2.1.10}$$

Since $\mathbb{F}$ is a unitary operator, $\mathbb{D}_0$ is self-adjoint in $L^2(\mathbb{R}^3,\mathbb{C}^4)$ and its spectral properties are identical to those of $\mathbb{M}_0$. Thus its spectrum is purely absolutely continuous and coincides with $\mathbb{R}\setminus(-1,1)$.

In view of (2.1.8), (1.5.5) and the Parseval formula (1.5.7), we have for $\hat{v} \in \mathcal{D}(\mathbb{M}_0)$ and $v = (v_1,\cdots,v_4)$,

$$\begin{aligned}
\|\mathbb{D}_0 v\|^2 &= \|\mathbb{M}_0\hat{v}\|^2 \\
&= \int_{\mathbb{R}^n} (|\mathbf{p}|^2 + 1)\, |\hat{v}(\mathbf{p})|^2 d\mathbf{p} \\
&= \sum_{k=1}^{4} \int_{\mathbb{R}^n} \left(|\nabla v_k(\mathbf{x})|^2 + |v_k(\mathbf{x})|^2\right) d\mathbf{x} \\
&=: \|v\|_{1,2}^2.
\end{aligned} \tag{2.1.11}$$

The norm $\|\cdot\|_{1,2}$ in (2.1.11) is the norm of the Sobolev space $H^1(\mathbb{R}^3,\mathbb{C}^4)$ of 4-spinors (i.e. $\mathbb{C}^4$-valued functions) $v = (v_1,\cdots,v_4) \in L^2(\mathbb{R}^3,\mathbb{C}^4)$ whose components have weak derivatives in $L^2(\mathbb{R}^3)$ and $\|v\|_{1,2}^2 < \infty$; in other words, each component lies in the space $H^1(\mathbb{R}^n)$. Thus $\mathbb{D}_0$ has domain

$$\mathcal{D}(\mathbb{D}_0) = H^1(\mathbb{R}^3,\mathbb{C}^4). \tag{2.1.12}$$

Since $C_0^\infty(\mathbb{R}^3,\mathbb{C}^4)$ is a dense subspace of $H^1(\mathbb{R}^3,\mathbb{C}^4)$, it follows from (2.1.12) that the operator $\mathbb{D}_0'$ in (2.1.3) on $C_0^\infty(\mathbb{R}^3,\mathbb{C}^4)$ is essentially self-adjoint and $\mathbb{D}_0$ is its closure.

For future reference, we also note that

$$\begin{aligned}
\|\mathbb{D}_0^2 u\|^2 &= \int_{\mathbb{R}^3} |\mathbb{M}_0^2 \hat{u}(\mathbf{p})|^2 d\mathbf{p} \\
&= \int_{\mathbb{R}^3} |[1+|\mathbf{p}|^2]\hat{u}(\mathbf{p})|^2 d\mathbf{p} = \int_{\mathbb{R}^3} |(1-\Delta)u(\mathbf{x})|^2 d\mathbf{x} \tag{2.1.13} \\
&\approx \|u\|_{H^2(\mathbb{R}^3,\mathbb{C}^4)}^2. \tag{2.1.14}
\end{aligned}$$

2.1.1 *Partial wave decomposition*

We shall show in this section that, when expressed in spherical polar co-ordinates, $\mathbb{D}_0$ is unitarily equivalent to the direct sum of first-order radial operators in $\mathbb{R}_+$. This representation will subsequently be used to analyse the operators $\mathbb{D}_0 + V$ with a spherically symmetric scalar potential V.

In the spherical polar co-ordinates $\mathbf{x} = (x_1,x_2,x_3) = (r,\theta,\varphi)$, where

$$x_1 = r\sin\theta\cos\varphi, \quad x_2 = r\sin\theta\sin\varphi, \quad x_3 = r\cos\theta,$$

$$r = |\mathbf{x}| \in (0,\infty), \quad \theta = \cos^{-1}(x_3/|\mathbf{x}|) \in [0,\pi), \quad \varphi \in [-\pi,\pi), \tag{2.1.15}$$

the following are orthonormal vectors in the directions of the polar co-ordinate lines:

$$\mathbf{e}_r = (\sin\theta\cos\varphi, \sin\theta\sin\varphi, \cos\theta) = \mathbf{x}/r,$$

$$\mathbf{e}_\theta = (\cos\theta\cos\varphi, \cos\theta\sin\varphi, -\sin\theta) = \frac{\partial}{\partial\theta}\mathbf{e}_r$$

$$\mathbf{e}_\varphi = (-\sin\varphi, \cos\varphi, 0) = \frac{1}{\sin\theta}\frac{\partial}{\partial\varphi}\mathbf{e}_r. \tag{2.1.16}$$

Thus $\mathbf{e}_r \times \mathbf{e}_\theta = \mathbf{e}_\varphi$, $\mathbf{e}_\theta \times \mathbf{e}_\varphi = \mathbf{e}_r$, $\mathbf{e}_\varphi \times \mathbf{e}_r = \mathbf{e}_\theta$, and it can be easily shown that

$$\nabla = \mathbf{e}_r\frac{\partial}{\partial r} + \mathbf{e}_\theta\frac{1}{r}\frac{\partial}{\partial\theta} + \mathbf{e}_\varphi\frac{1}{r\sin\theta}\frac{\partial}{\partial\varphi}. \tag{2.1.17}$$

The discussion that follows is brief, in that some details which require verification are omitted. It is similar to the treatment in [Thaller (1992)], Section 4.6, except that there are differences in the notation. We begin by introducing some of the operators which play a leading role.

(i) $-i\nabla$ is the *momentum* operator and

$$\mathbf{L} := \mathbf{x} \times (-i\nabla)$$

$$= i\left(\mathbf{e}_\theta\frac{1}{\sin\theta}\frac{\partial}{\partial\varphi} - \mathbf{e}_\varphi\frac{\partial}{\partial\theta}\right)$$

$$=: (L_1, L_2, L_3) \tag{2.1.18}$$

is the *orbital angular momentum*. Thus the expressions for $\mathbf{L}, L_1, L_2, L_3$ and $L^2 = L_1^2 + L_2^2 + L_3^2$ depend only on the angular variables θ and ϕ. It is readily shown that

$$L^2 = -\frac{1}{\sin\theta}\frac{\partial}{\partial\theta}\left(\sin\theta\frac{\partial}{\partial\theta}\right) - \frac{1}{\sin^2\theta}\frac{\partial^2}{\partial\varphi^2}. \tag{2.1.19}$$

(ii) The *spin angular momentum* $\mathbf{S}$ and *total angular momentum* $\mathbf{J}$ are respectively

$$\mathbf{S} := -\frac{i}{4}\alpha \times \alpha \tag{2.1.20}$$

$$= (1/2)\left\{\begin{pmatrix}\sigma_1 & 0_2 \\ 0_2 & \sigma_1\end{pmatrix} + \begin{pmatrix}\sigma_2 & 0_2 \\ 0_2 & \sigma_2\end{pmatrix} + \begin{pmatrix}\sigma_3 & 0_2 \\ 0_2 & \sigma_3\end{pmatrix}\right\}$$

$$=: (1/2)\begin{pmatrix}\sigma & 0_2 \\ 0_2 & \sigma\end{pmatrix}, \quad \sigma := (\sigma_1, \sigma_2, \sigma_3), \tag{2.1.21}$$

$$\mathbf{J} = \mathbf{L} + \mathbf{S} =: (J_1, J_2, J_3). \tag{2.1.22}$$

We may also consider $\mathbf{J}$ and its components J_1, J_2, J_3, as acting on $\mathbb{C}^2$-valued functions by setting

$$\mathbf{J} = \mathbf{L} + (1/2)\sigma. \tag{2.1.23}$$

On identifying $L^2(\mathbb{R}^3, \mathbb{C}^k)$ with the tensor product $L^2(\mathbb{R}_+; r^2 dr) \otimes L^2(\mathbb{S}^2, \mathbb{C}^k)$ we may view the angular momentum operators $\mathbf{L}, \mathbf{J}$, and their components, as acting in $L^2(\mathbb{S}^2, \mathbb{C}^k)$ with $k = 2$ or 4: $\mathbf{L}, L_1, L_2, L_3$ apply to each component of a $\mathbb{C}^k$-valued spinor. They are essentially self-adjoint on $C_0^\infty(\mathbb{S}^2, \mathbb{C}^k)$, their unique self-adjoint extensions being denoted by the same letter in each case. In $L^2(\mathbb{S}^2, \mathbb{C}^2)$, L^2, J^2 and J_3 have a discrete spectrum with common eigenvectors $\Omega_{l,m,s} = \Omega_{l,m,s}(\theta, \varphi)$: we have

$$L^2\Omega_{l,m,s} = l(l+1)\Omega_{l,m,s}$$
$$J^2\Omega_{l,m,s} = (l+s)(l+s+1)\Omega_{l,m,s}$$
$$J_3\Omega_{l,m,s} = m\Omega_{l,m,s}, \tag{2.1.24}$$

where $l \in \mathbb{N}_0$, $s = \pm 1/2$ and $m \in \{-l-1/2, \cdots, l+1/2\}$: l is referred to as the angular momentum channel, and s denotes the spin of the particle. The $\mathbb{C}^2$-valued functions $\Omega_{l,m,s}$ are called *spherical spinors* and are given by

$$\begin{pmatrix} \sqrt{\frac{l+s+m}{2(l+s)}} Y_{l,m-1/2}(\theta, \varphi) \\ \sqrt{\frac{l+s-m}{2(l+s)}} Y_{l,m+1/2}(\theta, \varphi) \end{pmatrix} \quad \text{if} \quad s = 1/2 \tag{2.1.25}$$

and

$$\begin{pmatrix} \sqrt{\frac{l+s-m+1}{2(l+s)+2}} Y_{l,m-1/2}(\theta, \varphi) \\ -\sqrt{\frac{l+s+m+1}{2(l+s)+2}} Y_{l,m+1/2}(\theta, \varphi) \end{pmatrix} \quad \text{if} \quad s = -1/2, \tag{2.1.26}$$

where the $Y_{l,k}(\theta, \varphi)$ are normalised spherical harmonics on $\mathbb{S}^2$, given in terms of the *associated Legendre polynomials*

$$P_l^k(x) = \frac{(-1)^k}{2^l l!}(1-x^2)^{k/2} \frac{d^{k+l}}{dx^{k+l}}(x^2-1)^l$$

by

$$Y_{l,k}(\theta, \varphi) = \sqrt{\frac{(2l+1)}{4\pi}\frac{(l-k)!}{(l+k)!}} e^{ik\varphi} P_l^k(\cos\theta), \quad k \geq 0,$$
$$Y_{l,-k}(\theta, \varphi) = (-1)^k \overline{Y_{l,k}(\theta, \varphi)}.$$

We adopt the convention that $Y_{l,k} = 0$ if $|k| > l$, and denote the set of admissible indices in (2.1.25) and (2.1.26) by $\mathfrak{J}$:

$$\mathfrak{J} := \{(l,m,s) : l \in \mathbb{N}_0, s = \pm 1/2, m = -l-1/2, \cdots, l+1/2, \Omega_{l,m,s} \neq 0\}; \tag{2.1.27}$$

note, in particular, that $l \geq 1$ when $s = -1/2$ since $\Omega_{l,m,s} = 0$ otherwise. The results in (2.1.24)–(2.1.26) can be verified by direct calculation from (2.1.18) and (2.1.22) on using the known facts that the $Y_{l,m}$ are eigenfunctions of L^2 and L_3 in $L^2(\mathbb{S}^2)$ corresponding to eigenvalues $l(l+1)$ and m, respectively, and that

$$(L_1 + iL_2)Y_{l,m} = \sqrt{(l+m+1)(l-m)}Y_{l,m+1}$$
$$(L_1 - iL_2)Y_{l,m} = \sqrt{(l-m+1)(l+m)}Y_{l,m-1}.$$

The 2-spinors $\Omega_{l,m,s}$, $(l,m,s) \in \mathfrak{J}$, constitute an orthonormal basis of $L^2(\mathbb{S}^2, \mathbb{C}^2)$, as a consequence of the well-known fact that the $Y_{l,k}$ form an orthonormal basis of $L^2(\mathbb{S}^2)$. Note that $\Omega_{l,m,s}$ and $\Omega_{l+2s,m,-s}$ are both eigenvectors of J^2 and J_3 corresponding to the eigenvalues $(l+s)(l+s+1)$ and m respectively. On using reduction formulae for the functions $Y_{l,m}$, it can be shown that

$$(\boldsymbol{\sigma} \cdot \mathbf{e}_r)\Omega_{l,m,s} = \Omega_{l+2s,m,-s}. \tag{2.1.28}$$

Set

$$\Phi^+_{l,m,s} = \begin{pmatrix} i\Omega_{l,m,s} \\ 0 \end{pmatrix}, \quad \Phi^-_{l,m,s} = \begin{pmatrix} 0 \\ \Omega_{l,m,s} \end{pmatrix}. \tag{2.1.29}$$

Then, from (2.1.4) and (2.1.28),

$$i(\boldsymbol{\alpha} \cdot \mathbf{e}_r)\Phi^+_{l,m,s} = -\Phi^-_{l+2s,m,-s}$$
$$i(\boldsymbol{\alpha} \cdot \mathbf{e}_r)\Phi^-_{l+2s,m,-s} = \Phi^+_{l,m,s}. \tag{2.1.30}$$

Also,

$$\beta\Phi^+_{l,m,s} = \Phi^+_{l,m,s}, \quad \beta\Phi^-_{l+2s,m,-s} = -\Phi^-_{l+2s,m,-s}. \tag{2.1.31}$$

Following [Thaller (1992)], Section 4.6.3, we now express $\mathbb{D}'_0$ as

$$\mathbb{D}'_0 = -i(\boldsymbol{\alpha} \cdot \mathbf{e}_r)\left(\frac{\partial}{\partial r} + \frac{1}{r} - \frac{1}{r}\beta K\right) + \beta \tag{2.1.32}$$

where K is the *spin–orbit* operator

$$K := \beta(2\mathbf{S} \cdot \mathbf{L} + 1), \tag{2.1.33}$$

which, by (2.1.22), satisfies

$$K := \beta(J^2 - L^2 + \frac{1}{4}). \tag{2.1.34}$$

To obtain (2.1.32), the first step is to verify the identity

$$(\boldsymbol{\sigma} \cdot \mathbf{a})(\boldsymbol{\sigma} \cdot \mathbf{b}) = \mathbf{a} \cdot \mathbf{b} + i\boldsymbol{\sigma} \cdot (\mathbf{a} \times \mathbf{b}) \tag{2.1.35}$$

for all vectors $\mathbf{a}, \mathbf{b} \in \mathbb{C}^3$, and that this yields

$$(\boldsymbol{\alpha} \cdot \mathbf{a})(2\mathbf{S} \cdot \mathbf{b}) = i\gamma_5(\mathbf{a} \cdot \mathbf{b}) - i\boldsymbol{\alpha} \cdot (\mathbf{a} \times \mathbf{b}), \tag{2.1.36}$$

where $\gamma_5 = -i\alpha_1\alpha_2\alpha_3$. From (2.1.17) and (2.1.18),

$$-i\boldsymbol{\nabla} = -i\mathbf{e}_r\frac{\partial}{\partial r} - \frac{1}{r}(\mathbf{e}_r \times \mathbf{L})$$

and so, from (2.1.36),

$$-i\boldsymbol{\alpha} \cdot \boldsymbol{\nabla} = -i(\boldsymbol{\alpha} \cdot \mathbf{e}_r)\frac{\partial}{\partial r} - \frac{1}{r}[\boldsymbol{\alpha} \cdot (\mathbf{e}_r \times \mathbf{L})]$$
$$= -i(\boldsymbol{\alpha} \cdot \mathbf{e}_r)\frac{\partial}{\partial r} + \frac{i}{r}(\boldsymbol{\alpha} \cdot \mathbf{e}_r)(2\mathbf{S} \cdot \mathbf{L}),$$

since $\mathbf{e}_r \cdot \mathbf{L} = 0$, and (2.1.32) follows.

The operators J^2, L^2 and K are essentially self-adjoint in $L^2(\mathbb{R}^3 \setminus \{0\}, \mathbb{C}^4)$. It follows from (2.1.24) that

$$(J^2 - L^2 + \frac{1}{4})\Omega_{l,m,s} = (2sl + s + \frac{1}{2})\Omega_{l,m,s} \tag{2.1.37}$$

and

$$(J^2 - L^2 + \frac{1}{4})\Omega_{l+2s,m,-s} = -(2sl + s + \frac{1}{2})\Omega_{l+2s,m,-s}, \tag{2.1.38}$$

whence, with $\kappa_{l,s} = 2sl + s + 1/2$,

$$\beta K \Phi^+_{l,m,s} = \kappa_{l,s}\Phi^+_{l,m,s}$$
$$\beta K \Phi^-_{l+2s,m,-s} = -\kappa_{l,s}\Phi^-_{l+2s,m,-s}. \tag{2.1.39}$$

There is a natural unitary isomorphism U between the Hilbert spaces $L^2(\mathbb{R}^3, \mathbb{C}^4)$ and $L^2(\mathbb{R}_+, r^2 dr) \otimes L^2(\mathbb{S}^2, \mathbb{C}^4)$, and $L^2(\mathbb{S}^2, \mathbb{C}^4)$ is the orthogonal sum of the 2-dimensional spaces $\mathcal{H}_{l,m,s}$ spanned by $\Phi^+_{l,m,s}$ and $\Phi^-_{l+2s,m,-s}$. Thus

$$UL^2(\mathbb{R}^3, \mathbb{C}^4) = \bigoplus_{l\in\mathbb{N}_0} \bigoplus_{m=\pm(l+1/2)} \bigoplus_{s=\pm 1/2} L^2(\mathbb{R}_+, r^2 dr) \otimes \mathcal{H}_{l,m,s} \tag{2.1.40}$$

and any $f \in L^2(\mathbb{R}^3, \mathbb{C}^4)$ has the representation

$$(Uf)(r, \theta, \varphi) = \sum_{l\in\mathbb{N}_0} \sum_{m=\pm(l+1/2)} \sum_{s=\pm 1/2} \left\{ r^{-1} f^+_{l,m,s}(r)\Phi^+_{l,m,s}(\theta, \varphi) \right.$$

$$\left. + r^{-1} f^-_{l+2s,m,-s}(r)\Phi^-_{l+2s,m,-s}(\theta, \varphi) \right\}, \tag{2.1.41}$$

where $f^+_{l,m,s}, f^-_{l+2s,m,-s} \in L^2(\mathbb{R}_+)$. If $g \in L^2(\mathbb{R}^3, \mathbb{C}^4)$ is similarly represented with coefficients $g^+_{l,m,s}, g^-_{l+2s,m,-s}$, then, in view of the orthonormality of the $\{\Phi^+_{l,m,s}, \Phi^-_{l+2s,m,-s}\}$, we have

$$(f, g) = \sum_{(l,m,s)\in\mathfrak{I}} \int_0^\infty \left(f^+_{l,m,s}\overline{g^+}_{l,m,s} + f^-_{l+2s,m,-s}\overline{g^+}_{l+2s,m,-s} \right) dr.$$

Also, $f \in C_0^\infty(\mathbb{R}^3, \mathbb{C}^4)$ if and only if the coefficients $f^+_{l,m,s}, f^-_{l+2s,m,-s}$ are in $C_0^\infty(\mathbb{R}_+)$.

From (2.1.30), (2.1.31), (2.1.32) and (2.1.39), it follows that for $f \in C_0^\infty(\mathbb{R}^3, \mathbb{C}^4)$,

$$\mathbb{D}'_0 Uf = \sum_{(l,m,s)\in\mathfrak{I}} \left\{ -i(\boldsymbol{\alpha}\cdot\boldsymbol{\epsilon}_r)\left(\frac{1}{r}\frac{d}{dr} - \frac{\kappa_{l,s}}{r^2}\right) f^+_{l,m,s}\Phi^+_{l,m,s} + \frac{1}{r}f^+_{l,m,s}\beta\Phi^+_{l,m,s} \right.$$

$$\left. - i(\boldsymbol{\alpha}\cdot\boldsymbol{\epsilon}_r)\left(\frac{1}{r}\frac{d}{dr} + \frac{\kappa_{l,s}}{r^2}\right) f^-_{l+2s,m,-s}\Phi^-_{l+2s,m,-s} + \frac{1}{r}f^-_{l+2s,m,-s}\beta\Phi^-_{l+2s,m,-s} \right\}$$

$$= \sum_{(l,m,s)\in\mathfrak{I}} \left\{ \left(\frac{1}{r}\frac{d}{dr} - \frac{\kappa_{l,s}}{r^2}\right) f^+_{l,m,s}\Phi^-_{l+2s,m,-s} + \frac{1}{r}f^+_{l,m,s}\Phi^+_{l,m,s} \right.$$

$$\left. - \left(\frac{1}{r}\frac{d}{dr} + \frac{\kappa_{l,s}}{r^2}\right) f^-_{l+2s,m,-s}\Phi^+_{l,m,s} - \frac{1}{r}f^-_{l+2s,m,-s}\Phi^-_{l+2s,m,-s} \right\}$$

and so

$$(U^{-1}\mathbb{D}'_0 Uf, g) = \sum_{(l,m,s)\in\mathfrak{I}} \int_0^\infty \left\{ \overline{g^+}_{l,m,s}(r)\left([-\frac{d}{dr} - \frac{\kappa_{l,s}}{r}]f^-_{l+2s,m,-s}(r) + f^+_{l,m,s}(r)\right) \right.$$

$$\left. + \overline{g^-}_{l+2s,m,-s}(r)\left([\frac{d}{dr} - \frac{\kappa_{l,s}}{r}]f^+_{l,m,s}(r) - f^-_{l+2s,m,-s}(r)\right) \right\}$$

$$= \sum_{(l,m,s)\in\mathfrak{I}} (\mathbb{D}'_{0;l,s}F_{l,m,s}, G_{l,m,s})_{L^2(\mathbb{R}_+, \mathbb{C}^2)}, \tag{2.1.42}$$

where $\mathbb{D}'_{0;l,s}$ is defined on $C_0^\infty(\mathbb{R}_+, \mathbb{C}^2)$ by

$$\mathbb{D}'_{0;l,s} := \begin{pmatrix} 1 & -\frac{d}{dr} - \frac{\kappa_{l,s}}{r} \\ \frac{d}{dr} - \frac{\kappa_{l,s}}{r} & -1 \end{pmatrix}$$

$$= -i\sigma_2 \frac{d}{dr} - \frac{\kappa_{l,s}}{r}\sigma_1 + \sigma_3 \qquad (2.1.43)$$

and

$$F_{l,m,s} = \begin{pmatrix} f^+_{l,m,s} \\ f^-_{l+2s,m,-s} \end{pmatrix}, \quad G_{l,m,s} = \begin{pmatrix} g^+_{l,m,s} \\ g^-_{l+2s,m,-s} \end{pmatrix}. \qquad (2.1.44)$$

We therefore have

$$U^{-1}\mathbb{D}'_0 U = \bigoplus_{l\in\mathbb{N}_0} \bigoplus_{m=\pm(l+1/2)} \bigoplus_{s=\pm 1/2} \mathbb{D}'_{0;l,s}.$$

Also

$$\|U^{-1}[\mathbb{D}'_0 - \lambda I]Uf\|^2 = \sum_{(l,m,s)\in\mathfrak{I}} \|(\mathbb{D}'_{0;l,s} - \lambda I_2)F_{l,m,s}\|^2_{L^2(\mathbb{R}_+,\mathbb{C}^2)}.$$

It will follow from Theorem 2.1.1 below that each $\mathbb{D}'_{0;l,s}$ is essentially self-adjoint in $L^2(\mathbb{R}_+, \mathbb{C}^2)$. Its self-adjoint extension $\mathbb{D}_{0;l,s}$, has domain

$$\mathcal{D}(\mathbb{D}_{0;l,s}) = \{u \in AC_{loc}(\mathbb{R}_+, \mathbb{C}^2) : u, \left(-i\sigma_2 \frac{d}{dr} - \frac{\kappa_{l,s}}{r}\sigma_1\right)u \in L^2(\mathbb{R}_+, \mathbb{C}^2)\},$$

$$(2.1.45)$$

where $AC_{loc}(\mathbb{R}_+, \mathbb{C}^2)$ denotes the set of 2-spinors whose components are absolutely continuous functions on $\mathbb{R}_+$. Moreover

$$U^{-1}\mathbb{D}_0 U = \bigoplus_{l\in\mathbb{N}_0} \bigoplus_{m=\pm(l+1/2)} \bigoplus_{s=\pm 1/2} \mathbb{D}_{0;l,s}. \qquad (2.1.46)$$

2.1.2   *Spherically symmetric electric potentials*

Let $V \in L^2_{loc}(\mathbb{R}^3)$ be such that $V(\mathbf{x}) = V(r), r = |\mathbf{x}|$. Then we have

$$\mathbb{D}'_0 + V \cong \bigoplus_{l\in\mathbb{Z}} \bigoplus_{m=\pm(l+1/2)} \bigoplus_{s=\pm 1/2} (\mathbb{D}'_{0;l,s} + V),$$

where $\cong$ indicates the unitary equivalence U of the last section. The operators $\mathbb{D}'_{0;l,s} + V$ are called the *Dirac partial wave operators*.

Theorem 2.1.1. *The operator $\mathbb{D}'_0 + V$ is essentially self-adjoint in $L^2(\mathbb{R}^3, \mathbb{C}^4)$ if and only if $\mathbb{D}'_{0;l,s} + V, l \in \mathbb{N}_0, s = \pm 1/2$, are essentially self-adjoint in $L^2(\mathbb{R}_+, \mathbb{C}^2)$. If $\mathbb{D}, \mathbb{D}_{l,s}$ denote the unique self-adjoint extensions of $\mathbb{D}'_0 + V, \mathbb{D}'_{0;l,s} + V$, respectively, then for $\lambda \in \mathbb{C} \setminus \mathbb{R}$,*

$$(\mathbb{D} - \lambda I)^{-1} \cong \bigoplus_{l\in\mathbb{N}_0} \bigoplus_{m=\pm(l+1/2)} \bigoplus_{s=\pm 1/2} (\mathbb{D}_{l,s} - \lambda I_2)^{-1} \qquad (2.1.47)$$

and

$$\sigma(\mathbb{D}) = \bigcup_{l \in \mathbb{N}_0, s = \pm 1/2} \sigma(\mathbb{D}_{l,s}). \tag{2.1.48}$$

Also

$$\|U^{-1}[\mathbb{D} - \lambda I]Uf\|^2 = \sum_{(l,m,s) \in \mathfrak{I}} \|(\mathbb{D}_{l,s} - \lambda I_2)F_{l,m,s}\|^2_{L^2(\mathbb{R}_+, \mathbb{C}^2)}, \tag{2.1.49}$$

in the notation of (2.1.41) and (2.1.44).

Proof. We shall use the criterion that a symmetric operator T in a Hilbert space $\mathcal{H}$ is essentially self-adjoint if and only if the range $\mathcal{R}(T - \lambda I)$ of $T - \lambda I$ is dense in $\mathcal{H}$ for some, and hence all, $\lambda \in \mathbb{C} \setminus \mathbb{R}$, see [Edmunds and Evans (1987)], Theorem III.4.2. Suppose that $\mathbb{D}'_{0;l,s} + V$ is not essentially self-adjoint for some l, s, say l', s'. Then, there exists a non-trivial $v \in L^2(\mathbb{R}_+, \mathbb{C}^2)$, such that

$$([\mathbb{D}'_{0;l',s'} + V]u, v)_{L^2(\mathbb{R}_+, \mathbb{C}^2)} = 0, \quad \text{for all } u \in C_0^\infty(\mathbb{R}_+, \mathbb{C}^2). \tag{2.1.50}$$

With $Ug = v^+ \Phi^+_{l',m',s'} + v^- \Phi_{l'+2s',m',-s'}$, where $v^\pm(r) := r \int_{\mathbb{S}^2} v(r, \omega) \Phi^\pm_{l',m',s'}(\omega) d\omega$, it follows from (2.1.42) that

$$([\mathbb{D}'_0 + V - \lambda I]Uf, Ug)_{L^2(\mathbb{R}^3, \mathbb{C}^4)} = 0 \quad \text{for all } f \in C_0^\infty(\mathbb{R}^3, \mathbb{C}^4). \tag{2.1.51}$$

Also,

$$\|Ug\|^2_{L^2(\mathbb{R}^3, \mathbb{C}^4)} = \|v^+\|^2_{L^2(0,\infty)} + \|v^-\|^2_{L^2(0,\infty)} \neq 0.$$

Hence $U^{-1}(\mathbb{D}'_0 + V)U$ is not essentially self-adjoint and, so, neither is $\mathbb{D}'_0 + V$.

Conversely, suppose there exists a non-trivial $g \in L^2(\mathbb{R}^3, \mathbb{C}^4)$ such that (2.1.51) holds. Then, for some $l', m', s', v = g^+ \Phi^+_{l',m',s'} + g^- \Phi_{l'+2s',m',-s'} \neq 0$ and, by (2.1.44), it follows that (2.1.50) holds. Hence $\mathbb{D}'_{0;l',s'} + V$ is not essentially self-adjoint. It follows that $\mathbb{D}$ is unitarily equivalent to the orthogonal sum of the partial wave operators and this yields (2.1.47) and (2.1.49). Thus a number λ lies in the resolvent set of $\mathbb{D}$ if and only if it lies in the intersection of the resolvent sets of the operators $\mathbb{D}_{l,s}$ and this is equivalent to (2.1.48). $\qquad \square$

The operators $\mathbb{D}'_{0;l',s'} + V$ are typified by an operator T'_0 defined on $C_0^\infty(\mathbb{R}_+, \mathbb{C}^2)$ by $T'_0 f = \tau f$, where

$$\tau := -i\sigma_2 \frac{d}{dr} - \frac{\kappa}{r}\sigma_1 + \sigma_3 + V, \quad \kappa \in \mathbb{R}. \tag{2.1.52}$$

It is densely defined and symmetric in $L^2(\mathbb{R}_+, \mathbb{C}^2)$ and its adjoint is defined by τ in the weak (or distributional) sense on the domain

$$\mathcal{D}^* := \{f : f, \tau f \in L^2(\mathbb{R}_+, \mathbb{C}^2)\}, \tag{2.1.53}$$

that is, for all $\varphi \in C_0^\infty(\mathbb{R}_+, \mathbb{C}^2)$

$$((T'_0)^* f, \varphi) := (f, \tau \varphi).$$

Since $C_0^\infty(\mathbb{R}_+, \mathbb{C}^2) \subset \mathcal{D}^*$, it follows that $(T_0')^*$ is densely defined, has an adjoint and T_0' has a closure $T_0 = ((T_0')^*)^*$; see [Edmunds and Evans (1987)], Theorem III.1.5. Thus $(T_0')^* = T_0^*$ and $T_0 = (T_0^*)^* =: T_0^{**}$. In common with all symmetric operators, T_0 satisfies

$$\|(T_0 - \lambda I)f\| \geq \nu\|f\|,$$

for all f in its domain and all $\lambda := \mu + i\nu \in \mathbb{C}$. It follows that the range $\mathcal{R}(T_0 - \lambda)$ is closed in $L^2(\mathbb{R}_+, \mathbb{C}^2)$ for $\lambda \in \mathbb{C} \setminus \mathbb{R}$ and this implies that for $\lambda \in \mathbb{C} \setminus \mathbb{R}$

$$\mathcal{R}(T_0 - \lambda I)^\perp = \mathcal{N}(T_0^* - \overline{\lambda} I); \tag{2.1.54}$$

here $\perp$ stands for the orthogonal complement and $\mathcal{N}$ the null space (or kernel). Recall the following terminology and facts from Section 1.1 in Chapter 1: the dimension of the space in (2.1.54) is called the *deficiency* of $T_0 - \lambda I$; it is constant for λ in each of the open half-planes $\mathbb{C}_\pm$; the numbers $m_+(T_0), m_-(T_0)$, where $m_\pm(T_0) := \mathrm{nul}(T_0^* \pm iI)$, are the deficiency indices of T_0.

Let $a > 0$, $|V(x)| \leq c_1|x|^{-1} + c_2$, for constants c_1, c_2, and consider the symmetric operator T_+' defined in $\mathcal{H}_+ := L^2((a, \infty), \mathbb{C}^2)$ by $T_+'f := \tau f$ on $C_0^\infty((a, \infty), \mathbb{C}^2)$. The adjoint of T_+' has domain

$$\begin{aligned}
\mathcal{D}_+^* &:= \{f : f, \tau f \in \mathcal{H}_+\} \\
&= \{f : f \in H_{loc}^1((a, \infty), \mathbb{C}^2),\ f, \tau f \in \mathcal{H}_+\} \\
&= \{f : f \in AC_{loc}((a, \infty), \mathbb{C}^2), f, df/dr \in \mathcal{H}_+\}; \tag{2.1.55}
\end{aligned}$$

see [Edmunds and Evans (1987)], Corollary V.3.12. The closure of T_+' is $T_+ = (T_+^*)^*$ and thus, since $T_+ \subseteq T_+^*$, we have that $T_+ \subseteq T_+^*$. The graph norm of T_+' is equivalent to the $H_0^1((a, \infty), \mathbb{C}^2)$ norm and hence the domain $\mathcal{D}_+$ of T_+ coincides with $H_0^1((a, \infty), \mathbb{C}^2)$. Moreover, for $f \in C_0^\infty((a, \infty), \mathbb{C}^2)$,

$$|f^2(x)| = 2\mathrm{Re} \int_a^x \langle \mathrm{f}(t), \mathrm{f}'(t)\rangle \mathrm{dt}$$

$$\leq 2\|f\|\|f'\| \leq \{\|f\|^2 + \|f'\|^2\},$$

from which it follows that $H_0^1((a, \infty), \mathbb{C}^2)$ is continuously embedded in $C([a, \infty), \mathbb{C}^2)$. This implies that

$$\mathcal{D}_+ \subseteq \{f : f \in \mathcal{D}_+^*, f(a) = 0\}. \tag{2.1.56}$$

We shall need Green's identity for $f, g \in \mathcal{D}_+^*$, namely,

$$\int_a^X \{\langle \tau f, g\rangle(t) - \langle f, \tau g\rangle(t)\}dt = [f, g](X) - [f, g](a), \quad f, g \in \mathcal{D}_+^*, \tag{2.1.57}$$

where, with $f = \begin{pmatrix} f_1 \\ f_2 \end{pmatrix}$, $g = \begin{pmatrix} g_1 \\ g_2 \end{pmatrix}$,

$$\begin{aligned}
[f, g](x) &:= -\langle i\sigma_2 f(x), g(x)\rangle \\
&= f_1(x)\overline{g}_2(x) - f_2(x)\overline{g}_1(x). \tag{2.1.58}
\end{aligned}$$

Since $\bar{g}\tau f - f\overline{\tau g} \in L^1(a, \infty)$, by the Cauchy–Schwarz inequality, (2.1.57) implies, in particular, that

$$[f, g](\infty) := \lim_{X \to \infty} [f, g](X) \quad \text{exists.}$$

Analogous to a celebrated result of H. Weyl in [Weyl (1909)] for singular Sturm–Liouville equations, the equation

$$(\tau - \lambda I)f = 0, \quad \lambda \in \mathbb{C} \setminus \mathbb{R} \tag{2.1.59}$$

has at least one solution $f \in \mathcal{H}_+$, i.e.,

$$\int_a^\infty \left(|f_1(r)|^2 + |f_2(r)|^2 \right) dr < \infty.$$

The equation (2.1.59) is said to be in the *limit-point* case at ∞ if there is precisely one solution in $\mathcal{H}_+$ and otherwise in the *limit-circle case*, when two linear independent, and hence all, solutions are in $\mathcal{H}_+$. This classification is independent of $\lambda \in \mathbb{C} \setminus \mathbb{R}$. For the number of solutions of (2.1.59) in $\mathcal{H}_+$ is equal to $\text{nul}(T_+^* - \lambda I) = \text{def}(T_+ - \bar{\lambda}I)$, which, as we noted in Section 1.1, is constant in $\mathbb{C}_+$ and $\mathbb{C}_-$. Furthermore, since $f \in \mathcal{N}(T_+^* - iI)$ implies that $\bar{f} \in \mathcal{N}(T_+^* + iI)$, it follows that the deficiency indices $m_\pm(T_+)$ of T_+ are equal. Hence Weyl's result can be expressed as $m(T_+) \geq 1$, where $m(T_+)$ is the common value of the deficiency indices. Note that $m(T_+) = 0$ would imply that T_+ is self-adjoint, which is impossible from (2.1.56).

From (1.1.1),

$$m(T_+) = \frac{1}{2}\dim\left\{ \mathcal{D}_+^* / \mathcal{D}_+ \right\}. \tag{2.1.60}$$

Hence, $m(T_+) \geq 1$ means that there exist functions $\varphi^{(j)}, j = 1, 2$, in $\mathcal{D}_+^*$, which are linearly independent modulo $\mathcal{D}_+$, that is, $c_1\varphi^{(1)} + c_2\varphi^{(2)} \in \mathcal{D}_+$ implies that $c_1 = c_2 = 0$. When $m(T_+) = 1$, equality holds in (2.1.56) and for any $c \in (a, \infty)$, functions satisfying the following conditions can be defined (see [Edmunds and Evans (1987)], Lemma III.10.4) and form a basis of $\mathcal{D}_+^* / \mathcal{D}_+$:

$$\varphi^{(j)} \in AC([a, c], \mathbb{C}^2), \, \varphi^{(j)}(t) = 0 \text{ for } t \geq c, (j = 1, 2),$$

$$\varphi^{(1)}(a) = \begin{pmatrix} 1 \\ 0 \end{pmatrix}, \, \varphi^{(2)}(a) = \begin{pmatrix} 0 \\ 1 \end{pmatrix}. \tag{2.1.61}$$

If all the solutions of (2.1.59) lie in $\mathcal{H}_+$ for some $\lambda = \lambda_0 \in \mathbb{C}$ it can be shown that this is true for all $\lambda \in \mathbb{C}$, cf. [Edmunds and Evans (1987)], Theorem III.10.10.

A similar analysis holds for the symmetric operator T_-' defined in $\mathcal{H}_- = L^2((0, a), \mathbb{C}^2)$ by $T_-'f = \tau f$ on $C_0^\infty((0, a), \mathbb{C}^2)$. It has a closure T_- whose adjoint T_-^* has domain

$$\mathcal{D}_-^* = \{f : f, \tau f \in \mathcal{H}_-\}$$
$$= \{f : f \in H_{loc}^1((0, a), \mathbb{C}^2), f, \tau f \in \mathcal{H}_-\}$$
$$= \{f : f \in AC_{loc}((0, a), \mathbb{C}^2), f, \tau f \in \mathcal{H}_-\}. \tag{2.1.62}$$

The graph norm of T'_- is equivalent to the $H^1_0((0,a),\mathbb{C}^2)$ norm on account of the Hardy inequality (1.7.5), $H^1_0((0,a),\mathbb{C}^2)$ is continuously embedded in $C((0,a],\mathbb{C}^2)$ and $T_- \subseteq T^*_-$ has domain

$$\mathcal{D}_- \subseteq \{f : f \in \mathcal{D}^*_-, f(a) = 0\}. \tag{2.1.63}$$

The limit $[f,g](0) := \lim_{x\to 0+}[f,g](x)$ exists for all $f,g \in \mathcal{D}^*_-$. The operator T_- has deficiency indices 1 or 2 determined by the number of linearly independent solutions of (2.1.59) in $\mathcal{H}_-$ for $\lambda \in \mathbb{C}\setminus\mathbb{R}$, the number being 2 for all $\lambda \in \mathbb{C}$ if it is so for one. The limit-point/limit-circle classification now relates to the singular point 0.

Lemma 2.1.2. *(i)* $m(T_+) = 1$ *if and only if* $[f,g](\infty) = 0$ *for all* $f,g \in \mathcal{D}^*_+$ *and in this case*

$$\mathcal{D}_+ = \{f : f \in \mathcal{D}^*_+, f(a) = 0\}. \tag{2.1.64}$$

(ii) $m(T_-) = 1$ *if and only if* $[f,g](0) = 0$ *for all* $f,g \in \mathcal{D}^*_-$ *and in this case*

$$\mathcal{D}_- = \{f : f \in \mathcal{D}^*_-, f(a) = 0\}. \tag{2.1.65}$$

Proof.   Suppose that $m(T_+) = 1$. Then $\dim\left(\mathcal{D}^*_+/\mathcal{D}_+\right) = 2$, by (2.1.60), and so the functions $\varphi^{(1)}, \varphi^{(2)}$ defined in (2.1.61) form a basis of $\mathcal{D}^*_+/\mathcal{D}_+$. Consequently, any $f \in \mathcal{D}^*_+$ can be written as

$$f = f_0 + c_1\varphi^{(1)} + c_2\varphi^{(2)}$$

for some $f_0 \in \mathcal{D}_+$ and constants c_1, c_2. Hence

$$[f,g](\infty) = [f_0,g](\infty).$$

But

$$[f_0,g](\infty) = (T_+f_0,g) - (f_0,T^*_+g) = 0$$

and so $[f,g](\infty) = 0$.

  Conversely, suppose $[f,g](\infty) = 0$ for all $f,g \in \mathcal{D}^*_+$. Let T_1 denote the restriction of T^*_+ to (2.1.64). Then, by Green's identity,

$$(T_1\varphi, \psi) = (\varphi, T^*_+\psi), \quad \text{for all } \phi \in \mathcal{D}(T_1),\ \psi \in \mathcal{D}^*_+,$$

whence $T_1 \subseteq T^{**}_+ = T_+$. Since $T_+ \subseteq T_1$ by (2.1.56), (2.1.64) is proved. If $\varphi^{(1)}, \varphi^{(2)}$ are the functions defined in (2.1.61), then $f \in \mathcal{D}^*_+$ if and only if

$$f - f_1(a)\varphi^{(1)} - f_2(a)\varphi^{(2)} \in \mathcal{D}_+.$$

Consequently $\dim\left(\mathcal{D}^*_+/\mathcal{D}_+\right) = 2$ and $m(T_+) = 1$ by (2.1.60).

  The proof is similar for (ii). $\square$

Lemma 2.1.3. $m(T_+) = 1.$

Proof.   Suppose that (2.1.59) is in the limit-circle case at ∞, that is, $m(T_+) = 2$. Then there exist linearly independent solutions φ, ψ of (2.1.59) in $\mathcal{H}_+$, and so $[\varphi, \psi](\cdot)$ is a non-zero constant. But this is not possible since $[\varphi, \psi](\cdot) \in L^1(a, \infty)$ by the Cauchy–Schwarz inequality. Hence we must have that $m(T_+) = 1$. $\square$

Lemma 2.1.4. T_0 *is self-adjoint (and so T_0' is essentially self-adjoint) if and only if $m(T_-) = 1$, or, equivalently, (2.1.59) is in the limit-point case at 0.*

Proof. Let $\tilde{T}_0$ be the orthogonal sum

$$\tilde{T}_0 = T_- \bigoplus T_+$$

in $\mathcal{H} = \mathcal{H}_- \bigoplus \mathcal{H}_+$. It is easily verified that $\tilde{T}_0$ is a closed symmetric operator with equal deficiency indices $m(\tilde{T}_0) = m_\pm(\tilde{T}_0)$ and

$$m(\tilde{T}_0) = m(T_-) + m(T_+);$$

see [Edmunds and Evans (1987)] (IX.5.1). Furthermore, $\tilde{T}_0 \subseteq T_0$ and by mimicking the proof of Theorem III.10.20 in [Edmunds and Evans (1987)] we have

$$\mathcal{D}(\tilde{T}_0) = \{f : f \in \mathcal{D}(T_0), f(a) = 0\},$$
$$\dim \left(\mathcal{D}(T_0)/\mathcal{D}(\tilde{T}_0) \right) = 2,$$

and

$$m(T_0) = m(T_-) + m(T_+) - 2.$$

Thus from Lemma 2.1.3, $m(T_0) = m(T_-) - 1$, whence the lemma. $\square$

Lemma 2.1.5. *Let V be the Coulomb potential $V(r) = \gamma/r$. Then T_0' is essentially self-adjoint in $L^2(\mathbb{R}_+, \mathbb{C}^2)$ if and only if $\gamma^2 \leq \kappa^2 - 1/4$.*

Proof. We noted above that if for some $\lambda \in \mathbb{C}$, all solutions of (2.1.59) are in $L^2((0, a), \mathbb{C}^2)$, then, for all $\lambda \in \mathbb{C}$, all solutions are in $L^2((0, a), \mathbb{C}^2)$. In view of Lemma 2.1.4 it is therefore sufficient to prove that for some value of $\lambda, \lambda = 0$ say, (2.1.59) has a solution that does not lie in $L^2((0, a), \mathbb{C}^2)$. The term σ_3 represents a bounded operator, and so we may consider

$$\left\{ -i\sigma_2 \frac{d}{dr} - \frac{\kappa}{r}\sigma_1 + \frac{\gamma}{r} \right\} f(r) = 0.$$

This has the solutions

$$f(r) = \binom{\kappa + s}{\gamma} r^s, \quad \text{where} \quad s = \pm\sqrt{\kappa^2 - \gamma^2}.$$

The solution with a negative s is not in $L^2((0, a), \mathbb{C}^2)$ if and only if

$$\gamma^2 \leq \kappa^2 - 1/4.$$

$\square$

On applying Lemmas 2.1.2–2.1.5 to Theorem 2.1.1, and noting that $\kappa_{l,s}^2 = (2sl + s + 1/2)^2 \geq 1$ since $l \geq 1$ when $s = -1/2$, we obtain

Theorem 2.1.6. *Let V be the Coulomb potential $V(r) = \gamma/r$ in Theorem 2.1.1. Then $\mathbb{D}' := \mathbb{D}_0' + V$ is essentially self-adjoint if and only if $|\gamma| \leq \sqrt{3}/2$.*

Remark 2.1.7. The essential self-adjointness of $\mathbb{D}' := \mathbb{D}_0' + V$ for any electric potential V satisfying $|V(\mathbf{x})| < \sqrt{3}/2$ was established in [Schmincke (1972b)]. The result for the spherically symmetric case of Theorem 2.1.6 was proved in [Evans (1970)] and also follows from results in [Weidmann (1971)].

Remark 2.1.8. For $|\gamma| \leq \sqrt{3}/2$, it follows from Theorem 2.1.6 that $\mathbb{D}$ is the closure of $\mathbb{D}_0 + V$ on $\mathcal{D}(\mathbb{D}_0)$. Let $\phi \in \mathcal{D}(\mathbb{D}_0)$ and let $\{\phi_j\}_{j \in \mathbb{N}}, \phi_j \in C_0^\infty(\mathbb{R}^3, \mathbb{C}^4)$, be such that $\phi_j \to \phi$ and $\mathbb{D}_0 \phi_j \to \mathbb{D}_0 \phi$ as $j \to \infty$. Since V is $\mathbb{D}_0$-bounded on account of the Hardy inequality and (2.1.11), we have that $V\phi_j \to V\phi$ and hence that $\mathbb{D}\phi_j = (\mathbb{D}_0 + V)\phi_j \to (\mathbb{D}_0 + V)\phi$. Consequently, $\phi \in \mathcal{D}(\mathbb{D})$ and $\mathbb{D}\phi = (\mathbb{D}_0 + V)\phi$, i.e., $(\mathbb{D}_0 + V) \subseteq \mathbb{D}$. This means, in particular, that $\mathcal{D}(\mathbb{D}_0) \subseteq \mathcal{D}(\mathbb{D})$. In fact, it is proved in [Landgren and Reitö (1979)] and [Landgren *et al.* (1980)] (see also [Arai and Yamada (1982)], [Klaus (1980)] and [Vogelsang (1987)]) that for $\mathbb{D} = \mathbb{D}_0 + V$ with $|V(\mathbf{x})| < (\sqrt{3}/2)|\mathbf{x}|^{-1}$, $\mathcal{D}(\mathbb{D}_0) = \mathcal{D}(\mathbb{D})$. This has important applications for us in Chapter 4, and we shall give a proof in the next section, based on that in [Vogelsang (1987)] for the case of matrix-valued potentials.

Remark 2.1.9. If $\sqrt{3}/2 < |\gamma| < 1$, $\mathbb{D}' = \mathbb{D}_0' + V, V(r) = \gamma/r$, is no longer essentially self-adjoint, but a *physically relevant* self-adjoint extension has been shown to exist: the literature on these *distinguished* self-adjoint extensions will be discussed in Remark 2.1.14 below. The range $|\gamma| \leq 1$ is significant, for it follows from Theorem 2 in [Schechter (1966)] that the spectrum of the distinguished self-adjoint extension is contained in $\mathbb{R} \setminus (-\sqrt{1 - \eta^2}, \sqrt{1 - \eta^2})$, for any $\eta \in (|\gamma|, 1)$.

Remark 2.1.10. The essence of Lemma 2.1.4 is that for spherically symmetric scalar potentials V, the essential self-adjointness of $\mathbb{D}_0' + V$ depends only on the local behaviour of V. This property was proved by Jörgens in [Jörgens (1972)] and Chernoff in [Chernoff (1973)], [Chernoff (1977)] to be satisfied by a wide class of operators of the form $\mathbb{D}_0' + P$, where P is a Hermitian 4×4-matrix-valued function, a class which includes operators $\boldsymbol{\alpha} \cdot (-i\boldsymbol{\nabla} + \mathbf{A}) + V$, where $\mathbf{A}$ is a vector (magnetic) potential and V a scalar potential. In [Thaller (1994)], Thaller gives an abstract result for operators exhibiting this type of behaviour which includes the results of Jörgens and Chernoff as special cases. In its application to the Dirac operator $\mathbb{D}_0' + P$, the essential self-adjointness of a symmetric operator T is a consequence of that of symmetric operators $A_n T A_n$ for each $n \in \mathbb{N}$, where, for instance, the A_n are the operators of multiplication by real functions ϕ_n, where $\phi_n(t) = \phi(t/n), \phi \in C_0^\infty(0, \infty)$. The crucial property which distinguishes the case when T is the Dirac operator from that of the Schrödinger operator is that the commutators $[T, A_n] = -i\phi_n'(\boldsymbol{\alpha} \cdot \mathbf{x}/r)$ are uniformly bounded operators in the Dirac case. If T is the Schrödinger operator the commutators are first-order differential operators.

2.1.3   *Matrix-valued potentials*

The main objective in this section is to establish a special case of the result in [Vogelsang (1987)] concerning the essential self-adjointness of

$$\mathbb{D}_P' := \mathbb{D}_0' + P, \quad \mathcal{D}(\mathbb{D}_P') = C_0^\infty(\mathbb{R}^3 \setminus \{0\}, \mathbb{C}^4), \tag{2.1.66}$$

where, in our case, P is a Hermitian 4×4-matrix-valued function with a Coulomb singularity at the origin. Our particular concern is to show that the domain of the unique self-adjoint extension of $\mathbb{D}_P'$ coincides with $\mathcal{D}(\mathbb{D}_0) = H^1(\mathbb{R}^3, \mathbb{C}^4)$.

   To prepare for the main theorem, we need some auxiliary results, based on ones in [Vogelsang (1987)], which are of intrinsic interest. We noted in Section 2.1.1 that the three components of the orbital angular momentum operator $\mathbf{L} = \mathbf{x} \times \mathbf{p}$ and total angular momentum operator $\mathbf{J} = \mathbf{L} + \mathbf{S}$ (where $\mathbf{S}$ is the spin angular momentum) are self adjoint in $L^2(\mathbb{S}^2, \mathbb{C}^k), k = 2, 4$. Also, L^2 and J^2 have a discrete spectrum with common eigenvectors $\Omega_{l,m,s}$ in the case $k = 2$, specifically, by (2.1.24),

$$L^2 \Omega_{l,m,s} = l(l+1)\Omega_{l,m,s}$$

$$J^2 \Omega_{l,m,s} = (l+s)(l+s+1)\Omega_{l,m,s}, \tag{2.1.67}$$

where $l \in \mathbb{N}_0$ when $s = 1/2$, $l \in \mathbb{N}$ when $s = -1/2$ and $m \in \{-l-1/2, \cdots, l+1/2\}$. It follows that the spectrum of $J^2 - L^2 + 1/4$ in $L^2(\mathbb{S}^2, \mathbb{C}^k), k = 2, 4$, consists of eigenvalues at $\{\pm 1, \pm 2, \cdots\}$.

   From

$$J^2 - L^2 + 1/4 = 1 + 2\mathbf{S} \cdot \mathbf{L} \tag{2.1.68}$$

and

$$2\mathbf{S} \cdot \mathbf{L} = -\frac{1}{2}(\boldsymbol{\alpha} \times \boldsymbol{\alpha}) \cdot (\mathbf{x} \times \boldsymbol{\nabla})$$

$$= -\sum_{i<j} \alpha_i \alpha_j (x_i \partial_j - x_j \partial_i) \tag{2.1.69}$$

it follows that $T := 1 + 2\mathbf{S} \cdot \mathbf{L}$ is a self-adjoint operator with domain $H^1(\mathbb{S}^2, \mathbb{C}^4)$ in $L^2(\mathbb{S}^2, \mathbb{C}^4)$ and

$$\|Tu\|_{L^2(\mathbb{S}^2, \mathbb{C}^4)} \geq \|u\|_{L^2(\mathbb{S}^2, \mathbb{C}^4)} \quad \text{for all} \quad u \in H^1(\mathbb{S}^2, \mathbb{C}^4). \tag{2.1.70}$$

Setting $\mathbf{x} = r\boldsymbol{\omega}, r = |\mathbf{x}|, \boldsymbol{\omega} = (\omega_1, \omega_2, \omega_3)$ and $\tilde{\alpha} := \sum_{i=1}^3 \alpha_i \omega_i$, we have by (2.1.69) and the Chain Rule that

$$r\tilde{\alpha} \sum_{j=1}^3 \alpha_j \partial_j = \sum_{i<j} \alpha_i \alpha_j (x_i \partial_j - x_j \partial_i) + \sum_{j=1}^3 x_j \partial_j$$

$$= (I - T) + r\partial_r.$$

Since $\tilde{\alpha}^2 = I_4$, we have with $u' = ru$,

$$\tilde{\alpha}(\partial_r - r^{-1}T)u' = r\sum_{j=1}^3 \alpha_j \partial_j u. \tag{2.1.71}$$

Lemma 2.1.11. *For all* $u \in H^1_{loc}(\mathbb{R}^3, \mathbb{C}^4)$,

$$\|r^{-1/2}u\| \leq \|r^{1/2}(\boldsymbol{\alpha} \cdot \mathbf{D})u\|, \quad \mathbf{D} := -i\boldsymbol{\nabla}, \tag{2.1.72}$$

whenever both sides are finite. In (2.1.72), $\|\cdot\|$ *is the* $L^2(\mathbb{R}^3, \mathbb{C}^4)$ *norm.*

Proof. Let $\varphi_N \in C_0^\infty(0,\infty)$ be such that $0 \leq \varphi_N(r) \leq 1$, $\varphi_N(r) = 1$ for $1/N \leq$ r $\leq$ N, and $\varphi_N'(r) = O(1/r)$. Suppose that (2.1.72) has been established for $\varphi_N u$. Then, with obvious notation,

$$
\begin{aligned}
\|r^{-1/2}u\|_{\{x:1/N<r<N\}} &\leq \|r^{-1/2}\varphi_N u\| \\
&\leq \|r^{1/2}\varphi_N(\boldsymbol{\alpha}\cdot\mathbf{D})u\| + \|r^{1/2}\varphi_N' u\|_{\{x:r<1/N\}\cup\{x:r>N\}} \\
&\leq \|r^{1/2}(\boldsymbol{\alpha}\cdot\mathbf{D})u\| + c\|r^{-1/2}u\|_{\{x:r<1/N\}\cup\{x:r>N\}}.
\end{aligned}
$$

On allowing $N \to \infty$, (2.1.72) follows for u in the prescribed sense. Hence, we may assume that u vanishes in neighbourhoods of zero and infinity.

From (2.1.71),

$$
\begin{aligned}
\|r^{1/2}(\boldsymbol{\alpha}\cdot\mathbf{D})u\|^2 &= \int_0^\infty \int_{\mathbb{S}^2} \left\{ |r(\boldsymbol{\alpha}\cdot\mathbf{D})u|^2 d\omega \right\} r\,dr \\
&= \int_0^\infty r\left\{ \|\tilde{\alpha}(\partial_r - r^{-1}T)u'\|_{L^2(\mathbb{S}^2,\mathbb{C}^4)}^2 \right\} dr \qquad (2.1.73)
\end{aligned}
$$

where

$$
\begin{aligned}
&\|\tilde{\alpha}(\partial_r - r^{-1}T)u'\|_{L^2(\mathbb{S}^2,\mathbb{C}^4)}^2 \\
&= \|\partial_r u'\|_{L^2(\mathbb{S}^2,\mathbb{C}^4)}^2 + \|r^{-1}Tu'\|_{L^2(\mathbb{S}^2,\mathbb{C}^4)}^2 - 2r^{-1}\mathrm{Re}\left[(\partial_r u', Tu')_{L^2(\mathbb{S}^2,\mathbb{C}^4)} \right].
\end{aligned}
$$

Since, T involves only the angular variables and therefore commutes with ∂_r, we have on changing the order of integration,

$$
\begin{aligned}
\int_0^\infty \int_{\mathbb{S}^2} \langle \partial_r u', Tu' \rangle dr\,d\omega &= -\int_{\mathbb{S}^2} \int_0^\infty \langle u', T\partial_r u' \rangle dr\,d\omega \\
&= -\int_0^\infty \int_{\mathbb{S}^2} \langle Tu', \partial_r u' \rangle d\omega\,dr
\end{aligned}
$$

and so

$$
\mathrm{Re} \int_0^\infty \left(\int_{\mathbb{S}^2} \langle \partial_r u', Tu' \rangle d\omega \right) dr = 0.
$$

Hence, from (2.1.73) and (2.1.70), we deduce that

$$
\|r^{1/2}(\boldsymbol{\alpha}\cdot\mathbf{D})u\|^2 = \int_0^\infty r\left(\|\partial_r u'\|_{L^2(\mathbb{S}^2,\mathbb{C}^4)}^2 + \|r^{-1}Tu'\|_{L^2(\mathbb{S}^2,\mathbb{C}^4)}^2 \right) dr \geq \|r^{-1/2}u\|^2
$$

and the proof is complete. $\qquad\square$

Lemma 2.1.12. *Let P be Hermitian-valued with $P\tilde{\alpha} = \tilde{\alpha}P$ and*

$$
|P(\mathbf{x})| \leq \mu/|\mathbf{x}|, \quad 0 \leq \mu < \sqrt{3}/2.
$$

Then, for all $u \in H^1(\mathbb{R}^3,\mathbb{C}^4)$,

$$
\|r^{-1}u\| \leq (1-a)^{-1}\|[(\boldsymbol{\alpha}\cdot\mathbf{D}) + P]u\|, \quad a^2 = \mu^2 + 1/4. \qquad (2.1.74)
$$

Proof. First note that $r^{-1}u \in L^2(\mathbb{R}^3, \mathbb{C}^4)$ by Hardy's inequality (1.7.3). On substituting $r^{-1/2}u$ for u in Lemma 2.1.11, we have

$$\|r^{-1}u\| \leq \|[(\boldsymbol{\alpha} \cdot \mathbf{D}) + P]u\| + \|(ir^{-1}\tilde{\alpha}/2 - P)u\|.$$

Furthermore, the assumptions on P imply that

$$\|(ir^{-1}\tilde{\alpha}/2 - P)u\|^2 = \|Pu\|^2 + \frac{1}{4}\|r^{-1}u\|^2 \leq a^2\|r^{-1}u\|^2,$$

whence the result. $\qquad\square$

The following theorem is a special case of Theorem 1 in [Vogelsang (1987)]; Vogelsang's theorem allows the matrix potential P to have Coulomb singularities at a sequence of distinct points without a finite limit point.

Theorem 2.1.13. *Let $P \in L^\infty_{loc}(\mathbb{R}^3 \setminus \{0\}, \mathbb{C}^4 \times \mathbb{C}^4)$ be Hermitian-valued with $P\tilde{\alpha} = \tilde{\alpha}P$ and*

$$|P(\mathbf{x})| \leq \mu/|\mathbf{x}|, \quad 0 \leq \mu < \sqrt{3}/2. \tag{2.1.75}$$

Then $\mathbb{D}'_P$ is essentially self-adjoint and its unique self-adjoint extension has domain $\mathcal{D}(\mathbb{D}_0) = H^1(\mathbb{R}^3, \mathbb{C}^4)$.

Proof. We first prove that $R := (\boldsymbol{\alpha} \cdot \mathbf{D}) + P$ is symmetric on

$$D_1 := \{u : u \in L^2(\mathbb{R}^3, \mathbb{C}^4) \cap H^1_{loc}(\mathbb{R}^3, \mathbb{C}^4), Ru \in L^2(\mathbb{R}^3, \mathbb{C}^4)\}.$$

Let $f, g \in D_1$ and $\phi_k(r) = \phi(r/k)$, where $\phi \in C_0^\infty[0, \infty)$, $0 \leq \phi \leq 1$, $\phi(r) = 1$ in $[0, 1)$, 0 for $r > 2$ and $\phi'(r) = O(1)$. Then

$$(\phi_k Rf, \phi_k g) - (\phi_k f, \phi_k Rg)$$
$$= i(\phi'_k \tilde{\alpha} f, \phi_k g) - i(\phi_k f, \phi'_k \tilde{\alpha} g)$$
$$= O\left(\frac{1}{k} \int_{k<r<2k} |fg| d\mathbf{x}\right).$$

That R is symmetric on D_1, follows on allowing $k \to \infty$.

The next step is to prove that R on D_1 is self-adjoint. This will follow if we show that $\mathcal{R}(R+iI) = L^2(\mathbb{R}^3, \mathbb{C}^2)$, i.e., for any $f \in L^2(\mathbb{R}^3, \mathbb{C}^4)$, there exists $u \in D_1$ such that

$$[(\boldsymbol{\alpha} \cdot \mathbf{D}) + P + i]u = f, \tag{2.1.76}$$

and similarly for $-i$.

Define

$$P_\lambda(\mathbf{x}) := \frac{r}{r+\lambda} P(\mathbf{x}), \quad (0 < \lambda \leq 1).$$

Then $(\boldsymbol{\alpha} \cdot \mathbf{D}) + P_\lambda$ is self-adjoint on $\mathcal{D}(\mathbb{D}_0) = H^1(\mathbb{R}^3, \mathbb{C}^4)$ by Theorem 1.1.1, and so there exists $u_\lambda \in H^1(\mathbb{R}^3, \mathbb{C}^4)$ such that

$$[(\boldsymbol{\alpha} \cdot \mathbf{D}) + P_\lambda + i]u_\lambda = f, \quad \|u_\lambda\| \leq \|f\|. \tag{2.1.77}$$

With the above $\phi_k \in C_0^\infty[0, \infty)$, we have

$$\|(\boldsymbol{\alpha} \cdot \mathbf{D})(\phi_k u_\lambda)\| = \|\phi_k(\boldsymbol{\alpha} \cdot \mathbf{D})u_\lambda - i\phi_k' \tilde{\alpha} u_\lambda\|$$
$$\leq \|\phi_k[(\boldsymbol{\alpha} \cdot \mathbf{D}) + P_\lambda + i]u_\lambda\| + \|\phi_k(P_\lambda + i)u_\lambda\| + \|\phi_k' u_\lambda\|$$
$$\leq C\|f\|, \tag{2.1.78}$$

where the constant C is independent of λ, on observing that the assumptions on P in Lemma 2.1.12 are satisfied by P_λ and consequently,

$$\|P_\lambda u_\lambda\| \leq \mu \|r^{-1}u_\lambda\| \leq \mu(1-a)^{-1}\|[(\boldsymbol{\alpha}\cdot\mathbf{D})+P_\lambda]u_\lambda\| \leq \mu(1-a)^{-1}\|f-iu_\lambda\| = O(\|f\|)$$

and

$$\|\phi_k' u_\lambda\| = O(\|u_\lambda\|) = O(\|f\|).$$

These imply that $\{\phi_k u_\lambda\}$ is bounded in $H_0^1(B_{2k})$, where B_{2k} is the ball $\{\mathbf{x} : |\mathbf{x}| < 2k\}$, and hence, by the Rellich–Kondrachov theorem (see property (v) in Section 1.6), there exists a sequence $\{\phi_k u_{\lambda_{n_k}} : n_k \in \mathbb{N}\}$, which, as $\lambda_{n_k} \to 0$, converges in $L^2(B_{2k}, \mathbb{C}^4)$ and, hence, $\{u_{\lambda_{n_k}}\}$ converges in $L^2(B_k, \mathbb{C}^4)$. On repeating the argument with $\{\phi_{2k} u_{\lambda_{n_k}}\}$, we see that $\{u_{\lambda_{n_k}}\}$ contains a subsequence $\{u_{\lambda_{2k}}\}$ which converges in $L^2(B_{2k}, \mathbb{C}^4)$. Continuing in this way and selecting the diagonal of the constructed subsequences, we obtain a sequence $\{u_{\lambda_n}\}$ which converges to a limit u, say, in $L_{loc}^2(\mathbb{R}^3, \mathbb{C}^4)$. Note also that, for all $k \in \mathbb{N}$,

$$\|u\|_{B_k} = \lim_{\lambda_n \to 0} \|u_{\lambda_n}\|_{B_k} \leq \|f\|$$

and so $u \in L^2(\mathbb{R}^3, \mathbb{C}^4)$.

We claim that $u \in D_1$ and satisfies (2.1.76). To see this, we first note that, for all $\psi \in C_0^\infty(\mathbb{R}^3, \mathbb{C}^4)$,

$$\int_{\mathbb{R}^3} \langle u_{\lambda_n}, (\boldsymbol{\alpha} \cdot \mathbf{D})\psi \rangle d\mathbf{x} = \int_{\mathbb{R}^3} \langle (\boldsymbol{\alpha} \cdot \mathbf{D})u_{\lambda_n}, \psi \rangle d\mathbf{x}$$
$$= \int_{\mathbb{R}^3} \langle f - (P_{\lambda_n} + i)u_{\lambda_n}, \psi \rangle d\mathbf{x}$$
$$= \int_{\mathbb{R}^3} \langle f - iu_{\lambda_n}, \psi \rangle d\mathbf{x} - \int_{\mathbb{R}^3} \langle u_{\lambda_n}, P_{\lambda_n}\psi \rangle d\mathbf{x}.$$

As $\lambda_n \to 0$, $u_{\lambda_n} \to u$ in $L_{loc}^2(\mathbb{R}^3, \mathbb{C}^4)$, and $P_{\lambda_n}\psi \to P\psi$ in $L^2(\mathbb{R}^3, \mathbb{C}^4)$ by dominated convergence, since $P_{\lambda_n} \to P$ and $|P_{\lambda_n}\psi|^2 \leq |P\psi|^2 \in L^1(\mathbb{R}^3, \mathbb{C}^4)$ by Hardy's inequality. It follows that

$$\int_{\mathbb{R}^3} \langle u, (\boldsymbol{\alpha} \cdot \mathbf{D})\psi \rangle d\mathbf{x} = \int_{\mathbb{R}^3} \langle f - (P + i)u, \psi \rangle d\mathbf{x}.$$

Thus $(\boldsymbol{\alpha} \cdot \mathbf{D})u = f - (P+i)u$ in the distributional sense in $\mathbb{R}^3$. Since $f - (P+i)u \in L_{loc}^2(\mathbb{R}^3 \setminus \{0\}, \mathbb{C}^4)$, we also have that $(\boldsymbol{\alpha} \cdot \mathbf{D})u = f - (P+i)u$ in the sense of a weak derivative in $\mathbb{R}^3 \setminus \{0\}$. This implies that $u \in D_1$ and (2.1.76) is satisfied. The same argument works when i is replaced by $-i$ in (2.1.76). Hence R is self-adjoint on D_1.

The operator R is the closure of $\mathbb{D}_P'$ (and hence $\mathbb{D}_P'$ is essentially self-adjoint) if and only if any $u \in D_1$ can be approximated in the graph norm $\|\cdot\|_R$ of R by functions in $C_0^\infty(\mathbb{R}^3 \setminus \{0\}, \mathbb{C}^4)$. From the identity $R[(1 - \phi_k)u] = (1 - \phi_k)Ru + i\phi'_k\tilde{\alpha}u$ we have that $\|u - \phi_k u\|_R \to 0$ as $k \to \infty$. Moreover, $\|\phi_k u\|_R$ is easily seen to be equivalent to $\|\phi_k u\|_{H^1(\mathbb{R}^3, \mathbb{C}^4)}$ on using Lemma 2.1.12 and Hardy's inequality. Since $C_0^\infty(\mathbb{R}^3 \setminus \{0\}, \mathbb{C}^4)$ is dense in $H^1(\mathbb{R}^3, \mathbb{C}^4)$ (see [Edmunds and Evans (1987)], Corollary VIII.6.4), it follows that $R = \overline{\mathbb{D}'_P}$.

Finally, for $u \in D_1, \phi_k u \in H^1(\mathbb{R}^3, \mathbb{C}^4)$ and by Lemma 2.1.12,

$$\begin{aligned}
\|Pu\|_{\{\mathbf{x}:r \le k\}} &\le \|\phi_k Pu\| \le \mu \|r^{-1}\phi_k u\| \\
&\le \mu(1-a)^{-1}\|[(\boldsymbol{\alpha} \cdot \mathbf{D}) + P](\phi_k u)\| \\
&\le C\{\|[(\boldsymbol{\alpha} \cdot \mathbf{D}) + P]u\| + k^{-1}\|u\|_{\mathbf{x}:k<r<2k}\}.
\end{aligned}$$

It follows that $Pu \in L^2(\mathbb{R}^3, \mathbb{C}^4)$ and, hence, $u \in H^1(\mathbb{R}^3, \mathbb{C}^4)$. The proof of the theorem is therefore complete. $\qquad\square$

In [Arai (1975)] and [Arai (1983)], it is shown that there are matrix-valued potentials P satisfying

$$|P(\mathbf{x})| \le \frac{1/2 + \varepsilon}{|\mathbf{x}|},$$

with $\varepsilon > 0$ arbitrary, for which $\mathbb{D}'_P$ is not essentially self-adjoint. Thus, the condition $P\tilde{\alpha} = \tilde{\alpha}P$ is necessary in Theorem 2.1.13. Hardy's inequality ensures that if

$$|P(\mathbf{x})| \le a/2|\mathbf{x}| + b$$

for some constants $b \ge 0$ and $a < 1$, then P has $\mathbb{D}'_0$-bound $a < 1$ and hence $\mathbb{D}'_P$ is essentially self-adjoint with $\mathcal{D}(\mathbb{D}_P) = \mathcal{D}(\mathbb{D}_0)$, by the Kato–Rellich theorem 1.1.1. Therefore $a < 1$ is optimal for general matrix-valued potentials.

Remark 2.1.14. For

$$\gamma = \sup_{\mathbf{x} \ne 0} |\mathbf{x}||V(\mathbf{x})| < 1, \tag{2.1.79}$$

where V is now a scalar potential, Schmincke in [Schmincke (1972a)] and Wüst in [Wüst (1973, 1975, 1977)] proved the existence of a distinguished self-adjoint extension $\tilde{T}$ of $T := \mathbb{D}' := \mathbb{D}'_0 + V \upharpoonright C_0^\infty(\mathbb{R}^3 \setminus \{0\}, \mathbb{C}^4)$ as a certain limit of Dirac operators with cut-off potentials: Schmincke removed Wüst's assumption that T be semi-bounded. The extension is characterised by its domain $\mathcal{D}(\tilde{T})$ being contained in $\mathcal{D}(T^*) \cap \mathcal{D}(r^{-1/2})$. This was shown by Klaus and Wüst in [Klaus and Wüst (1979)] to coincide with the self-adjoint extension constructed by Nenciu in [Nenciu (1976, 1977)] as the unique self-adjoint extension of $\mathbb{D}'_0 + V$ with domain in $\mathcal{D}(|\mathbb{D}_0|^{1/2}) = H^{1/2}(\mathbb{R}^3, \mathbb{C}^4)$. In [Nenciu (1976)], $\tilde{T}$ is defined as the self-adjoint operator satisfying the following:

$$(\tilde{T}\phi, \psi) = (|\mathbb{D}_0|^{1/2}\phi, U|\mathbb{D}_0|^{1/2}\psi) + (V\phi, \psi), \quad V(\mathbf{x}) = \gamma/|\mathbf{x}|, \tag{2.1.80}$$

for $\phi \in \mathcal{D}(\tilde{T}), \psi \in \mathcal{D}(\mathbb{D}_0)$, where U is the partial isometry in the polar decomposition of $\mathbb{D}_0$; see Section 1.2. We shall see in Section 2.2.3 below that $U = \Lambda_+ \oplus \Lambda_-$, where $\Lambda_\pm$ are the projections of $L^2(\mathbb{R}^3, \mathbb{C}^4)$ onto the positive and negative spectral subspaces of $\mathbb{D}_0$ and $|\mathbb{D}_0| = \sqrt{-\Delta + 1}I_4$.

The pivotal role played by Hardy-type inequalities in establishing the essential self-adjointness of T is clear from the proof of Theorem 2.1.13, and their fundamental importance in proving the existence of a distinguished self-adjoint extension of T is demonstrated in [Esteban and Loss (2007)] and [Esteban and Loss (2008)]. A special case of the inequality in Theorem 1 of [Esteban and Loss (2007)], which is adequate for dealing with Coulomb potentials, is the following inequality proved originally in [Dolbeault *et al.* (2000a)]:

$$\int_{\mathbb{R}^3} \left(\frac{|(\boldsymbol{\sigma} \cdot \boldsymbol{\nabla})\varphi(\mathbf{x})|^2}{1 + 1/|\mathbf{x}|} + |\varphi(\mathbf{x})|^2 \right) d\mathbf{x} \geq \int_{\mathbb{R}^3} \frac{|\varphi(\mathbf{x})|^2}{|\mathbf{x}|^2} d\mathbf{x}, \qquad (2.1.81)$$

where $\boldsymbol{\sigma} \cdot \boldsymbol{\nabla} = \sum_{j=1}^{3} \sigma_j \partial_j$, with $\sigma_j, j = 1, 2, 3$ denoting the Pauli matrices (2.1.5). The powers of $|\mathbf{x}|$ and the constants are optimal in (2.1.81). By scaling, replacing $\varphi(\cdot)$ by $\varepsilon^{-1}\varphi(\varepsilon^{-1}\cdot)$ and allowing $\varepsilon \to 0$, (2.1.81) yields

$$\int_{\mathbb{R}^3} |\mathbf{x}||(\boldsymbol{\sigma} \cdot \boldsymbol{\nabla})\varphi(\mathbf{x})|^2 d\mathbf{x} \geq \int_{\mathbb{R}^3} \frac{|\varphi(\mathbf{x})|^2}{|\mathbf{x}|} d\mathbf{x}. \qquad (2.1.82)$$

The analysis in [Esteban and Loss (2007)] applies to the critical case $\gamma = 1$ in (2.1.79). It is shown that when $V(\mathbf{x}) = -1/|\mathbf{x}|$, the domain of the distinguished self-adjoint extension is not contained in $H^{1/2}(\mathbb{R}^3, \mathbb{C}^4)$.

2.2   The quasi-relativistic operator

In [Herbst (1977)] and [Weder (1974, 1975)], a comprehensive description is given of the spectral properties of the operator

$$\mathbb{H} := (-\Delta + m^2)^{1/2} - Ze^2/|\mathbf{x}|, \qquad (2.2.1)$$

which serves as a model for the interaction of a relativistic spin-zero particle of charge e and mass m in the Coulomb field of an infinitely heavy nucleus of charge Z. Unlike the Dirac operator, this operator has the virtue of being non-negative for a range of values of Ze^2, and provides an interesting comparison with the Brown–Ravenhall operator which will be investigated in the next section. To simplify notation again, we consider the operator

$$\mathbb{H} := \mathbb{H}_0 + V, \quad \mathbb{H}_0 := \sqrt{-\Delta + 1}, \quad V(\mathbf{x}) = -\gamma/|\mathbf{x}|, \quad \gamma > 0. \qquad (2.2.2)$$

The operator $\mathbb{H}_0$ has domain $\mathcal{D}(\mathbb{H}_0) = H^1(\mathbb{R}^3)$ and form domain $H^{1/2}(\mathbb{R}^3)$; see Section 1.6. We first need some preliminary results, which will also be needed in the next section.

2.2.1 The Kato inequality

In this section we give a proof of Kato's inequality (1.7.7) in the cases $n = 2, 3$, as a consequence of technical lemmas which will be needed elsewhere in this chapter. In the first lemma, the normalised spherical harmonics $Y_{l,m}$ feature and also the Legendre functions of the second kind, namely,

$$Q_l(z) = \frac{1}{2} \int_{-1}^1 \frac{P_l(t)}{z-t} dt, \tag{2.2.3}$$

where P_l are the Legendre polynomials.

Lemma 2.2.1. Let $\mathbf{p} = p\,\boldsymbol{\omega}_{\mathbf{p}}, \mathbf{p}' = p'\boldsymbol{\omega}_{\mathbf{p}'} \in \mathbb{R}^3$. Then

$$\int_{\mathbb{S}^2} \int_{\mathbb{S}^2} \frac{1}{|\mathbf{p}-\mathbf{p}'|^2} Y_{l',m'}(\boldsymbol{\omega}_{\mathbf{p}'}) Y_{l,m}(\boldsymbol{\omega}_{\mathbf{p}}) d\boldsymbol{\omega}_{\mathbf{p}'} d\boldsymbol{\omega}_{\mathbf{p}} = \frac{2\pi}{pp'} Q_l\left(\frac{p'^2+p^2}{2pp'}\right) \delta_{ll'}\delta_{mm'}. \tag{2.2.4}$$

Proof. On setting $z = (p^2 + p'^2)/2pp'$ and $\mathbf{p} \cdot \mathbf{p}' = pp' \cos\gamma$, we have from [Whittaker and Watson (1940)], Chapter XV that

$$\frac{1}{|\mathbf{p}-\mathbf{p}'|^2} = \frac{1}{2pp'(z - \cos\gamma)}$$

$$= \frac{1}{2pp'} \sum_{l''=0}^\infty (2l''+1) Q_{l''}(z) P_{l''}(\cos\gamma)$$

$$= \frac{4\pi}{2pp'} \sum_{l''=0}^\infty Q_{l''}(z) \sum_{m''=-l''}^{l''} \overline{Y}_{l'',m''}(\boldsymbol{\omega}_{\mathbf{p}}) Y_{l'',m''}(\boldsymbol{\omega}_{\mathbf{p}'}). \tag{2.2.5}$$

The lemma now follows from (2.2.5) and the orthonormality of the spherical harmonics. $\qquad\square$

The following analogue of Lemma 2.2.1 for the spherical spinors of (2.1.25) and (2.1.26) will be used in the partial wave decomposition of the Brown–Ravenhall operator in the next section.

Lemma 2.2.2. Let $\mathbf{p} = p\,\boldsymbol{\omega}_{\mathbf{p}}, \mathbf{p}' = p'\boldsymbol{\omega}_{\mathbf{p}'} \in \mathbb{R}^3$. Then

$$\int_{\mathbb{S}^2} \int_{\mathbb{S}^2} \frac{1}{|\mathbf{p}-\mathbf{p}'|^2} \langle \Omega_{l',m',s'}(\boldsymbol{\omega}_{\mathbf{p}'}), \Omega_{l,m,s}(\boldsymbol{\omega}_{\mathbf{p}}) \rangle d\boldsymbol{\omega}_{\mathbf{p}'} d\boldsymbol{\omega}_{\mathbf{p}}$$

$$= \frac{2\pi}{pp'} Q_l\left(\frac{p'^2+p^2}{2pp'}\right) \delta_{ll'}\delta_{mm'}\delta_{ss'}. \tag{2.2.6}$$

Proof. For $s = s' = 1/2$, we derive from (2.1.25) and (2.2.4)

$$I := \int_{\mathbb{S}^2} \int_{\mathbb{S}^2} \frac{1}{|\mathbf{p}-\mathbf{p}'|^2} \langle \Omega_{l',m',s'}(\boldsymbol{\omega}_{\mathbf{p}'}), \Omega_{l,m,s}(\boldsymbol{\omega}_{\mathbf{p}}) \rangle d\boldsymbol{\omega}_{\mathbf{p}'} d\boldsymbol{\omega}_{\mathbf{p}}$$

$$= \int_{\mathbb{S}^2} \int_{\mathbb{S}^2} \frac{1}{|\mathbf{p}-\mathbf{p}'|^2} \left\{ \sqrt{\frac{l'+m'+1/2}{2(l'+1/2)}} \sqrt{\frac{l+m+1/2}{2(l+1/2)}} Y_{l',m'-1/2}(\boldsymbol{\omega}_{\mathbf{p}'}) Y_{l,m-1/2}(\boldsymbol{\omega}_{\mathbf{p}}) \right.$$

$$\left. + \sqrt{\frac{l'-m'+1/2}{2(l'+1/2)}} \sqrt{\frac{l-m+1/2}{2(l+1/2)}} Y_{l',m'+1/2}(\boldsymbol{\omega}_{\mathbf{p}'}) Y_{l,m+1/2}(\boldsymbol{\omega}_{\mathbf{p}}) \right\} d\boldsymbol{\omega}_{\mathbf{p}'} d\boldsymbol{\omega}_{\mathbf{p}}$$

$$= \frac{2\pi}{pp'} Q_l\left(\frac{p'^2+p^2}{2pp'}\right) \delta_{ll'}\delta_{mm'}.$$

Similarly for $s = s' = -1/2$ on using (2.1.26).

When $s = 1/2$, $s' = -1/2$, (2.1.25), (2.1.26) and (2.2.4) yield

$$I = \left\{ \sqrt{\frac{l+m+1/2}{2(l+1/2)}} \sqrt{\frac{l-m+1/2}{2(l+1/2)}} \right.$$

$$\left. - \sqrt{\frac{l-m+1/2}{2(l+1/2)}} \sqrt{\frac{l+m+1/2}{2(l+1/2)}} \right\} \frac{2\pi}{pp'} Q_l \left(\frac{p'^2 + p^2}{2pp'} \right) \delta_{ll'} \delta_{mm'}$$

$$= 0.$$

$\square$

The next lemma is a consequence of Hilbert's inequality in Theorem 1.7.2 and will be used repeatedly later.

Lemma 2.2.3. *For $u, v \in L^2(\mathbb{R}_+; x dx)$ and $l \in \mathbb{N}_0$,*

$$I_l := \int_0^\infty \int_0^\infty \overline{u}(x) v(y) Q_l \left(\frac{1}{2} \left[\frac{x}{y} + \frac{y}{x} \right] \right) dx dy$$

$$\leq C_l \left(\int_0^\infty x|u(x)|^2 dx \right) \left(\int_0^\infty y|v(y)|^2 dy \right), \qquad (2.2.7)$$

where

$$C_l = \int_0^\infty Q_l \left(\frac{1}{2} \left[x + \frac{1}{x} \right] \right) x^{-1} dx$$

is sharp. In particular,

$$C_l = \begin{cases} \pi^2/2, & \text{if } l = 0 \\ 2, & \text{if } l = 1 \end{cases} \qquad (2.2.8)$$

and $C_l \leq 2$ for $l > 2$.

Proof. The functions $x^{-1/2} y^{-1/2} Q_l \left(\frac{1}{2} \left[\frac{x}{y} + \frac{y}{x} \right] \right)$ are homogeneous of degree -1. Hence (2.2.7) follows by Hilbert's inequality.

Next we note that it is shown in [Whittaker and Watson (1940)], Chapter XV, Section 32, that for $t > 1$, the Legendre functions $Q_l(t)$ have the integral representation

$$Q_l(t) = \int_{t+\sqrt{t^2-1}}^\infty \frac{z^{-l-1}}{\sqrt{1 - 2tz + z^2}} dz \qquad (2.2.9)$$

and so, for $t > 1$,

$$Q_0(t) \geq Q_1(t) \geq \cdots \geq Q_l(t) \geq \cdots \geq 0. \qquad (2.2.10)$$

Thus $C_l \leq C_0$ if $l \in \mathbb{N}_0$ and $C_l \leq C_1$ if $l \in \mathbb{N}$. Furthermore,

$$
\begin{aligned}
C_0 &= \int_0^\infty Q_0\left(\frac{1}{2}\left[x+\frac{1}{x}\right]\right) x^{-1}dx = \int_0^\infty \ln\left|\frac{x+1}{x-1}\right|\frac{dx}{x} \\
&= 2\int_0^1 \ln\left|\frac{x+1}{x-1}\right|\frac{dx}{x} \\
&= 4\int_0^1 \left(\sum_{k=0}^\infty \frac{x^{2k}}{2k+1}\right)dx \\
&= 4\sum_{k=0}^\infty \frac{1}{(2k+1)^2} \\
&= \frac{\pi^2}{2}
\end{aligned}
\tag{2.2.11}
$$

and

$$
\begin{aligned}
C_1 &= \int_0^\infty Q_1\left(\frac{1}{2}\left[x+\frac{1}{x}\right]\right) x^{-1}dx \\
&= 2\int_0^1 Q_1\left(\frac{1}{2}\left[x+\frac{1}{x}\right]\right) x^{-1}dx \\
&= 2\int_0^1 \left\{\frac{1}{2}\left(x+\frac{1}{x}\right)\ln\left|\frac{x+1}{x-1}\right|-1\right\}\frac{dx}{x} \\
&= 2\lim_{\varepsilon\to 0+,\delta\to 1-}\left[\frac{1}{2}\left(x-\frac{1}{x}\right)\ln\left|\frac{x+1}{x-1}\right|\right]_\varepsilon^\delta \\
&= 2
\end{aligned}
\tag{2.2.12}
$$

as asserted. □

Kato's inequality in $\mathbb{R}^n$ is now shown (in the cases $n = 2, 3$) to be a consequence of Lemmas 2.2.1 and 2.2.3. We use the notation

$$
D^{1/2}(\mathbb{R}^n) := \{u : (\cdot)^{1/2}\hat{u} \in L^2(\mathbb{R}^n)\}.
\tag{2.2.13}
$$

Equivalently, in configuration space, $D^{1/2}(\mathbb{R}^n)$ is the completion of $\mathbf{S}(\mathbb{R}^n)$ with respect to the norm

$$
\|u\|_{D^{1/2}(\mathbb{R}^n)} := \|\{-\Delta\}^{1/4}u\|_{L^2(\mathbb{R}^n)} := \|\{|\cdot|\}^{1/2}\hat{u}\|_{L^2(\mathbb{R}^n)}.
\tag{2.2.14}
$$

Note that $D^{1/2}(\mathbb{R}^n)$ is not a subspace of $L^2(\mathbb{R}^n)$. Its intersection with $L^2(\mathbb{R}^n)$ is the Bessel potential space

$$
H^{1/2}(\mathbb{R}^n) := \{u : (|\cdot|^2 + 1)^{1/4}\hat{u} \in L^2(\mathbb{R}^n)\};
\tag{2.2.15}
$$

see (1.6.4). Equivalently, in configuration space, $H^{1/2}(\mathbb{R}^n)$ is the completion of $\mathbf{S}(\mathbb{R}^n)$ with respect to the norm

$$
\|u\|_{H^{1/2}(\mathbb{R}^n)} := \|(-\Delta+1)^{1/4}u\|_{L^2(\mathbb{R}^n)} := \|(|\cdot|^2+1)^{1/4}\hat{u}\|_{L^2(\mathbb{R}^n)};
\tag{2.2.16}
$$

see Section 1.6 in Chapter 1.

Theorem 2.2.4. *For all $f \in D^{1/2}(\mathbb{R}^3)$,*

$$\int_{\mathbb{R}^3} |\mathbf{x}|^{-1} |f(\mathbf{x})|^2 d\mathbf{x} \leq \frac{\pi}{2} \int_{\mathbb{R}^3} |\mathbf{p}| |\hat{f}(\mathbf{p})|^2 d\mathbf{p}. \qquad (2.2.17)$$

The constant $\pi/2$ is sharp.

Proof.   It is sufficient to prove the theorem for $f \in \mathbf{S}(\mathbb{R}^3)$.

By Parseval's formula (1.5.7), and (1.5.9), we have

$$\int_{\mathbb{R}^3} |\mathbf{x}|^{-1} |f(\mathbf{x})|^2 d\mathbf{x} = \int_{\mathbb{R}^3} \mathbb{F}(|\cdot|^{-1} f)(\mathbf{p}) \mathbb{F}(\overline{f})(\mathbf{p}) d\mathbf{p}$$

$$= \int_{\mathbb{R}^3} \left(\mathbb{F}(|\cdot|^{-1}) * \hat{f} \right)(\mathbf{p}) \overline{\hat{f}(\mathbf{p})} d\mathbf{p}$$

$$= \frac{1}{(2\pi)^{3/2}} \sqrt{\frac{2}{\pi}} \int_{\mathbb{R}^3} |\mathbf{p} - \mathbf{p}'|^{-2} \hat{f}(\mathbf{p}') \overline{\hat{f}(\mathbf{p})} d\mathbf{p} d\mathbf{p}' =: I$$

$$(2.2.18)$$

since

$$\mathbb{F}(|\cdot|^{-1})(\mathbf{p}) = \sqrt{\frac{2}{\pi}} |\mathbf{p}|^{-1}$$

by (1.5.14). Since the spherical harmonics $\{Y_{l,m}\}$ form an orthonormal basis of $L^2(\mathbb{S}^2)$, we can write, with $\mathbf{p} = p\boldsymbol{\omega}_\mathbf{p}$ in polar co-ordinates,

$$\hat{f}(\mathbf{p}) = \sum_{l,m} c_{l,m}(p) Y_{l,m}(\boldsymbol{\omega}_\mathbf{p}), \quad c_{l,m}(p) = \int_{\mathbb{S}^2} \hat{f}(p\boldsymbol{\omega}_\mathbf{p}) Y_{l,m}(\boldsymbol{\omega}_\mathbf{p}) d\boldsymbol{\omega}_\mathbf{p},$$

where the summation is over $m = -l, -l+1, \cdots, l-1, l, \ l \in \mathbb{N}_0$. On substituting in the integral I of (2.2.18) and using Lemma 2.2.1, we get

$$I = \frac{1}{2\pi^2} \sum_{l,m} \sum_{l',m'} \int_0^\infty \int_0^\infty p^2 p'^2 c_{l',m'}(p') \overline{c_{l,m}(p)}$$

$$\times \int_{\mathbb{S}^2} \int_{\mathbb{S}^2} \frac{1}{|\mathbf{p} - \mathbf{p}'|^2} Y_{l',m'}(\boldsymbol{\omega}_{\mathbf{p}'}) \overline{Y_{l,m}(\boldsymbol{\omega}_\mathbf{p})} d\boldsymbol{\omega}_\mathbf{p} d\boldsymbol{\omega}_{\mathbf{p}'} dp dp'$$

$$= \frac{1}{\pi} \sum_{l,m} \int_0^\infty \int_0^\infty p p' Q_l(\frac{p^2 + p'^2}{2pp'}) c_{l,m}(p') \overline{c_{l,m}(p)} dp dp'$$

$$= \frac{1}{\pi} \sum_{l,m} \int_0^\infty \int_0^\infty Q_l \left(\frac{1}{2} \left[\frac{p}{p'} + \frac{p'}{p} \right] \right) \left(\overline{p c_{l,m}(p)} \right) (p' c_{l,m}(p')) \, dp dp'. \quad (2.2.19)$$

It now follows from Lemma 2.2.3 and (2.2.10) that

$$I \leq \frac{\pi}{2} \sum_{l,m} \int_0^\infty p^3 |c_{l_0,m}(p)|^2 dp$$

$$= \frac{\pi}{2} \int_{\mathbb{R}^3} |\hat{f}(\mathbf{p})|^2 |\mathbf{p}| d\mathbf{p}$$

with sharp constant $\pi/2$, and so the proof is complete. $\square$

The following proof of Kato's inequality in $\mathbb{R}^2$ is taken from [Bouzouina (2002)] where a method that is analogous to that above is used.

Theorem 2.2.5. *For all $f \in D^{1/2}(\mathbb{R}^2)$*

$$\int_{\mathbb{R}^2} \frac{|f(\mathbf{x})|^2}{|\mathbf{x}|} d\mathbf{x} \leq \left(\frac{\Gamma(1/4)}{\sqrt{2}\Gamma(3/4)}\right)^2 \int_{\mathbb{R}^2} |\mathbf{p}||\hat{f}(\mathbf{p})|^2 d\mathbf{p}, \qquad (2.2.20)$$

where the constant is sharp.

Proof. As in the proof of Theorem 2.2.4, but noting that in $\mathbb{R}^2, \mathbb{F}(1/|\cdot|)(\mathbf{p}) = |\mathbf{p}|^{-1}$, by (1.5.14), it follows that

$$\int_{\mathbb{R}^2} \frac{|f(\mathbf{x})|^2}{|\mathbf{x}|} d\mathbf{x} = \int_{\mathbb{R}^2} \left\{\mathbb{F}\left(\frac{1}{|\cdot|}\right) * \hat{f}\right\}(\mathbf{p})\overline{\hat{f}}(\mathbf{p})d\mathbf{p}$$

$$= \frac{1}{2\pi}\int_{\mathbb{R}^2}\int_{\mathbb{R}^2} \frac{1}{|\mathbf{p}-\mathbf{p}'|}\hat{f}(\mathbf{p}')\overline{\hat{f}}(\mathbf{p})d\mathbf{p}'d\mathbf{p}$$

$$=: I \qquad (2.2.21)$$

say. Since $\{e^{ik\theta}/\sqrt{2\pi} : k \in \mathbb{Z}\}$ is an orthonormal basis in $L^2(0, 2\pi)$, we can set, in polar co-ordinates,

$$\hat{f}(\mathbf{p}) = \frac{1}{\sqrt{2\pi}}\sum_{k\in\mathbb{Z}}p^{-1/2}v_k(p)e^{ik\theta},$$

where

$$\sum_{k\in\mathbb{Z}}\int_0^\infty p|v_k(p)|^2 dp = \int_{\mathbb{R}^2} |\mathbf{p}||\hat{f}(\mathbf{p})|^2 d\mathbf{p}.$$

On substituting in (2.2.21), we derive

$$I = \frac{1}{(2\pi)^2}\sum_{k,k'\in\mathbb{Z}}\int_0^\infty\int_0^\infty v_{k'}(p')\overline{v_k(p)}J_{k,k'}(p,p')(pp')^{1/2}dpdp', \qquad (2.2.22)$$

where

$$J_{k,k'}(p'p') := \int_0^{2\pi}\int_0^{2\pi} [p^2 + p'^2 - 2pp'\cos(\theta - \theta')]^{-1/2}e^{ik'\theta'}e^{-ik\theta}d\theta d\theta'.$$

Let $a := 2pp'/(p^2 + p'^2)$ and suppose that $k \neq k'$. Then, on setting $\theta' = \phi' + \alpha, \theta = \phi + \alpha$, we obtain

$$J_{k,k'}(p,p') = e^{i(k'-k)\alpha}J_{k,k'}(p,p')$$

and hence $J_{k,k'}(p,p') = 0$. When $k = k'$,

$$J_{k,k}(p,p') = \frac{1}{\sqrt{p^2 + p'^2}}\int_0^{2\pi}\int_0^{2\pi} \frac{e^{ik(\theta'-\theta)}}{\sqrt{1 - a\cos(\theta' - \theta)}}d\theta' d\theta.$$

The change of variables $\theta' - \theta = u, \theta = v$ yields

$$J_{k,k}(p,p') = \frac{1}{\sqrt{p^2 + p'^2}}\int_0^{2\pi} dv \int_{-v}^{2\pi-v} \frac{e^{iku}}{\sqrt{1 - a\cos u}}du$$

$$= \frac{2\pi}{\sqrt{p^2 + p'^2}}I_k(a), \qquad (2.2.23)$$

where

$$I_k(a) := \int_0^{2\pi} \frac{e^{iku}}{\sqrt{1 - a\cos u}} du$$

$$= -\sqrt{\frac{2}{a}} \int_{|z|=1} \frac{z^{k-1/2}}{\sqrt{(z - z_-)(z - z_+)}} dz \qquad (2.2.24)$$

on setting $z = e^{iu}$ and with $z_\pm = (1/a)\{1 \pm \sqrt{1 - a^2}\}$: note that $I_{-k}(a) = \overline{I_k(a)} = I_k(a)$. The function

$$\frac{z^{k-1/2}}{\sqrt{(z - z_-)(z - z_+)}}$$

is analytic in $\mathbb{C} \setminus [0, \infty)$ and it follows by Cauchy's theorem that $I_k = I_k^1 + I_k^2$, where

$$I_k^1 = \sqrt{\frac{2}{a}} \int_0^{z_-} \frac{x^{k-1/2}}{\sqrt{(x - z_-)(x - z_+)}} dx + \sqrt{\frac{2}{a}} \int_{z_-}^1 \frac{x^{k-1/2}}{i\sqrt{(x - z_-)(z_+ - x)}} dx$$

and

$$I_k^2 = \sqrt{\frac{2}{a}} \int_0^{z_-} \frac{x^{k-1/2}}{\sqrt{(x - z_-)(x - z_+)}} dx + \sqrt{\frac{2}{a}} \int_{z_-}^1 \frac{x^{k-1/2}}{-i\sqrt{(x - z_-)(z_+ - x)}} dx.$$

Hence, on setting $t = 1/a$, we have

$$I_{|k|}(a) = 2\sqrt{2t} \int_0^{t-\sqrt{t^2-1}} \frac{x^{|k|-1/2}}{\sqrt{x^2 - 2tx + 1}} dx$$

$$= 2\sqrt{2t} \int_{t+\sqrt{t^2-1}}^\infty \frac{x^{-|k|-1/2}}{\sqrt{1 - 2tx + x^2}} dx$$

$$= 2\sqrt{2t} Q_{|k|-1/2}(t)$$

by (2.2.9). Note that this remains true with the non-integer index $|k| - 1/2$. For the substitution $t = (1/2)(z + 1/z)$ gives

$$t + \sqrt{t^2 - 1} = \max(z, 1/z), \quad x^2 - 2tx + 1 = (x - z)(x - 1/z),$$

and so

$$I_{|k|}(1/t) = 2\sqrt{2t} \int_{\max(z,1/z)} \frac{x^{-|k|-1/2}}{\sqrt{(x - z)(x - 1/z)}} dx$$

$$= 2\sqrt{2t} Q_{|k|-1/2}((1/2)(z + 1/z)) = 2\sqrt{2t} Q_{|k|-1/2}(t),$$

by (1.5.24) with $l = |k| - 1/2$. We therefore have in (2.2.22)

$$I = \frac{1}{\pi} \sum_{k \in \mathbb{Z}} \int_0^\infty \int_0^\infty v_k(p') \overline{v_k(p)} Q_{|k|-1/2} \left(\frac{1}{2} \left[\frac{p}{p'} + \frac{p'}{p} \right] \right) dp dp'.$$

The function $Q_{|k|-1/2} \left(\frac{1}{2} \left[\frac{p}{p'} + \frac{p'}{p} \right] \right) (p'p)^{-1/2}$ is homogeneous of degree -1, and thus, on using Hilbert's inequality (1.7.10) and the fact that $Q_{|k|-1/2}$ decreases with $|k|$, it follows that

$$I \leq \frac{1}{\pi} \left(\int_0^\infty Q_{-1/2} \left(\frac{1}{2} \left[p + \frac{1}{p} \right] \right) p^{-1} dp \right) \sum_{k \in \mathbb{Z}} \int_0^\infty p |v_k(p)|^2 dp$$

$$= \sqrt{\frac{2}{\pi}} \left((\mathcal{M}Q_{-1/2}) \left(\frac{1}{2} \left[\cdot + \frac{1}{\cdot} \right] \right) \right) (-i/2) \int_{\mathbb{R}^2} |\mathbf{p}| |\hat{f}(\mathbf{p})|^2 d\mathbf{p}$$

where $\mathcal{M}$ denotes the Mellin transform. Hence by (1.5.21) and (1.5.22),

$$I \leq \left(\frac{\Gamma(1/4)}{\sqrt{2}\Gamma(3/4)}\right)^2 \int_{\mathbb{R}^2} |\mathbf{p}||\hat{f}(\mathbf{p})|^2 d\mathbf{p}.$$

The lemma follows from (2.2.21). □

2.2.2 The self-adjoint realisation

In view of Kato's inequality (2.2.17), a natural self-adjoint realisation of the operator (2.2.2) in $L^2(\mathbb{R}^3)$ is that given by the first representation theorem associated with the form

$$h[f] := \int_{\mathbb{R}^3} \left(|(\mathbb{H}_0^{1/2}f)(\mathbf{x})|^2 + V(\mathbf{x})|f(\mathbf{x}|^2\right) d\mathbf{x}, \quad V(\mathbf{x}) = -\gamma/|\mathbf{x}|, \qquad (2.2.25)$$

on $H^{1/2}(\mathbb{R}^3)$, for γ not exceeding the critical value $2/\pi$. The complete picture is provided by the following result from [Herbst (1977)].

Theorem 2.2.6.

(1) If $\gamma \leq 2/\pi$, the form (2.2.25) is non-negative on $H^{1/2}(\mathbb{R}^3)$. The operator $\mathbb{H}$ in (2.2.2) is defined to be the associated self-adjoint operator, having form domain $H^{1/2}(\mathbb{R}^3)$; it is the Friedrichs extension of the restriction of $\mathbb{H}_0 + V$ to $\mathcal{D}(\mathbb{H}_0) = H^1(\mathbb{R}^3)$. If $\gamma < 2/\pi, \mathbb{H}$ coincides with the form sum $\mathbb{H}_0 \dotplus V$.

(2) If $\gamma > 2/\pi$, the form (2.2.25) is unbounded below.

(3) If $\gamma < 1/2, \mathbb{H}$ is self-adjoint on $\mathcal{D}(\mathbb{H}_0)$, being the operator sum of $\mathbb{H}_0$ and V; it is essentially self-adjoint if $\gamma = 1/2$. For all γ and $f \in \mathcal{D}(\mathbb{H}_0)$, $Vf \in L^2(\mathbb{R}^3)$.

Proof.

(1) This is an immediate consequence of Kato's inequality (2.2.17) and the stability theorem for forms, Theorem 1.2.1.

(2) Suppose that $\mathbb{H}_0 + V$ is bounded below, $\mathbb{H}_0 + V \geq -E$, say, that is,

$$\int_{\mathbb{R}^3} \left\{\overline{f(\mathbf{x})}\sqrt{-\Delta + 1}f(\mathbf{x}) + [V(\mathbf{x}) + E]|f(\mathbf{x})|^2\right\} d\mathbf{x} \geq 0, \qquad (2.2.26)$$

for all f in the form domain of $\mathbb{H}_0 + V$, and, in particular, for all $f \in \mathbf{S}(\mathbb{R}^3)$. Let Λ denote the dilation $(\Lambda f)(\mathbf{x}) = f(\lambda\mathbf{x}), \lambda > 0$. Then

$$0 \leq \int_{\mathbb{R}^3} \{\sqrt{p^2 + 1}|[\mathbb{F}(\Lambda f)](\mathbf{p})|^2 d\mathbf{p} + \int_{\mathbb{R}^3}[V(\mathbf{x}) + E]|\Lambda f(\mathbf{x})|^2\}d\mathbf{x}$$

$$= \int_{\mathbb{R}^3} \sqrt{p^2 + \lambda^{-2}}|\hat{f}(\mathbf{p})|^2 d\mathbf{p} + \int_{\mathbb{R}^3}[V(\mathbf{x}) + E\lambda^{-1}]|f(\mathbf{x})|^2 d\mathbf{x}.$$

On allowing $\lambda \to \infty$, we obtain

$$\int_{\mathbb{R}^3} p|\hat{f}(\mathbf{x})|^2 d\mathbf{x} \geq \gamma \int_{\mathbb{R}^3} \frac{|f(\mathbf{x})|^2}{|\mathbf{x}|} d\mathbf{x},$$

which, from Theorem 2.2.4, can only hold for $\gamma \leq 2/\pi$.

(3) By Hardy's inequality (1.7.3), we have
$$\|Vf\| \le (2\gamma)\|\mathbb{H}_0 f\|, \quad f \in H^1(\mathbb{R}^3).$$

Hence by Theorem 1.1.1, $\mathbb{H}_0 + V$ is self-adjoint with domain $\mathcal{D}(\mathbb{H}_0)$ if $\gamma < 1/2$ and is the operator sum of $\mathbb{H}_0$ and V. If $\gamma = 1/2$, the essential self-adjointness of $\mathbb{H}_0 + V$ follows from Wüst's Theorem 1.1.2.

□

The following lower bound for the operator $\mathbb{H} = \mathbb{H}_0 + V$ in the first part of Theorem 2.2.6 is obtained by Herbst in [Herbst (1977)], Theorem 2.2.

Theorem 2.2.7. *If $\gamma < 2/\pi$, then*
$$\mathbb{H} \ge \sqrt{1 - (\pi\gamma/2)^2}.$$

Proof.   Let $\kappa := \pi\gamma/2 < 1$ and suppose that $0 < \lambda < 1$, so that $\lambda \in \rho(\mathbb{H}_0)$. Define
$$B := (\mathbb{H}_0 - \lambda)^{-1/2} V (\mathbb{H}_0 - \lambda)^{-1/2}$$
$$= -A^*A,$$

where
$$A = \left(|V|^{1/2}(-\Delta)^{-1/4}\right)\left((-\Delta)^{1/4}(\mathbb{H}_0 - \lambda)^{-1/2}\right).$$

From Kato's inequality (2.2.17) we have
$$\||V|^{1/2}(-\Delta)^{-1/4}\|^2 \le \kappa.$$

Also, on using the Spectral Theorem for $(-\Delta)^{1/2}$, we have
$$\|(-\Delta)^{1/4}(\mathbb{H}_0 - \lambda)^{-1/2}f\|^2 \le \int_{\mathbb{R}_+} x\left(\sqrt{x^2+1} - \lambda\right)^{-1} d\|E_x f\|^2$$
$$\le \sup_{x \ge 0}\left\{x\left(\sqrt{x^2+1} - \lambda\right)^{-1}\right\}\|f\|^2$$
$$= (1 - \lambda^2)^{-1/2}\|f\|^2.$$

Hence
$$\|B\| \le (1 - \lambda^2)^{-1/2}\kappa < 1$$

if $\lambda < \sqrt{1 - \kappa^2}$. For $u \in \mathcal{D}(\mathbb{H}_0)$ and $v \in \mathcal{Q}(\mathbb{H}_0) = \mathcal{Q}(\mathbb{H})$,
$$h[u, v] - \lambda(u, v) = ([\mathbb{H}_0 + V - \lambda]u, v)$$
$$= ([\mathbb{H}_0 - \lambda]^{1/2}[1 + B][\mathbb{H}_0 - \lambda]^{1/2}u, v),$$

while if $v \in \mathcal{D}(\mathbb{H})$, the left-hand side is equal to $(u, [\mathbb{H} - \lambda]v)$. It follows that
$$\mathbb{H} - \lambda = [\mathbb{H}_0 - \lambda]^{1/2}[1 + B][\mathbb{H}_0 - \lambda]^{1/2}$$

and so
$$[\mathbb{H} - \lambda]^{-1} = [\mathbb{H}_0 - \lambda]^{-1/2}[1 + B]^{-1}[\mathbb{H}_0 - \lambda]^{-1/2}$$

for $\lambda < \sqrt{1 - \kappa^2}$. Thus the resolvent set of $\mathbb{H}$ contains $(-\infty, \sqrt{1 - \kappa^2})$ and this implies the theorem.

□

In order to investigate further the range $1/2 < \gamma < 2/\pi$, we decompose $\mathbb{H}$ in terms of the spherical harmonics $Y_{l,m}$ and mimic some of the analysis of Section 2.2.1. We have

$$\mathbb{F}(\mathbb{H}_0 f)(\mathbf{p}) = \sqrt{p^2 + 1}\hat{f}(\mathbf{p})$$

and from (1.5.11) and (1.5.16)

$$\mathbb{F}(\frac{1}{|\cdot|}f)(\mathbf{p}) = \left(\mathbb{F}\frac{1}{|\cdot|} * \hat{f}\right)(\mathbf{p})$$

$$= \frac{1}{2\pi^2}\int_{\mathbb{R}^3}\frac{1}{|\mathbf{p} - \mathbf{p}'|^2}\hat{f}(\mathbf{p}')d\mathbf{p}'.$$

Since $|\cdot|^{-1}f \in L^2(\mathbb{R}^3)$ for all $f \in H^1(\mathbb{R}^3)$, by Hardy's inequality, $\mathbb{H}$ is defined on $H^1(\mathbb{R}^3)$ and

$$\mathbb{F}(\mathbb{H}f)(\mathbf{p}) = \sqrt{p^2 + 1}\hat{f}(\mathbf{p}) - \frac{\gamma}{2\pi^2}\int_{\mathbb{R}^3}\frac{1}{|\mathbf{p} - \mathbf{p}'|^2}\hat{f}(\mathbf{p}')d\mathbf{p}'.$$

Let

$$\hat{f}(\mathbf{p}) = \sum_{l,m}p^{-1}c_{l,m}(p)Y_{l,m}(\boldsymbol{\omega}_{\mathbf{p}}),$$

where

$$p^{-1}c_{l,m}(p) = \int_{\mathbb{S}^2}\hat{f}(p\boldsymbol{\omega}_{\mathbf{p}})Y_{l,m}(\boldsymbol{\omega}_{\mathbf{p}})d\boldsymbol{\omega}_{\mathbf{p}},$$

and

$$\sum_{l,m}\int_0^\infty(1 + p^2)|c_{l,m}(p)|^2dp = \int_{\mathbb{R}^3}(1 + |\mathbf{p}|^2)|\hat{f}(\mathbf{p})|^2d\mathbf{p} = \|f\|_{H^1(\mathbb{R}^3)}^2.$$

Then, on using the identity (2.2.5) and the orthonormality of the spherical harmonics, we obtain

$$\mathbb{F}(\mathbb{H}f)(\mathbf{p}) = \sum_{l,m}\left\{\sqrt{p^2 + 1}p^{-1}c_{l,m}(p)\right.$$

$$\left. - \frac{\gamma}{\pi}\int_0^\infty p^{-1}c_{l,m}(p')Q_l\left(\frac{1}{2}\left[\frac{p}{p'} + \frac{p'}{p}\right]\right)dp'\right\}Y_{l,m}(\boldsymbol{\omega}_{\mathbf{p}})$$

and so

$$(f, \mathbb{H}f) = \sum_{l,m}\left\{\int_0^\infty\sqrt{p^2 + 1}|c_{l,m}(p)|^2dp\right.$$

$$\left. - \frac{\gamma}{\pi}\int_0^\infty\int_0^\infty Q_l\left(\frac{1}{2}\left[\frac{p}{p'} + \frac{p'}{p}\right]\right)c_{l,m}(p')\overline{c_{l,m}(p)}dpdp'\right\}$$

$$= \sum_{l,m}(c_{l,m}, \mathfrak{h}_l c_{l,m})_{L^2(\mathbb{R}_+)}, \tag{2.2.27}$$

say. Formally, the operator $\mathfrak{h}_l$ is given by

$$\mathfrak{h}_l = E - K_l \tag{2.2.28}$$

where E is the operator of multiplication by $\sqrt{p^2 + 1}$ and

$$(K_l u)(p) = \frac{\gamma}{\pi} \int_0^\infty Q_l \left(\frac{1}{2} \left[\frac{p}{p'} + \frac{p'}{p} \right] \right) u(p') dp'. \tag{2.2.29}$$

Lemma 2.2.8. *For $u \in L^2(\mathbb{R}_+; p^2 dp)$,*

$$\|K_l u\|_{L^2(\mathbb{R}_+)} \le \frac{\gamma}{\pi} I_{l,1/2} \|u\|_{L^2(\mathbb{R}_+, p^2 dp)}, \tag{2.2.30}$$

where

$$I_{l,1/2} = \frac{\pi}{\sqrt{2}} \int_{-1}^1 \frac{P_l(t)}{\sqrt{1-t}} dt.$$

In particular

$$\|K_0 u\|_{L^2(\mathbb{R}_+)} \le 2\gamma \|u\|_{L^2(\mathbb{R}_+, p^2 dp)}$$

$$\|K_1 u\|_{L^2(\mathbb{R}_+)} \le \frac{2}{3}\gamma \|u\|_{L^2(\mathbb{R}_+, p^2 dp)},$$

and for $l \ge 1$,

$$\|K_l u\|_{L^2(\mathbb{R}_+)} \le \|K_1 u\|_{L^2(\mathbb{R}_+)}.$$

Proof.   We may clearly assume that u is real. We have

$$\|K_l u\|_{L^2(\mathbb{R}_+)}^2 = \left(\frac{\gamma}{\pi} \right)^2 \int_0^\infty \int_0^\infty k_l(p', p'')(p'p'')u(p')u(p'')dp'dp'',$$

where

$$k_l(p', p'') = (p'p'')^{-1} \int_0^\infty Q_l \left(\frac{1}{2} \left[\frac{p}{p'} + \frac{p'}{p} \right] \right) Q_l \left(\frac{1}{2} \left[\frac{p}{p''} + \frac{p''}{p} \right] \right) dp.$$

Note that $k_l(p', p'')$ is homogeneous of degree -1. Hence, by Hilbert's inequality (1.7.10),

$$\|K_l u\|_{L^2(\mathbb{R}_+)}^2 \le \left(\frac{\gamma}{\pi} \right)^2 C_l \|u\|_{L^2(\mathbb{R}_+, p^2 dp)}^2,$$

where

$$\begin{aligned}
C_l &= \int_0^\infty k_l(p', 1) p'^{-3/2} dp' \\
&= \int_0^\infty p'^{-3/2} \left(\int_0^\infty Q_l \left(\frac{1}{2} \left[\frac{p'}{p} + \frac{p}{p'} \right] \right) Q_l \left(\frac{1}{2} \left[p + \frac{1}{p} \right] \right) dp \right) dp' \\
&= \int_0^\infty Q_l \left(\frac{1}{2} \left[p + \frac{1}{p} \right] \right) p^{-1/2} dp \int_0^\infty Q_l \left(\frac{1}{2} \left[x + \frac{1}{x} \right] \right) x^{-3/2} dx.
\end{aligned}$$

Let

$$I_{l,\beta} := \int_0^\infty Q_l \left(\frac{1}{2} \left[x + \frac{1}{x} \right] \right) x^{-\beta} dx$$

with $\beta = 1/2$ or $3/2$. It follows that $I_{l,1/2} = I_{l,3/2}$, and from (2.2.3)

$$I_{l,1/2} = \int_0^\infty \left(\frac{1}{2} \int_{-1}^1 \frac{P_l(t)}{\frac{1}{2}[x + 1/x] - t} dt \right) x^{-1/2} dx$$

$$= \int_{-1}^1 \left\{ \int_0^\infty \frac{\sqrt{x}}{(x^2 + 1 - 2tx)} dx \right\} P_l(t) dt. \tag{2.2.31}$$

The inequality (2.2.30) follows since

$$\int_0^\infty \frac{\sqrt{x}}{(x^2 + 1 - 2tx)} dx = \frac{\pi}{\sqrt{2(1-t)}}.$$

On substituting $P_0(t) = 1, P_1(t) = t$, it is readily shown that $I_{0,1/2} = 2\pi, I_{1,1/2} = 2\pi/3$ and from these, and the fact that the Q_l decrease with l, we deduce the rest of the lemma. □

The following theorem is now an immediate consequence of Lemma 2.2.8, Theorem 1.1.1 and Theorem 1.1.2.

Theorem 2.2.9.

(1) For $\gamma < 1/2, \mathfrak{h}_0$ is self-adjoint with domain $L^2(\mathbb{R}_+; (p^2 + 1)dp)$.

(2) For $\gamma = 1/2, \mathfrak{h}_0$ on $L^2(\mathbb{R}_+; (p^2 + 1)dp)$ is essentially self-adjoint.

(3) For $l \geq 1, \mathfrak{h}_l$ with domain $L^2(\mathbb{R}_+; (p^2 + 1)dp)$ is self-adjoint for all $\gamma < 3/2$ and in particular throughout the critical range $\gamma \leq 2/\pi$.

This provides additional information to that in the third part of Theorem 2.2.6 since the operator $\mathbb{H}$ in (2.2.2) is self-adjoint or essentially-self-adjoint if and only if the same is true for all the operators $\mathfrak{h}_l$; cf. Theorem 2.1.1. Thus it is $\mathfrak{h}_0$ which determines the range of self-adjointness and essential-self-adjointness. We examine the behaviour outside the range $\gamma \leq 1/2$ next.

2.2.3 *The range $1/2 < \gamma < 2/\pi$ for $\mathfrak{h}_0$*

To analyse $\mathfrak{h}_0$ in this range, we follow the development in [Yaouanc *et al.* (1997)] using Mellin transforms. Recall from Section 1.5 that the Mellin transform $\mathcal{M}$ is a unitary map from $L^2(\mathbb{R}_+)$ onto $L^2(\mathbb{R})$ defined by

$$\psi^\sharp(s) := \mathcal{M}\psi(s) := \frac{1}{\sqrt{2\pi}} \int_0^\infty p^{-\frac{1}{2}-is} \psi(p) dp$$

with inverse

$$\mathcal{M}^{-1}\psi^\sharp(s) := \frac{1}{\sqrt{2\pi}} \int_{-\infty}^\infty p^{-\frac{1}{2}+is} \psi^\sharp(s) ds.$$

Also, with the convolution defined by

$$(\psi \star \phi)(s) := \frac{1}{\sqrt{2\pi}} \int_0^\infty \psi(p) \phi \left(\frac{s}{p} \right) p^{-1} dp$$

we have

$$(\mathcal{M}(\psi \star \phi))(s) = \psi^{\#}(s)\phi^{\#}(s).$$

Since the multiplication operator $\sqrt{p^2 + 1} - p$ has a bounded self-adjoint extension on $L^2(\mathbb{R}_+)$, the properties of $\mathfrak{h}_0$ we seek will follow from those of the operator

$$\mathfrak{h} = p - K_0, \quad \mathcal{D}(\mathfrak{h}) = L^2(\mathbb{R}_+; (p^2 + 1)dp). \tag{2.2.32}$$

Set $\tilde{Q}_0(x) := Q_0(\frac{1}{2}[x + \frac{1}{x}])$. Then, by (2.2.29),

$$\frac{\pi}{\gamma}(K_0\psi)(p) = \int_0^\infty Q_0\left(\frac{1}{2}\left[\frac{p}{p'} + \frac{p'}{p}\right]\right)\psi(p')dp'$$

$$= \sqrt{2\pi}\left(\tilde{Q}_0 \star [(\cdot)\psi(\cdot)]\right)(p)$$

and hence

$$\frac{\pi}{\gamma}[\mathcal{M}(K_0\psi)](s) = \sqrt{2\pi}\mathcal{M}(\tilde{Q}_0)(s)(\mathcal{M}[(\cdot)\psi(\cdot)])(s).$$

Moreover,

$$(\mathcal{M}[(\cdot)\psi(\cdot)])(s) = \psi^{\#}(s + i)$$

and

$$\sqrt{2\pi}\left[\mathcal{M}(\tilde{Q}_0)\right](s) = \pi V(s + i/2),$$

where

$$V(z) := \int_0^\infty x^{-1-iz}\tilde{Q}_0(x)dx.$$

This is the case $l = 0$ of the function evaluated in (1.5.22), thus

$$V(z) = \frac{1}{2}\frac{\Gamma\left(\frac{1-iz}{2}\right)\Gamma\left(\frac{1+iz}{2}\right)}{\Gamma\left(\frac{2-iz}{2}\right)\Gamma\left(\frac{2+iz}{2}\right)}.$$

On using the identities

$$\Gamma(w)\Gamma(1 - w) = \pi/\sin(\pi w)$$
$$\Gamma(w - 1) = w\Gamma(w),$$

we obtain

$$V(z) = \frac{\tanh(\pi z/2)}{z}. \tag{2.2.33}$$

The operator $\mathfrak{h}^{\#} := \mathcal{M}\mathfrak{h}\mathcal{M}^{-1}$ therefore satisfies

$$\mathfrak{h}^{\#}\psi^{\#}(s) = [1 - \gamma V(s + i/2)]\psi^{\#}(s + i). \tag{2.2.34}$$

The domain of $\mathfrak{h}$ consists of functions $\psi \in L^2(\mathbb{R}_+)$ which are such that $p\psi \in L^2(\mathbb{R}_+)$; hence $\psi^{\#}, \psi^{\#}(\cdot + i) \in L^2(\mathbb{R})$. Since $V(\cdot + i/2)$ is bounded on $\mathbb{R}$ we have that $\mathfrak{h}^{\#}\psi^{\#}$ is defined for all $\psi \in \mathcal{D}(\mathfrak{h})$ and

$$\mathcal{D}(\mathfrak{h}^{\#}) = \mathcal{D}(p^{\#}) = \{\psi^{\#} : \psi^{\#}, (p^{\#}\psi^{\#}) \in L^2(\mathbb{R})\}$$

where

$$(p^\sharp \psi^\sharp)(s) = \psi^\sharp(s+i).$$

The following theorem from [Titchmarsh (1986)] gives an important equivalent characterisation of functions in $\mathcal{D}(p^\sharp)$.

Theorem 2.2.10. *A function $\psi^\sharp \in \mathcal{D}(p^\sharp)$ if and only if there exists a unique function $\Psi^\sharp$ that is holomorphic in the strip $S(1) := \{z \in \mathbb{C} : 0 < \mathrm{Im}[z] < 1\}$ and is such that:*

$$\lim_{y \to 0} \int_{\mathbb{R}} |\Psi^\sharp(s+iy) - \psi^\sharp(s)|^2 ds = 0,$$

$$\lim_{y \to 1} \int_{\mathbb{R}} |\Psi^\sharp(s+iy) - \psi^\sharp(s+i)|^2 ds = 0,$$

$$\sup_{y \in (0,1)} \int_{\mathbb{R}} |\Psi^\sharp(s+iy)|^2 ds < \infty.$$

We shall identify the function $\Psi^\sharp$ with $\psi^\sharp$ in what follows.

Lemma 2.2.11. *The operator $\mathfrak{h}^\sharp$ is closed in $L^2(\mathbb{R})$ if and only if $\gamma \notin \Delta_0 := \{V(s+i/2)^{-1} : s \in \mathbb{R}\}$.*

Proof. If $\gamma \notin \Delta_0$, $[1 - \gamma V(s+i/2)]^{-1}$ is bounded on $\mathbb{R}$, since $V(s+i/2) \to 0$ as $s \to \pm\infty$. Let $\psi_n^\sharp \in \mathcal{D}(\mathfrak{h}^\sharp), n \in \mathbb{N}$, be such that $\psi_n^\sharp \to \psi^\sharp$ and $\{\mathfrak{h}^\sharp \psi_n^\sharp\}$ is convergent in $L^2(\mathbb{R})$. Then $\{p^\sharp \psi_n^\sharp\} = \{\psi_n^\sharp(\cdot + i)\}$ is convergent in $L^2(\mathbb{R})$ by (2.2.34). Since $\mathcal{M}$ is unitary and p is closed in $L^2(\mathbb{R}_+)$, it follows that $p^\sharp$ is closed, and consequently, $p^\sharp \psi_n^\sharp \to p^\sharp \psi^\sharp$ in $L^2(\mathbb{R})$. In view of the boundedness of $V(\cdot + i/2)$ on $\mathbb{R}$, we therefore have that $\mathfrak{h}^\sharp \psi_n^\sharp \to \mathfrak{h}^\sharp \psi^\sharp$ in $L^2(\mathbb{R})$ and, hence, $\mathfrak{h}^\sharp$ is closed. This implies that $\mathfrak{h}$ is closed and hence that $\mathfrak{h}_0$ is closed since $\mathfrak{h}_0 - \mathfrak{h}$ is bounded on $L^2(\mathbb{R}_+)$.

To prove the converse, suppose $\gamma = V(s_0 + i/2)^{-1}$ for some $s_0 \in \mathbb{R}$, and set

$$\psi_n^\sharp = (s - s_0 - i - i/n)^{-1}, \quad \varphi(s) = (s - s_0 - i)^{-1}.$$

Then, it is easy to see that $\psi_n^\sharp \to \varphi$ in $L^2(\mathbb{R})$ and $\mathfrak{h}^\sharp \psi_n^\sharp(s) = \{[1 - \gamma V(s+i/2)]\psi_n^\sharp(s+i)$ converges in $L^2(\mathbb{R}^n)$. If $\mathfrak{h}^\sharp$ is closed, the previous sentence would imply that $\varphi \in \mathcal{D}(\mathfrak{h}^\sharp)$ and $\mathcal{M}^{-1}\varphi \in \mathcal{D}(\mathfrak{h})$. But, by (1.5.25), $[\mathcal{M}^{-1}\varphi](p) = i\sqrt{2\pi}\theta(p-1)/p^{3/2 - is_0}$, which does not lie in $\mathcal{D}(p)$ and, hence, $\varphi \notin \mathcal{D}(\mathfrak{h}^\sharp)$. Therefore $\mathfrak{h}^\sharp$ is not closed. This in turn implies that $\mathfrak{h}$ and $\mathfrak{h}_0$ are not closed if $\gamma \in \Delta_0$. $\square$

Corollary 2.2.12. *The operator $\mathfrak{h}^\sharp$ (and consequently $\mathfrak{h}$ and $\mathfrak{h}_0$) is closed if and only if $\gamma \neq 1/2$.*

Proof. From the expansion (1.5.23) with $l = 0$,

$$V(z) = \frac{4}{\pi} \sum_{n=0}^{\infty} \frac{1}{(2n+1)^2 + z^2}, \tag{2.2.35}$$

it follows that $V(s+i/2)$ is real if and only if $s = 0$. Since

$$[V(i/2)] = \frac{\tan(\pi/4)}{1/2} = 2,$$

the corollary follows. $\square$

The set Δ_0 is a curve, which separates the complex plane into connected components Δ_1 and Δ_2, where $\Delta_1 = \{V(z)^{-1} : |\mathrm{Im}[z]| < 1/2\}$ and $\Delta_2 = \mathbb{C} \setminus (\Delta_0 \cup \Delta_1)$. For $\alpha \in \Delta_1$, $1 - \alpha V(z)$ has precisely two zeros in the strip $|\mathrm{Im}[z]| < 1/2$. To see this, consider the image of the closed path $C : i/2 - X \to i/2 + X \to -i/2 + X \to -i/2 - X$ under the map $z \mapsto V(z)$. This is a closed curve $V(C)$, which is covered twice in view of (2.2.35), and the image of the interior of C is the interior of $V(C)$. Therefore the values of $V(z)$ for $|\mathrm{Im}[z]| < 1/2$ are attained twice. Moreover, if $z_0 \neq 0$ is one zero of $1 - \alpha V(z)$, the other one must be $-z_0$ since $V(z)$ is quadratic in z. If $z_0 = 0$ it is a double zero. The interval $(1/2, \infty)$ clearly lies in Δ_1.

Theorem 2.2.13. *For $1/2 < \gamma \leq 2/\pi$, the adjoint of $\mathfrak{h}^\sharp$ is an extension of $\mathfrak{h}^\sharp$ and given by*

$$\left[(\mathfrak{h}^\sharp)^* \psi^\sharp \right](s) = [1 - \gamma V(s + i/2)] \, \psi^\sharp(s + i) \tag{2.2.36}$$

with domain

$$\mathcal{D}((\mathfrak{h}^\sharp)^*) = \{ \psi^\sharp \in L^2(\mathbb{R}) : [1 - \gamma V(\cdot - i/2)]\psi^\sharp \in \mathcal{D}(p^\sharp) \}. \tag{2.2.37}$$

The solutions $\pm z_0$ of $1 - \gamma V(z) = 0$ in $|\mathrm{Im}[z]| < 1/2$ are such that $z_0 \neq 0$ corresponds to $\gamma < 2/\pi$ and $z_0 = 0$ to $\gamma = 2/\pi$. The domain of the adjoint $(\mathfrak{h}^\sharp)^*$ is the set of functions $\varphi^\sharp$ of the form

$$\varphi^\sharp(s) = \psi^\sharp(s) + \frac{c_1}{\left(s - \frac{i}{2} - z_0\right)} + \frac{c_2}{\left(s - \frac{i}{2} + z_0\right)} \quad \text{if} \quad z_0 \neq 0 \ \ (\gamma < 2/\pi) \tag{2.2.38}$$

and

$$\varphi^\sharp(s) = \psi^\sharp(s) + \frac{c_1}{\left(s - \frac{i}{2}\right)} + \frac{c_2}{\left(s - \frac{i}{2}\right)^2} \quad \text{if} \quad z_0 = 0 \ \ (\gamma = 2/\pi), \tag{2.2.39}$$

where $\psi^\sharp \in \mathcal{D}(\mathfrak{h}^\sharp)$ and $c_1, c_2 \in \mathbb{C}$.

Proof. The multiplication operator $A := 1 - \gamma V(\cdot + i/2)$ is bounded on $L^2(\mathbb{R})$ and has adjoint $A^* = 1 - \gamma V(\cdot - i/2)$. Since $\mathfrak{h}^\sharp = Ap^\sharp$, we have that $\varphi \in \mathcal{D}((\mathfrak{h}^\sharp)^*)$ if and only if $A^*\varphi \in \mathcal{D}(p^\sharp)$, and then, $(\mathfrak{h}^\sharp)^*\varphi = p^\sharp A^*\varphi = Ap^\sharp\varphi$, as asserted in (2.2.36) and (2.2.37). The function $V(z - i/2)$ is analytic and bounded in the strip $-1/2 + \varepsilon < \mathrm{Im}[z] < 3/2 + \varepsilon$, $\varepsilon > 0$, and from this we can infer that $A^*\varphi \in \mathcal{D}(p^\sharp)$ if $\varphi \in \mathcal{D}(p^\sharp)$. Hence $(\mathfrak{h}^\sharp)^*$ is an extension of $\mathfrak{h}^\sharp$, i.e., $\mathfrak{h}^\sharp$ is a symmetric operator.

Consider the case $z_0 \neq 0$; the case $z_0 = 0$ can be treated similarly. It is easily verified that the functions $[1 - \gamma V(s - i/2)](s - i/2 \pm z_0)^{-1}$ belong to $\mathcal{D}(p^\sharp)$ and, hence, any $\varphi^\sharp$ given by (2.2.38) belongs to $\mathcal{D}((\mathfrak{h}^\sharp)^*)$.

Conversely, suppose that $\varphi^\sharp \in \mathcal{D}((\mathfrak{h}^\sharp)^*)$, so that $[1 - \gamma V((\cdot) - i/2)]\varphi^\sharp \in \mathcal{D}(p^\sharp)$ and $\varphi^\sharp \in L^2(\mathbb{R})$. We are required to show that for some constants c_1, c_2,

$$\psi^\sharp(s) := \varphi^\sharp(s) - \frac{c_1}{\left(s - \frac{i}{2} - z_0\right)} - \frac{c_2}{\left(s - \frac{i}{2} + z_0\right)}$$

belongs to $\mathcal{D}(p^\sharp)$. Since $1 - \gamma V(z)$ has simple zeros at $\pm z_0$, then, on using the characterisation of $\mathcal{D}(p^\sharp)$ in Theorem 2.2.10, $\varphi^\sharp$ may have simple poles in the strip

$0 < \text{Im}[z] < 1$ at $z = i/2 + z_0$ and $z = i/2 - z_0$, which would mean that $\varphi^\sharp \notin \mathcal{D}(p^\sharp)$. We can choose constants c_1, c_2 such that

$$\psi^\sharp(z) := \phi^\sharp(z) - \frac{c_1}{z - i/2 - z_0} - \frac{c_2}{s - i/2 + z_0}$$

is holomorphic in $0 < \text{Im}[z] < 1$ and it remains to prove that

$$\sup_{0 < y < 1} \int_{\mathbb{R}} |\psi^\sharp(s + iy)|^2 ds < \infty.$$

For suitable $\varepsilon, X, i/2 \pm z_0$ are interior points of the rectangle $R := \{z \in \mathbb{C} : \varepsilon \leq \text{Im}[z] \leq 1 - \varepsilon, |\text{Re}[z]| \leq X\}$. Let C_1 be an upper bound for $|\psi^\sharp(z)|$ in R and $C_2 > 0$ a lower bound of $|1 - \gamma V(z - i/2)|$ in $\{z \in \mathbb{C} : 0 \leq \text{Im}[z] \leq 1\} \setminus R$. Then, for $0 \leq y \leq 1$,

$$\sup_{0 \leq y \leq 1} \int_{\mathbb{R}} |\psi^\sharp(s + iy)|^2 ds \leq \sup_{\varepsilon \leq y \leq 1 - \varepsilon} \int_{|s| < X} |\psi^\sharp(s + iy)|^2 ds$$

$$+ \left(\sup_{y \in [0,\varepsilon) \cup (1-\varepsilon,1]} \int_{|s| < X} + \sup_{y \in [0,1]} \int_{|s| > X} \right) \left| \frac{[1 - \gamma V(s - i/2 + iy)]}{[1 - \gamma V(s - i/2 + iy)]} \psi^\sharp \right|^2 ds$$

$$\leq C_1^2 X + C_2^{-2} \sup_{0 \leq y \leq 1} \int_{\mathbb{R}} |[1 - \gamma V(s - i/2 + iy)] \psi^\sharp(s + iy)|^2 ds.$$

Since $[1 - \gamma V(\cdot - i/2)] \varphi^\sharp$ is assumed to belong to $\mathcal{D}(p^\sharp)$, the boundedness of the second term follows from Theorem 2.2.10. The theorem is therefore proved. $\quad\square$

Corollary 2.2.14. *If* $1/2 < \gamma \leq 2/\pi$, $\mathfrak{h}^\sharp$ *is a closed symmetric operator in* $L^2(\mathbb{R})$ *with deficiency indices* $(1, 1)$.

Proof. It follows from Theorem 2.2.13 that $\dim \left(\mathcal{D}([\mathfrak{h}^\sharp]^*) / \mathcal{D}(\mathfrak{h}^\sharp) \right) = 2$. Hence by (1.1.1), the deficiency indices (n_+, n_-) of $\mathfrak{h}^\sharp$ satisfy $n_+ + n_- = 2$. The operator $\mathfrak{h}$ is real, that is $\overline{\mathfrak{h}\psi} = \mathfrak{h}\overline{\psi}$, and hence

$$(\overline{\mathfrak{h}^\sharp \psi^\sharp})(s) = \overline{[\mathcal{M}(\mathfrak{h}\psi)](s)}$$
$$= [\mathcal{M}(\overline{\mathfrak{h}\psi})](-s) = [\mathcal{M}(\mathfrak{h}\overline{\psi})](-s)$$
$$= [\mathfrak{h}^\sharp \overline{\psi^\sharp}](s).$$

Thus $\mathfrak{h}^\sharp$ is real and has equal deficiency indices, whence $n_+ = n_- = 1$. $\quad\square$

Corollary 2.2.15. *If* $1/2 < \gamma \leq 2/\pi$, $\mathfrak{h}$ *is a closed symmetric operator in* $L^2(\mathbb{R}_+)$ *with deficiency indices* $(1, 1)$.

Proof. The result follows immediately from Corollary 2.2.14 since $\mathcal{M}$ is unitary.

$\quad\square$

Remark 2.2.16. In [Yaouanc *et al.* (1997)] it is shown that for $1/2 < \gamma \leq 2/\pi$, the Friedrichs extension of $\mathfrak{h}^\sharp$ is the restriction of its adjoint to the set of functions of the form

$$\varphi^\sharp = \psi^\sharp + \frac{c}{s - i/2 + i\beta}$$

where $\varphi^\sharp \in \mathcal{D}(p^\sharp), c \in \mathbb{C}$, and β satisfies

$$V(i\beta)^{-1} = \beta / \tan(\pi\beta/2) = \gamma, \quad -1/2 < \beta \leq 0.$$

Thus the domain of the Friedrichs extension consists of functions that are mero-morphic in $0 < \text{Im}[z] < 1$ with, possibly, a pole at $z = i/2 - i\beta$. As γ increases to $2/\pi, \beta$ increases to 0 and the pole approaches $i/2$. On increasing β further, the corresponding values of $\gamma = V(i\beta)^{-1}$ decrease from $2/\pi$ and the pole reaches the real axis only when $\beta = 1/2$, and hence $\gamma = 1/2$.

For any complex $\beta \neq 2n, n \in \mathbb{Z} \setminus \{0\}$, a family $\mathfrak{h}(\beta)^\sharp$ of operators in $L^2(\mathbb{R})$ is defined by

$$\left(\mathfrak{h}(\beta)^\sharp \varphi^\sharp\right)(s) := \left[1 - \frac{V(s+i/2)}{V(i\beta)}\right] \varphi^\sharp(s+i)$$

on the set of functions $\varphi^\sharp$ which are such that

$$\psi^\sharp(s) = \frac{s - i/2 + i\beta}{s - iA} \varphi^\sharp \in \mathcal{D}(p^\sharp),$$

where $A > 1$ is a fixed constant. For $-1/2 \leq \beta < 1/2, \mathfrak{h}(\beta)^\sharp$ is a proper extension of $\mathfrak{h}^\sharp$, being the Friedrichs extension for $-1/2 < \beta \leq 0$, the closure of $\mathfrak{h}^\sharp$ when $\beta = -1/2$ and another self-adjoint extension for $0 < \beta < 1/2$. When $|\beta| > 1/2, \mathfrak{h}(\beta)^\sharp$ coincides with $\mathfrak{h}^\sharp$.

In [Yaouanc *et al.* (1997)], Theorem 3, it is shown in particular that, for β in $\{z \in \mathbb{C} : -1/2 \leq \text{Re } z < 1/2\}$, the $\mathfrak{h}(\beta)^\sharp$ form a self-adjoint holomorphic family in the sense of Kato. Also, in [Yaouanc *et al.* (1997)], Theorem 4, for $-1/2 \leq \beta < 1/2$, it is proved that the operator

$$\mathfrak{h}_0(\beta)^\sharp := \left(\sqrt{p^2 + 1} - p\right)^\sharp + \mathfrak{h}(\beta)^\sharp$$

is positive, has essential spectrum $[1, \infty)$ and with only isolated eigenvalues of finite multiplicity in $(0, 1)$. The smallest eigenvalue tends to 0 as $\beta \to 1/2$.

On reverting to the original representations in $L^2(\mathbb{R}_+)$, and using (1.5.25), it follows that for $1/2 < \gamma \leq 2/\pi$ and $\gamma = \beta/\tan(\pi\beta/2), 0 \leq \beta < 1/2$, self-adjoint extensions of $\mathfrak{h}_0$ are given by $\mathfrak{h}_0(\beta)$ defined on the set of functions of the form

$$\varphi(p) = \psi(p) + c\theta(p-1)p^{-1\pm\beta}$$

where $\psi \in \mathcal{D}(p) = L^2(\mathbb{R}_+; (1+p^2)dp)$ and $c \in \mathbb{C}$, the negative sign in $p^{-1\pm\beta}$ giving the Friedrichs extension.

We refer to [Yaouanc *et al.* (1997)] for a detailed analysis.

2.3   The Brown–Ravenhall operator

The Brown–Ravenhall operator describes an electron under a Coulomb force and subject to relativity. It was initially proposed as an alternative to the Dirac oper-ator for analysing multi-particle relativistic systems: even in the case of two non-interacting electrons in the electric field of the nucleus the spectrum of the Dirac operator coincides with the whole real line, and this situation is not improved when the interaction between electrons is taken into account. The Brown–Ravenhall operator is defined in terms of the associated Dirac operator $\mathbb{D}$ and the projection Λ_+ onto the positive spectral subspace of the free Dirac operator $\mathbb{D}_0$. Specifically, $\Lambda_+ = \chi_{(0,\infty)}(\mathbb{D}_0)$, where $\chi_{(0,\infty)}$ denotes the characteristic function of $(0,\infty)$. From the definition of the free Dirac operator $\mathbb{D}_0$ in (2.1.10), we have

$$\mathbb{F}\mathbb{D}_0\mathbb{F}^{-1} = \mathbb{M}_0, \tag{2.3.1}$$

where $\mathbb{M}_0$ is the operator of multiplication by $\sqrt{p^2+1}\Lambda(\mathbf{p})$ and

$$\Lambda(\mathbf{p}) = \frac{\boldsymbol{\alpha}\cdot\mathbf{p} + \beta}{\sqrt{p^2+1}}. \tag{2.3.2}$$

The 4×4 matrix $\Lambda(\mathbf{p})$ is a unitary involution, i.e., $\Lambda(\mathbf{p})^2 = I_4$ and, hence, has double eigenvalues at ± 1. The projections $\Lambda_\pm(\mathbf{p})$ onto the eigenspaces at ± 1 are given by

$$\Lambda_\pm(\mathbf{p}) = \frac{1}{2}\left\{ I_4 \pm \frac{\boldsymbol{\alpha}\cdot\mathbf{p} + \beta}{\sqrt{p^2+1}} \right\}. \tag{2.3.3}$$

The operator $\Lambda_+ = \chi_{(0,\infty)}(\mathbb{D}_0)$, which is known as the *Casimir projection*, is then given by

$$\Lambda_+ = \mathbb{F}^{-1}\Lambda_+(\mathbf{p})\mathbb{F}. \tag{2.3.4}$$

Note that, with $E(\mathbf{p}) := \sqrt{p^2+1}, p = |\mathbf{p}|$,

$$\begin{aligned}
\mathbb{D}_0\Lambda_+ &= \mathbb{F}^{-1}E(\mathbf{p})\Lambda(\mathbf{p})\Lambda_+(\mathbf{p})\mathbb{F} \\
&= \mathbb{F}^{-1}E(\mathbf{p})\Lambda_+(\mathbf{p})\mathbb{F} \\
&= \mathbb{F}^{-1}E(\mathbf{p})\mathbb{F}\mathbb{F}^{-1}\Lambda_+(\mathbf{p})\mathbb{F} \\
&= \sqrt{-\Delta+1}\Lambda_+,
\end{aligned} \tag{2.3.5}$$

since $\sqrt{-\Delta+1} = \mathbb{F}^{-1}\sqrt{p^2+1}\mathbb{F}$.

The *absolute value* $|\mathbb{D}_0|$, i.e., the positive square root of $\mathbb{D}_0^2$, is $\sqrt{-\Delta+1}I_4$. Also, if Λ_- denotes the projection onto the negative spectral subspace of $\mathbb{D}_0$ and $U := \Lambda_+ \oplus \Lambda_-$, then

$$\mathbb{D}_0 = \sqrt{-\Delta+1}\, U = U\sqrt{-\Delta+1}$$

is the *polar decomposition* of $\mathbb{D}_0$.

In terms of the Pauli matrices, $\Lambda_+(\mathbf{p})$ has the form

$$\Lambda_+(\mathbf{p}) = \frac{1}{2E(\mathbf{p})}\begin{pmatrix} E(p)+1 & \boldsymbol{\sigma}\cdot\mathbf{p} \\ \boldsymbol{\sigma}\cdot\mathbf{p} & E(\mathbf{p})-1 \end{pmatrix}. \tag{2.3.6}$$

Corresponding to its eigenvalue $1, \Lambda_+(\mathbf{p})$ has the two orthonormal eigenvectors

$$\frac{1}{N(\mathbf{p})}\begin{pmatrix} [E(\mathbf{p})+1]u_j \\ (\boldsymbol{\sigma}\cdot\mathbf{p})u_j \end{pmatrix}, \; j = 1,2, \quad N(\mathbf{p}) = \sqrt{2E(\mathbf{p})[E(\mathbf{p})+1]}, \tag{2.3.7}$$

where u_1, u_2 are orthonormal vectors in $\mathbb{C}^2$, say $u_1 = (1,0)^t, u_2 = (0,1)^t$, where t denotes the transpose of the vector. Thus any eigenvector of $\Lambda_+(\mathbf{p})$ has the form

$$\widehat{\psi}(\mathbf{p}) = \frac{1}{N(\mathbf{p})} \begin{pmatrix} [E(\mathbf{p})+1]u(\mathbf{p}) \\ (\boldsymbol{\sigma} \cdot \mathbf{p})u(\mathbf{p}) \end{pmatrix}, \tag{2.3.8}$$

where $u(\mathbf{p}) \in \mathbb{C}^2$: u is called a 2-spinor or *Pauli spinor*. Furthermore,

$$\int_{\mathbb{R}^3} |\widehat{\psi}(\mathbf{p})|^2 d\mathbf{p} = \int_{\mathbb{R}^3} |u(\mathbf{p})|^2 d\mathbf{p}.$$

A Dirac spinor in the positive spectral subspace of $\mathbb{D}_0$ is, therefore, of the form $\psi = \mathbb{F}^{-1}\widehat{\psi}$ for $u \in L^2(\mathbb{R}^3, \mathbb{C}^2)$. Conversely, any such spinor must lie in the positive spectral subspace of $\mathbb{D}_0$ since

$$\left[\mathbb{F}\Lambda_+ \mathbb{F}^{-1}\widehat{\psi} \right](\mathbf{p}) = \Lambda_+(\mathbf{p})\widehat{\psi}(\mathbf{p}) = \widehat{\psi}(\mathbf{p})$$

and so $\Lambda_+ \mathbb{F}^{-1}\widehat{\psi} = \mathbb{F}^{-1}\widehat{\psi}$. Consequently, there is a one-one correspondence between Dirac spinors ψ in the positive spectral subspace of $\mathbb{D}_0$ and $\mathbb{F}^{-1}\widehat{\psi}$, where $\widehat{\psi}(\mathbf{p})$ is of the form (2.3.8), with u a Pauli spinor in $L^2(\mathbb{R}^3, \mathbb{C}^2)$. Note that if $\widehat{\psi}^{(j)}(\mathbf{p})$ is of the form (2.3.8) for $u = u^{(j)}, j = 1, 2$, then

$$< \widehat{\psi}^{(1)}(\mathbf{p}), \widehat{\psi}^{(2)}(\mathbf{p}) > = < u^{(1)}(\mathbf{p}), u^{(2)}(\mathbf{p}) > \tag{2.3.9}$$

so that $\widehat{\psi}^{(1)}(\mathbf{p}), \widehat{\psi}^{(2)}(\mathbf{p})$ are orthogonal in $\mathbb{C}^4$ if and only if $u^{(1)}(\mathbf{p}), u^{(2)}(\mathbf{p})$ are orthogonal in $\mathbb{C}^2$. In (2.3.9), as is clear from the context, the inner products $\langle \cdot, \cdot \rangle$ on the left- and right-hand sides are those of $\mathbb{C}^4$ and $\mathbb{C}^2$ respectively.

2.3.1   *Coulomb potentials*

We saw in Theorem 2.1.6 that with $\mathbb{D}_0'$ the restriction of $\mathbb{D}_0$ to $C_0^\infty(\mathbb{R}^3, \mathbb{C}^4)$,

$$\mathbb{D}' := \mathbb{D}_0' + V, \quad V(\mathbf{x}) = -\gamma/|\mathbf{x}|$$

is essentially self-adjoint if and only if $|\gamma| \leq \sqrt{3}/2$. Also, by Hardy's inequality (1.7.3) its unique self-adjoint extension (and closure) $\mathbb{D}$ is the operator sum $\mathbb{D} = \mathbb{D}_0 + V$ if $|\gamma| < 1/2$ since then the multiplication operator V is bounded relative to $\mathbb{D}_0'$ with relative bound < 1. The Brown–Ravenhall operator is given formally as

$$\mathbb{B} = \Lambda_+ (\mathbb{D}_0 + V) \Lambda_+ \tag{2.3.10}$$

acting in $L^2(\mathbb{R}^3, \mathbb{C}^4)$, or, equivalently,

$$\mathbb{B} = \Lambda_+ (\mathbb{D}_0 + V) \tag{2.3.11}$$

acting in $\mathcal{H} = \Lambda_+ L^2(\mathbb{R}^3, \mathbb{C}^4)$. The self-adjoint operator it generates in $\mathcal{H}$ will be defined by the quadratic form $(\psi, \mathbb{B}\psi)$ on the Schwartz space $\mathbf{S}^4 := \Lambda_+ \mathbf{S}(\mathbb{R}^3, \mathbb{C}^4)$ of

Dirac spinors $\psi \in \mathcal{H}$ with rapidly decreasing components. By the Parseval identity, for $\psi \in \mathcal{H}$,

$$
\begin{aligned}
(\psi, \mathbb{B}\psi) &= \int_{\mathbb{R}^3} \langle \mathbb{F}\psi, \mathbb{F}(\mathbb{B}\psi) \rangle(\mathbf{p}) d\mathbf{p} \\
&= \int_{\mathbb{R}^3} \langle \hat{\psi}, \mathbb{F}[\Lambda_+(\mathbb{D}_0 + V)\psi] \rangle(\mathbf{p}) d\mathbf{p} \\
&= \int_{\mathbb{R}^3} \langle \hat{\psi}, \mathbb{F}[\sqrt{-\Delta + 1}\Lambda_+\psi] \rangle(\mathbf{p}) d\mathbf{p} \\
&\quad + \int_{\mathbb{R}^3} \langle \hat{\psi}, \mathbb{F}[\Lambda_+ V\psi] \rangle(\mathbf{p}) d\mathbf{p} \\
&=: I_1 + I_2
\end{aligned}
\tag{2.3.12}
$$

since Λ_+ commutes with $\mathbb{D}_0$ and on using (2.3.5). In I_1, since $\Lambda_+\psi = \psi$, we have

$$
\mathbb{F}[\sqrt{-\Delta + 1}\Lambda_+\psi](\mathbf{p}) = \sqrt{p^2 + 1}\hat{\psi}(\mathbf{p}).
$$

Hence, from (2.3.9), it follows that

$$
I_1 = \int_{\mathbb{R}^3} E(\mathbf{p})|\psi(\mathbf{p})|^2 d\mathbf{p} = \int_{\mathbb{R}^3} E(\mathbf{p})|u(\mathbf{p})|^2 d\mathbf{p}.
$$

In I_2,

$$
\begin{aligned}
\int_{\mathbb{R}^3} \langle \hat{\psi}, \mathbb{F}[\Lambda_+ V\psi] \rangle(\mathbf{p}) d\mathbf{p} &= \int_{\mathbb{R}^3} \langle \hat{\psi}(\mathbf{p}), \Lambda_+(\mathbf{p})[\mathbb{F}(V\psi)](\mathbf{p}) \rangle d\mathbf{p} \\
&= \int_{\mathbb{R}^3} \langle \hat{\psi}(\mathbf{p}), [\mathbb{F}(V\psi)](\mathbf{p}) \rangle d\mathbf{p}
\end{aligned}
$$

and by (1.5.9) and (1.5.16),

$$
\begin{aligned}
[\mathbb{F}(V\psi)](\mathbf{p}) &= (\hat{V} * \hat{\psi})(\mathbf{p}) \\
&= \frac{\gamma}{2\pi^2} \int_{\mathbb{R}^3} |\mathbf{p} - \mathbf{p}'|^{-2}\hat{\psi}(\mathbf{p}') d\mathbf{p}'.
\end{aligned}
$$

On substituting in (2.3.12), we get

$$
\begin{aligned}
(\psi, \mathbb{B}\psi) &= \int_{\mathbb{R}^3} E(\mathbf{p})|\hat{\psi}(\mathbf{p})|^2 d\mathbf{p} - \int_{\mathbb{R}^3} \langle \hat{\psi}(\mathbf{p}), \mathbb{F}(V\psi)(\mathbf{p}) \rangle d\mathbf{p} \\
&= \int_{\mathbb{R}^3} E(\mathbf{p})|u(\mathbf{p})|^2 d\mathbf{p} - \frac{\gamma}{2\pi^2} \int_{\mathbb{R}^3} \int_{\mathbb{R}^3} u(\mathbf{p}')^* K(\mathbf{p}', \mathbf{p}) u(\mathbf{p}) d\mathbf{p} d\mathbf{p}' \\
&=: (u, \mathfrak{b}u)
\end{aligned}
\tag{2.3.13}
$$

where $*$ denotes the Hermitian conjugate, and

$$
K(\mathbf{p}', \mathbf{p}) = \frac{(E(\mathbf{p}') + 1)(E(\mathbf{p}) + 1)I_2 + (\mathbf{p}' \cdot \boldsymbol{\sigma})(\mathbf{p} \cdot \boldsymbol{\sigma})}{N(\mathbf{p}')|\mathbf{p}' - \mathbf{p}|^2 N(\mathbf{p})}.
\tag{2.3.14}
$$

Note that (2.3.13) defines the quadratic form $(u, \mathfrak{b}u)$ on the space $\mathbf{S}(\mathbb{R}^3, \mathbb{C}^2)$ of rapidly decreasing Pauli spinors u: we shall in fact prove in the next subsection that it is defined on $H^{1/2}(\mathbb{R}^3, \mathbb{C}^2)$. It is bounded below in $L^2(\mathbb{R}^3, \mathbb{C}^2)$ if and only if $(\psi, \mathbb{B}\psi)$ is bounded below in $\mathcal{H}$, and in this case both associated self-adjoint operators will be called a Brown–Ravenhall operator. The map $\hat{\psi} \to u$ defined by (2.3.8) determines a unitary equivalence between the two operators, and, hence, they have the same spectral properties.

To establish the aforementioned facts about $(u, \mathfrak{b}u)$ we use the spherical spinors $\Omega_{l,m,s}$ of (2.1.25) and (2.1.26) to decompose the form. Since the $\Omega_{l,m,s}$, $(l, m, s) \in \mathfrak{I}$ form an orthonormal basis of $L^2(\mathbb{S}^2, \mathbb{C}^2)$, any $u \in L^2(\mathbb{R}^3, \mathbb{C}^2)$ can be written as

$$u(\mathbf{p}) = \sum_{(l,m,s)\in\mathfrak{I}} p^{-1}a_{l,m,s}(p)\Omega_{l,m,s}(\boldsymbol{\omega}), \qquad (2.3.15)$$

where $p = |\mathbf{p}|, \boldsymbol{\omega} = \boldsymbol{\omega}_{\mathbf{p}} = p^{-1}\mathbf{p}$ and

$$\sum_{(l,m,s)\in\mathfrak{I}} \int_0^\infty |a_{l,m,s}(p)|^2 dp = \int_{\mathbb{R}^3} |u(\mathbf{p})|^2 d\mathbf{p}. \qquad (2.3.16)$$

From (2.1.28) and (2.2.6)

$$(\boldsymbol{\sigma}\cdot\mathbf{p})\Omega_{l,m,s}(\boldsymbol{\omega}_{\mathbf{p}}) = p\Omega_{l+2s,m,-s}(\boldsymbol{\omega}_{\mathbf{p}}) \qquad (2.3.17)$$

and

$$\int_{\mathbb{S}^2}\int_{\mathbb{S}^2} \frac{1}{|\mathbf{p}-\mathbf{p}'|^2}\Omega_{l,m,s}(\boldsymbol{\omega}_{\mathbf{p}})\Omega_{l',m',s'}(\boldsymbol{\omega}_{\mathbf{p}'})d\boldsymbol{\omega}_{\mathbf{p}}d\boldsymbol{\omega}_{\mathbf{p}'} =$$

$$= \frac{2\pi}{|\mathbf{p}||\mathbf{p}'|}Q_l\left(\frac{1}{2}\left[\frac{p}{p'}+\frac{p'}{p}\right]\right)\delta_{ll'}\delta_{mm'}\delta_{ss'}.$$

On using these identities after substituting (2.3.15) in (2.3.13) we obtain

$$(u, \mathfrak{b}u) = \sum_{(l,m,s)\in\mathfrak{I}} (a_{l,m,s}, \mathfrak{b}_{l,s}a_{l,m,s})$$

$$:= \sum_{(l,m,s)\in\mathfrak{I}} \left(\int_0^\infty e(p)|a_{l,m,s}(p)|^2 dp - \right. \qquad (2.3.18)$$

$$\left. - \frac{\gamma}{\pi}\int_0^\infty\int_0^\infty \overline{a_{l,m,s}}(p')k_{l,s}(p',p)a_{l,m,s}(p)dp'dp\right),$$

with $e(p) = E(\mathbf{p}) = \sqrt{p^2+1}$, $n(p) = N(\mathbf{p}) = \sqrt{2e(p)[e(p)+1]}$, and

$$k_{l,s}(p',p) = \frac{(e(p')+1)Q_l\left(\frac{1}{2}\left[\frac{p'}{p}+\frac{p}{p'}\right]\right)(e(p)+1)+p'Q_{l+2s}\left(\frac{1}{2}\left[\frac{p'}{p}+\frac{p}{p'}\right]\right)p}{n(p)n(p')}.$$

$$(2.3.19)$$

The forms $(a_{l,m,s}, \mathfrak{b}_{l,s}a_{l,m,s})$, $(l, m, s) \in \mathfrak{I}$, in (2.3.18) will next be shown to be defined on $L^2(\mathbb{R}_+; e(p)dp) := \{u : e|u|^2 \in L^1(\mathbb{R}_+)\}$, to be bounded below and to define self-adjoint operators $\mathfrak{b}_{l,s}$ in $L^2(\mathbb{R}_+)$.

2.3.2 *Lower semi-boundedness*

It follows from Lemma 2.2.3 that the forms $(\cdot, \mathfrak{b}_{l,s}\cdot)$ have domain $L^2(\mathbb{R}_+; e(p)dp)$. The next lemma shows that the angular momentum channel defined by $l = 0, s = 1/2$ yields the lowest energy.

Lemma 2.3.1. *For all $l, s \in \mathfrak{I}$,*

$$\inf\{(u, \mathfrak{b}_{l,s}u) : u \in L^2(\mathbb{R}_+; e(p)dp), \|u\| = 1\}$$
$$= \inf\{(u, \mathfrak{b}_{0,1/2}u) : u \in L^2(\mathbb{R}_+; e(p)dp), \|u\| = 1\}, \qquad (2.3.20)$$

where $(\cdot, \cdot)$ and $\| \cdot \|$ are the $L^2(\mathbb{R}_+)$ inner product and norm respectively.

Proof. It follows from (2.2.10) that

$$(u, \mathfrak{b}_{l,s}u) = \int_0^\infty e(p)|u(p)|^2 dp - \frac{\gamma}{\pi} \int_0^\infty \int_0^\infty \overline{u}(p')k_{l,s}(p', p)u(p)dpdp'$$
$$\geq \int_0^\infty e(p)|u(p)|^2 dp - \frac{\gamma}{\pi} \int_0^\infty \int_0^\infty |u(p')|k_{l,s}(p', p)|u(p)|dpdp'$$
$$= (|u|, \mathfrak{b}_{l,s}|u|).$$

Hence, when minimising $(u, \mathfrak{b}_{l,s}u)$, we need only consider non-negative functions, and in view of (2.2.10) and the definition of the index set $\mathfrak{I}$, take (l, s) to be either $(0, 1/2)$ or $(1, -1/2)$ since $l = 0, 1$ are, respectively, the least values of l when $s = 1/2, -1/2$. The lemma will be proved if we show that $k_{1,-1/2}(p', p) \leq k_{0,1/2}(p', p)$. But, with $Q_l \equiv Q_l \left(\frac{1}{2}\left[\frac{p}{p'} + \frac{p'}{p}\right]\right)$ to simplify notation,

$$n(p)n(p') \left[k_{1,-1/2}(p', p) - k_{0,1/2}(p', p)\right] = [(e(p') + 1)(e(p) + 1) - p'p] (Q_1 - Q_0)$$
$$\leq 0$$

by (2.2.10), whence the lemma. □

The main result in this section is Theorem 2.3.2 below, which establishes that the forms $(\cdot, \mathfrak{b}_{l,s}\cdot)$ are all positive and, hence, define in the form sense, positive self-adjoint operators formally given by

$$\mathfrak{b}_{l,s}u(p) = e(p)u(p) - \frac{\gamma}{\pi} \int_0^\infty k_{l,s}(p', p)u(p')dp' \qquad (2.3.21)$$

with form domain $L^2(\mathbb{R}_+; e(p)dp)$. A corollary is Theorem 2.3.7, which gives the optimal value γ_c for the Brown–Ravenhall operator $\mathfrak{b}$ to be bounded below, and indeed positive, for $\gamma \leq \gamma_c$. The proof we give is from [Burenkov and Evans (1998)]. This proves that the forms are positive, but, in fact, they are strictly positive (i.e., $\mathfrak{b}_{l,s} \geq \delta$ for some positive δ) as was proved by Tix in [Tix (1998)]. Tix's result and approach will be discussed in Remark 2.3.8 below. In Section 2.3.3 we analyse the self-adjoint operators generated by (2.3.21) with domain $L^2(\mathbb{R}_+; e^2(p)dp)$.

In view of Lemma 2.3.1, to prove that $b_{l,s} > 0$ for all $l, s \in \mathfrak{I}$, it is sufficient to prove that

$$\int_0^\infty \int_0^\infty \phi(p') k_{0,1/2}(p', p) \phi(p) dp' dp \; < \; \frac{\pi}{\gamma_c} \int_0^\infty e(p) \phi^2(p) dp \qquad (2.3.22)$$

for all positive, non-trivial $\phi \in L^2((0, \infty); e(p) dp)$. It is convenient to make some notational changes, which set the problem in an equivalent form. Let

$$p = x, \; p' = y, \; \psi(x) = \sqrt{e(x)} \phi(x),$$

and for $l = 0, 1$,

$$g_l(r) := Q_l \left(\frac{1}{2} \left[r + \frac{1}{r} \right] \right) = \begin{cases} \ln \left| \frac{r+1}{r-1} \right|, & \text{if } l = 0, \\ \frac{1}{2} \left(r + \frac{1}{r} \right) \ln \left| \frac{r+1}{r-1} \right| - 1, & \text{if } l = 1. \end{cases}$$

Then (2.3.22) becomes

$$\int_0^\infty \int_0^\infty t(x, y) \psi(x) \psi(y) dx dy \; < \; \frac{\pi}{\gamma_c} \int_0^\infty \psi^2(x) dx \qquad (2.3.23)$$

where

$$t(x, y) = \frac{1}{2} \left\{ \sqrt{\frac{\sqrt{x^2 + 1} + 1}{x^2 + 1}} g_0(x/y) \sqrt{\frac{\sqrt{y^2 + 1} + 1}{y^2 + 1}} \right. $$
$$\left. + \sqrt{\frac{\sqrt{x^2 + 1} - 1}{x^2 + 1}} g_1(x/y) \sqrt{\frac{\sqrt{y^2 + 1} - 1}{y^2 + 1}} \right\}. \qquad (2.3.24)$$

The integral operator T defined by

$$(T\psi)(x) \; = \; \int_0^\infty t(x, y) \psi(y) dy \qquad (2.3.25)$$

is bounded and symmetric on $L^2(0, \infty)$ and so

$$\sup \left\{ \frac{(T\phi, \phi)}{\|\phi\|^2} : \phi \in L^2(0, \infty), \phi \neq 0 \right\} = \|T : L^2(0, \infty) \to L^2(0, \infty)\| =: \|T\|.$$

Therefore (2.3.23) holds with an optimal value of γ_c if and only if the following theorem is valid.

Theorem 2.3.2. *Let T be the bounded symmetric operator defined by (2.3.25).* *Then*

$$\|T\| \; = \; \frac{\pi}{\gamma_c}, \; \gamma_c = \frac{4\pi}{\pi^2 + 4}, \qquad (2.3.26)$$

and T has no extremal function, i.e., a non-trivial function $\phi \in L^2(\mathbb{R}_+)$ such that $\|T\| = |(T\phi, \phi)|/\|\phi\|^2$.

The theorem will follow from a series of lemmas which break up the proof into digestible pieces.

Lemma 2.3.3.

$$\|T\| \geq \pi/\gamma_c. \tag{2.3.27}$$

Proof. Let $f_\delta(x) = 1/\sqrt{x}$ for $x \in (1, \delta), 1 < \delta < \infty$, and 0 otherwise. Then, by l'Hôpital's rule

$$\|T\| \geq \lim_{\delta \to \infty} \frac{(Tf_\delta, f_\delta)}{\|f_\delta\|^2}$$

$$= \lim_{\delta \to \infty} \left\{ (\ln \delta)^{-1} \int_1^\delta \left(\int_1^\delta t(x, y) \frac{dy}{\sqrt{y}} \right) \frac{dx}{\sqrt{x}} \right\}$$

$$= \lim_{\delta \to \infty} \left\{ \sqrt{\delta} \int_1^\delta t(\delta, y) \frac{dy}{\sqrt{y}} + \sqrt{\delta} \int_1^\delta t(x, \delta) \frac{dx}{\sqrt{x}} \right\}$$

$$= 2 \lim_{\delta \to \infty} \int_{1/\delta}^1 \delta \, t(\delta, \delta u) \frac{du}{\sqrt{u}}$$

$$= 2 \int_0^1 \lim_{\delta \to \infty} [\delta \, t(\delta, \delta u)] \frac{du}{\sqrt{u}} = \int_0^1 [g_0(u) + g_1(u)] \frac{du}{u}$$

$$= \frac{1}{2} \int_0^\infty [g_0(u) + g_1(u)] \frac{du}{u} = \frac{\pi}{\gamma_c}$$

by (2.2.11) and (2.2.12). In deriving this lower bound, we have used that

$$\frac{\delta \, t(\delta, \delta u)}{\sqrt{u}} \leq \frac{g_0(u) + g_1(u)}{u} \in L^1(0, 1)$$

to justify the use of the Dominated Convergence Theorem for taking the limits under the integral sign, and

$$\lim_{\delta \to \infty} \frac{\delta \, t(\delta, \delta u)}{\sqrt{u}} = \frac{g_0(u) + g_1(u)}{2u}. \qquad \square$$

Lemma 2.3.4. *For all functions h_0, h_1 that are positive and measurable on $(0, \infty)$,*

$$\|T\| \leq A(h_0, h_1) = \frac{1}{2} \sup_{0 < x < \infty} F(x; h_0, h_1), \tag{2.3.28}$$

where

$$F(x; h_0, h_1) = (\sqrt{x^2 + 1} + 1) \frac{x}{x^2 + 1} \int_0^\infty \frac{h_0(xz)}{h_0(x)} g_0(z) dz$$

$$+ (\sqrt{x^2 + 1} - 1) \frac{x}{x^2 + 1} \int_0^\infty \frac{h_1(xz)}{h_1(x)} g_1(z) dz.$$

The operator T has an extremal function ϕ if and only if

$$\phi(x) = A_0 h_0(x)\sqrt{\frac{x^2+1}{\sqrt{x^2+1}+1}} = A_1 h_1(x)\sqrt{\frac{x^2+1}{\sqrt{x^2+1}-1}} \qquad (2.3.29)$$

and

$$F(x; h_0, h_1) = A_3, \qquad (2.3.30)$$

for some non-zero constants A_0, A_1 and A_3.

Proof. A key step in determining the upper bound of $\|T\|$ is the following simple consequence of the Cauchy–Schwarz inequality: for a positive measurable function h and real positive measurable functions f, g with $g(1/u) = g(u)$ on $(0, \infty)$

$$\int_0^\infty\int_0^\infty f(x)g(x/y)f(y)\,dxdy = \int_0^\infty\int_0^\infty f(x)\sqrt{g(x/y)\frac{h(y)}{h(x)}}f(y)\sqrt{g(y/x)\frac{h(x)}{h(y)}}\,dxdy$$

$$\leq \left(\int_0^\infty\int_0^\infty f^2(x)g(x/y)\frac{h(y)}{h(x)}\,dxdy\right)^{\frac{1}{2}}\left(\int_0^\infty\int_0^\infty f^2(y)g(y/x)\frac{h(x)}{h(y)}\,dxdy\right)^{\frac{1}{2}}$$

$$= \int_0^\infty f^2(x)\left(\int_0^\infty g(y/x)\frac{h(y)}{h(x)}\,dy\right)dx.$$

Equality holds if and only if, for some constants λ and μ,

$$\mu f(x)\sqrt{g(x/y)\frac{h(y)}{h(x)}} = \lambda f(y)\sqrt{g(y/x)\frac{h(x)}{h(y)}}$$

a.e. on $\mathbb{R}_+ \times \mathbb{R}_+$, which is equivalent to $f(x) = Ah(x)$ a.e. on $\mathbb{R}_+$ for some constant A. It follows that for all positive measurable functions h_0, h_1 on $(0, \infty)$

$$\int_0^\infty\int_0^\infty t(x,y)\phi(x)\phi(y)\,dxdy$$

$$= \frac{1}{2}\Bigg\{\int_0^\infty\int_0^\infty \sqrt{\frac{\sqrt{x^2+1}+1}{x^2+1}}\phi(x)g_0(x/y)\sqrt{\frac{\sqrt{y^2+1}+1}{y^2+1}}\phi(y)\,dxdy$$

$$+ \int_0^\infty\int_0^\infty \sqrt{\frac{\sqrt{x^2+1}-1}{x^2+1}}\phi(x)g_1(x/y)\sqrt{\frac{\sqrt{y^2+1}-1}{y^2+1}}\phi(y)\,dxdy\Bigg\}$$

$$\leq \frac{1}{2}\int_0^\infty\Bigg(\frac{\sqrt{x^2+1}+1}{x^2+1}\int_0^\infty\frac{h_0(y)}{h_0(x)}g_0(y/x)\,dy$$

$$+ \frac{\sqrt{x^2+1}-1}{x^2+1}\int_0^\infty\frac{h_1(y)}{h_1(x)}g_1(y/x)\,dy\Bigg)\phi^2(x)\,dx$$

$$\leq A(h_0, h_1)\|\phi\|^2. \qquad (2.3.31)$$

The first inequality in the derivation of (2.3.31) becomes an equality if and only if

$$\sqrt{\frac{\sqrt{x^2+1}+1}{x^2+1}}\phi(x)=A_0h_0(x), \qquad \sqrt{\frac{\sqrt{x^2+1}-1}{x^2+1}}\phi(x)=A_1h_1(x), \qquad (2.3.32)$$

a.e. on $\mathbb{R}_+$, for some constants A_0, A_1. The second inequality in (2.3.31) becomes an equality if and only if $\phi(x)=0$ a.e. on the set of all $x \in \mathbb{R}_+$ for which $F(x;h_0,h_1)<A(h_0,h_1)$. Hence if ϕ is an extremal function we must have that $F(x;h_0,h_1)=A_3$ a.e. on $\mathbb{R}_+$, for some constant A_3. This completes the proof. $\qquad\square$

Lemma 2.3.5. *For all $\phi \in L^2(0,\infty)$, $\phi(x) \neq 0$ a.e., we have*

$$\int_0^\infty \int_0^\infty t(x,y)\phi(x)\phi(y)dxdy < C\int_0^\infty \phi^2(x)dx, \qquad (2.3.33)$$

where $C=\sup_{0<x<\infty}F(x)$ and

$$F(x):=\frac{\pi}{2}(\sqrt{x^2+1}+1)\frac{\arctan x}{x}+\frac{(\sqrt{x^2+1}-1)x}{x^2+1}. \qquad (2.3.34)$$

Proof. We first claim that, for all positive measurable functions h_0, h_1 on $\mathbb{R}_+$,

$$\liminf_{x\to\infty}\frac{1}{2}F(x;h_0,h_1)\geq\frac{\pi}{\gamma_c}=\frac{\pi^2}{4}+1. \qquad (2.3.35)$$

To see this, let $\hat{h}_j(\xi):=\liminf_{x\to\infty}\frac{h_j(\xi x)}{h_j(x)}$, $0<\xi<\infty$, $j=0,1$, where the lim inf can be infinite. Then

$$\hat{h}_j(1/\xi)=\liminf_{x\to\infty}\frac{h_j(x/\xi)}{h_j(x)}$$

$$=\liminf_{y\to\infty}\frac{h_j(y)}{h_j(\xi y)}=\frac{1}{\hat{h}_j(\xi)}.$$

By Fatou's theorem,

$$\frac{1}{2}\liminf_{x\to\infty}F(x)\geq\frac{1}{2}\left(\int_0^\infty \liminf_{x\to\infty}\left[\frac{h_0(ux)}{h_0(x)}\right]g_0(u)du+\int_0^\infty \liminf_{x\to\infty}\left[\frac{h_1(ux)}{h_1(x)}\right]g_1(u)du\right)$$

$$=\frac{1}{2}\left(\int_0^\infty \hat{h}_0(u)g_0(u)du+\int_0^\infty \hat{h}_0(u)g_0(u)du\right).$$

Furthermore, on substituting $u=v-\sqrt{v^2-1}$ when $0<u<1$ and $u=v+\sqrt{v^2-1}$ when $u>1$, we have

$$\int_0^\infty \hat{h}_0(u)g_j(u)du=\int_1^\infty \left\{\hat{h}_j(v-\sqrt{v^2-1})(v-\sqrt{v^2-1})\right.$$

$$\left.+\hat{h}_j(v+\sqrt{v^2+1})(v+\sqrt{v^2+1})\right\}g_j(v)\frac{dv}{\sqrt{v^2-1}}$$

$$=\int_1^\infty \left\{\frac{1}{\hat{h}_j(v+\sqrt{v^2-1})(v+\sqrt{v^2-1})}\right.$$

$$\left.+\hat{h}_j(v+\sqrt{v^2-1})(v+\sqrt{v^2-1})\right\}g_j(v)\frac{dv}{\sqrt{v^2-1}}$$

$$\geq 2\int_1^\infty g_j(v)\frac{dv}{\sqrt{v^2-1}}$$

$$=\int_0^\infty g_j(u)\frac{du}{u},$$

whence (2.3.35) by (2.2.11) and (2.2.12). We also note that equality holds if and only if $\hat{h}_j(u) = 1/u$ a.e. on $\mathbb{R}_+$. Thus to prove that $A(h_0, h_1) \leq (\pi^2/4) + 1$, and thus complete the proof of the first part of Theorem 2.3.2, we must choose h_0 and h_1 in such a way that $\hat{h}_0(u) = \hat{h}_1(u) = 1/u$ a.e. on $\mathbb{R}_+$.

Guided by the last comment, we now choose

$$h_0(x) = \frac{x}{x^2 + 1}, \quad h_1(x) = \frac{1}{x} \tag{2.3.36}$$

so that

$$\int_0^\infty h_0(xz)g_0(z)dz = x \int_0^\infty \frac{z}{x^2 z^2 + 1} g_0(z)dz$$

and

$$\int_0^\infty h_1(xz)g_1(z)dz = x \int_0^\infty g_1(z)\frac{dz}{z} = 2x,$$

by (2.2.12). On using Cauchy's Residue Theorem, we obtain

$$\int_0^\infty \frac{z g_0(z)}{x^2 z^2 + 1} dz = \int_0^\infty \frac{z}{x^2 z^2 + 1} \ln\left|\frac{z+1}{z-1}\right| dz$$

$$= \frac{1}{2} \int_{-\infty}^\infty \frac{z}{x^2 z^2 + 1} \ln\left|\frac{z+1}{z-1}\right| dz$$

$$= \frac{1}{2}\mathrm{Re}\left[\int_{-\infty}^\infty \frac{z}{x^2 z^2 + 1} \ln\left(\frac{z+1}{z-1}\right) dz\right]$$

$$= \frac{\pi}{x^2}\mathrm{Re}\left[\frac{i}{2} \ln\left(\frac{x+i}{x-i}\right)\right]$$

$$= \frac{\pi}{x^2} \arctan x.$$

This completes the proof of (2.3.33). The inequality is strict in view of Lemma 2.3.4, since the equality (2.3.29) with the current choice of h_0, h_1, is not satisfied by a non-trivial $\phi \in L^2(0, \infty)$. □

The final link in the chain of results to establish Theorem 2.3.2 is

Lemma 2.3.6. *The constant C in (2.3.33) is given by*

$$C = \frac{\pi}{\gamma_c} = \frac{\pi^2}{4} + 1. \tag{2.3.37}$$

Proof. It is significant to note that

$$\lim_{x \to \infty} \frac{1}{2} F(x) = \int_0^\infty g_0(z)\frac{dz}{z} + \int_0^\infty g_1(z)\frac{dz}{z}$$

$$= \frac{\pi}{\gamma_c}$$

by (2.2.11) and (2.2.12). It therefore suffices to prove that F attains its supremum at infinity. On substituting $x = \tan 2v$ we obtain

$$\frac{1}{2} F(\tan 2v) = \frac{\pi v + 4 \sin^4 v}{\tan v}, \quad 0 \le v \le \pi/4.$$

The final step is to prove that

$$f(v) := \pi v + 4 \sin^4 v - \left(\frac{\pi^2}{4} + 1\right) \tan v \le 0, \quad 0 \le v \le \pi/4.$$

This is more difficult than it seems!

We start with the following identities which are easily verified:

$$f'(v) = \pi + 16 \sin^3 v \cos v - \left(\frac{\pi^2}{4} + 1\right) \sec^2 v$$

$$f''(v) = 2 \sin v \sec^3 v \, g(v)$$

where

$$g(v) = 3 \sin 2v + 3 \sin 4v + \sin 6v - \left(\frac{\pi^2}{4} + 1\right).$$

Since $g(0) < 0, g(\pi/8) > 0, g(\pi/4) < 0, g'(v) > 0$ on $[0, \pi/8)$ and $g'(v) < 0$ on $(\pi/8, \pi/4]$, there exist v_1, v_2 such that $0 < v_1 < v_2 < \pi/4, g(v_1) = g(v_2) = 0, g(v) < 0$ on $[0, v_1)$ and $(v_2, \pi/4]$, and $g(v) > 0$ on (v_1, v_2). Thus $f(0) = 0, f'(0) < 0, f''(0) = 0, f(\pi/4) = 0, f'(\pi/4) > 0$ and $f''(\pi/4) < 0$. Moreover, f'' vanishes at v_1 and v_2, is negative on $(0, v_1)$ and (v_2, π), and positive on (v_1, v_2). In particular, it follows that f' is negative on $[0, v_1]$ and positive on $(v_2, \pi/4]$.

Suppose that $f(\xi) = 0$ for some $\xi \in (0, \pi/4)$. Then we have $f(0) = f(\xi) = f(\pi/4) = 0$, and from the last sentence of the last paragraph it follows that there exist η_1, η_2 such that $f'(\eta_1) = f'(\eta_2) = 0$ and $v_1 < \eta_1 < \eta_2 < v_2$. But this implies that there exists $v_3 \in (v_1, v_2)$ such that $f''(v_3) = 0$, contrary to what was established in the last paragraph. Thus $f(v) \ne 0$ on $(0, \pi/4)$. Since $f(0) = f(\pi/4) = 0$ and $f'(0) < 0, f'(\pi/4) > 0$, we conclude that $f(v) \le 0$ on $[0, \pi/4]$. This proves the lemma and completes the proof of Theorem 2.3.2. $\qquad\square$

Theorem 2.3.7. *Let $\mathfrak{b}$ be the Brown–Ravenhall operator defined by the form in (2.3.21). Then, with $\gamma_c = 2(\pi/2 + 2/\pi)^{-1}$,*
(i) if $\gamma \le \gamma_c$, $\mathfrak{b} \ge (1 - \gamma/\gamma_c)$;
(ii) if $\gamma = \gamma_c$, $\mathfrak{b}$ has no eigenvalue at 0;
(iii) if $\gamma > \gamma_c$, $\mathfrak{b}$ is unbounded below.

Proof.   (i) From (2.3.16), (2.3.25) and Lemma 2.3.1

$$(\psi, \mathfrak{b}\, \psi) \geq \sum_{(l,m,s)\in\mathfrak{J}} \left\{ \int_0^\infty e(p)|a_{l,m,s}(p)|^2 dp \right.$$

$$\left. - \frac{\gamma}{\pi} \int_0^\infty \int_0^\infty |a_{l,m,s}(p')||k_{0,1/2}(p'p)||a_{l,m,s}(p)|dpdp' \right\}$$

$$\geq \sum_{(l,m,s)\in\mathfrak{J}} \int_0^\infty \left(1 - \frac{\gamma}{\gamma_c}\right) e(p)|a_{l,m,s}(p)|^2 dp$$

$$\geq \left(1 - \frac{\gamma}{\gamma_c}\right) \|\psi\|^2.$$

(ii) Suppose that 0 is an eigenvalue of $\mathfrak{b}$ with corresponding eigenfunction ψ. Since all the summands on the right-hand side of (2.3.21) with $u = \psi$ are non-negative, they must be zero. Also (2.3.16) implies that at least one of the functions $a_{l,m,s}$, a_{l_0,m_0,s_0}, say, must be non-trivial. But this would imply that there is equality in (2.3.22) with $\phi = a_{l_0,m_0,s_0}$, contrary to Theorem 2.3.2.

(iii) Choose

$$u(p) := \begin{cases} 1/p, & p \in (\tilde{c}, \tilde{d}) \\ 0, & p \notin (\tilde{c}, \tilde{d}), \end{cases}$$

where $\tilde{c}, \tilde{d}$ are positive numbers to be selected later. For $p, p' \in (\tilde{c}, \tilde{d})$,

$$2k_{1,-1/2}(p',p) \geq Q_0 \left(\frac{1}{2}\left[\frac{p'}{p} + \frac{p}{p'}\right]\right) + Q_1 \left(\frac{1}{2}\left[\frac{p'}{p} + \frac{p}{p'}\right]\right) (1 - O(1/\tilde{c}))$$

and

$$\int_0^\infty \int_0^\infty u(p')k_{1,-1/2}(p',p)u(p)dpdp' \geq$$

$$\frac{1}{2}\int_{\tilde{c}}^{\tilde{d}} \int_{\tilde{c}}^{\tilde{d}} \frac{1}{p'p} \left\{ Q_0\left(\frac{1}{2}\left[\frac{p'}{p} + \frac{p}{p'}\right]\right) + Q_1\left(\frac{1}{2}\left[\frac{p'}{p} + \frac{p}{p'}\right]\right)(1 - O(1/\tilde{c}) \right\} dpdp'.$$

Next we estimate the integrals

$$I(\nu) := \int_{\tilde{c}}^{\tilde{d}} \int_{\tilde{c}}^{\tilde{d}} \frac{1}{p'p} \left(\frac{p}{p'}\right)^\nu \left\{ Q_0\left(\frac{1}{2}\left[\frac{p'}{p} + \frac{p}{p'}\right]\right) \right\} dpdp', \quad \nu = 0, \pm1$$

by considering the regions where $p > p'$ and $p' > p$ in turn. As $\tilde{c}, \tilde{d}$ and $\tilde{d}/\tilde{c}$ tend to infinity, it is readily shown that

$$I(0) = \frac{\pi^2}{2} \ln\left(\frac{\tilde{d}}{\tilde{c}}\right) + O(1),$$

$$I(\pm1) = \ln^2\left(\frac{\tilde{d}}{\tilde{c}}\right) + 2\ln\left(\frac{\tilde{d}}{\tilde{c}}\right) + O(1)$$

and these yield

$$\frac{\gamma}{\pi} \int_0^\infty \int_0^\infty u(p') k_{1,-1/2}(p',p) u(p) dp' dp$$

$$\geq \frac{\gamma}{4\pi} \left\{ 2I(0) + \left(I(-1) + I(1) - 2\ln^2\left(\frac{\tilde{d}}{\tilde{c}}\right) \right) \left(1 + O(1/\tilde{d}) \right) \right\}$$

$$\geq \frac{\gamma}{\gamma_c} \ln\left(\frac{\tilde{d}}{\tilde{c}}\right) (1 + O(1/\tilde{c})).$$

Moreover,

$$\int_0^\infty e(p) |u(p)|^2 dp = \ln\left(\frac{\tilde{d}}{\tilde{c}}\right) + O(1)$$

and, hence,

$$(u, \mathfrak{b}_{1,-1/2} u) \leq \left[1 - \frac{\gamma}{\gamma_c} (1 - O(1/\tilde{c})) \right] \ln\left(\frac{\tilde{d}}{\tilde{c}}\right) + O(1)$$

$$\to -\infty$$

if $\gamma > \gamma_c$, on choosing $\tilde{c}, \tilde{d}/\tilde{c} \to \infty$. Also

$$\|u\|^2 = \frac{1}{\tilde{c}} - \frac{1}{\tilde{d}} \leq 1$$

if $\tilde{c} \geq 1$. The proof of (iii), and, hence, of the theorem, is concluded. $\square$

Remark 2.3.8. (i) If Z denotes the nuclear charge, $\gamma = \alpha Z$, where α is Sommerfeld's fine-structure constant. On taking $\alpha = 1/137$, its physical value, we have that the largest value Z_c of Z for which $\mathfrak{b}$ is bounded below is $Z_c = \gamma_c/\alpha \approx 124$, a value larger than the 111 for the heaviest known element. This is to be compared with $Z_c \approx 87$ given by the critical value $\gamma_c = 2/\pi$ for the quasi-relativistic operator of Section 2.2.

(ii) In [Tix (1998)], Tix proves that for $\gamma \leq \gamma_c, \mathfrak{b} \geq (1 - \gamma_c) > 0.09$, and in fact derives a larger lower bound numerically in [Tix (1997a)].

(iii) It follows from Theorem 2.3.2 that for all values of l, s, the form

$$(u, k_{l,s} u) := \int_0^\infty \int_0^\infty \overline{u}(p) k_{l,s}(p',p) u(p') dp' dp, \quad u \in L^2(\mathbb{R}_+; e(p) dp),$$

is such that $\gamma/\pi (\cdot, k_{l,s} \cdot)$ is bounded relative to $(\cdot, e\cdot)$, with relative bound γ/γ_c. Hence, if $\gamma < \gamma_c, \mathfrak{b}_{l,s}$ is a closed positive form with domain $L^2(\mathbb{R}_+; e(p) dp)$, which is the form domain of the associated self-adjoint operator given by the first representation theorem. If $\gamma = \gamma_c$, the self-adjoint operator associated with $\mathfrak{b}_{l,s}$ is the Friedrichs extension of the symmetric operator (2.3.21) on

$L^2(\mathbb{R}_+; e^2(p)dp)$, its form domain being the completion of $L^2(\mathbb{R}_+; e^2(p)dp)$ with respect to $\{(\cdot, \mathfrak{b}_{l,s} \cdot) + \| \cdot \|^2\}^{1/2}$, and hence, equivalently, with respect to $(\cdot, \mathfrak{b}_{l,s} \cdot)^{1/2}$ in view of Tix's result in [Tix (1998)].

(iv) The critical value γ_c is relevant because of Lemma 2.3.1, which shows that the angular momentum channel $(l, s) = (0, 1/2)$ yields the lowest energy. In [Balinsky and Evans (2002a)], the critical value of γ for the lower semi-boundedness of (2.3.21) was determined for all channels. Subject to numerical estimates to assist the complicated analysis, it was shown that $\mathfrak{b}_{l,s}$ is positive for $\gamma \leq \gamma_c(l, s)$, where

$$\gamma_c(l, s) = 4 \left\{ \frac{\Gamma^2(\frac{1}{2}[l+1])}{\Gamma^2(\frac{1}{2}[l+2])} + \frac{\Gamma^2(\frac{1}{2}[l+2s+1])}{\Gamma^2(\frac{1}{2}[l+2s+2])} \right\}^{-1}. \tag{2.3.38}$$

For $\gamma > \gamma_c(l, s)$, $\mathfrak{b}_{l,s}$ is unbounded below.

(v) We have already seen that the map $\hat{\psi} \mapsto u$ defined by (2.3.8) determines a unitary equivalence between the Brown–Ravenhall operator $\mathbb{B}$ in (2.3.10) and $\mathfrak{b}$. Hence, if $\gamma < \gamma_c$, $\mathbb{B}$ is the form sum of $\Lambda_+ \mathbb{D}_0 \Lambda_+$ and $\Lambda_+ V \Lambda_+$, with form domain $\mathbb{Q}(\Lambda_+ \mathbb{D}_0 \Lambda_+) = \Lambda_+ H^{1/2}(\mathbb{R}^3, \mathbb{C}^4)$, and is the Friedrichs extension of $\Lambda_+ \mathbb{D}' \Lambda_+$ when $\gamma = \gamma_c$. Moreover, if $\gamma < \gamma_c$, $\mathfrak{b}$ is the form sum of E and $-K$ where E is the operator of multiplication by $e(p) = \sqrt{p^2 + 1}$ and

$$(Ku)(\mathbf{p}) = \frac{\gamma}{2\pi^2} \int_{\mathbb{R}^3} K(\mathbf{p}', \mathbf{p}) u(\mathbf{p}') d\mathbf{p}'$$

with kernel (2.3.14). The form domain of $\mathfrak{b}$ is $L^2(\mathbb{R}^3; \sqrt{p^2+1}d\mathbf{p}) \otimes \mathbb{C}^2$, the space of 2-spinors u with $|u| \in L^2(\mathbb{R}^3; \sqrt{p^2+1}d\mathbf{p})$.

(vi) From (2.3.13), for $\psi \in \mathbf{S}^4 = \Lambda_+ \mathbf{S}(\mathbb{R}^3, \mathbb{C}^4)$,

$$(\psi, \mathbb{B}\psi) = \int_{\mathbb{R}^3} e(p) |\hat{\psi}(\mathbf{p})|^2 d\mathbf{p} - \int_{\mathbb{R}^3} \langle \hat{\psi}(\mathbf{p}), \mathbb{F}(V\psi)(\mathbf{p}) \rangle d\mathbf{p}$$

$$= \int_{\mathbb{R}^3} e(p) |\hat{\psi}(\mathbf{p})|^2 d\mathbf{p} - \gamma \int_{\mathbb{R}^3} \frac{|\psi(\mathbf{x})|^2}{|\mathbf{x}|} d\mathbf{x}.$$

This yields the Kato-type inequality

$$\int_{\mathbb{R}^3} \frac{|\psi(\mathbf{x})|^2}{|\mathbf{x}|} d\mathbf{x} < \frac{1}{2} \left(\frac{\pi}{2} + \frac{2}{\pi} \right) \int_{\mathbb{R}^3} e(p) |\hat{\psi}(\mathbf{p})|^2 d\mathbf{p}$$

$$= \frac{1}{2} \left(\frac{\pi}{2} + \frac{2}{\pi} \right) \int_{\mathbb{R}^3} \langle \sqrt{-D+1}\psi, \psi \rangle(\mathbf{x}) d\mathbf{x}, \tag{2.3.39}$$

for $\psi \in \Lambda_+ H^{1/2}(\mathbb{R}^3, \mathbb{C}^4)$. The constant is sharp and equality is only attained if $\psi = 0$. The inequality also holds for $\psi \in \Lambda_- H^{1/2}(\mathbb{R}^3, \mathbb{C}^4)$, and has an important application in [Dolbeault *et al.* (2000b)].

2.3.3 The Brown–Ravenhall operators $\mathfrak{b}_{l,s}$

From Theorem 2.3.2 for $\gamma \leq \gamma_c$ and the first representation theorem for semi-bounded quadratic forms, it follows that non-negative self-adjoint operators $\mathfrak{b}_{l,s}$ are associated with the forms $(\cdot, \mathfrak{b}_{l,s}\cdot)$ having form domain $L^2(\mathbb{R}_+; e(p)dp)$. Formally we have that

$$\mathfrak{b}_{l,s}u(p) = e(p)u(p) - \frac{\gamma}{\pi}\int_0^\infty k_{l,s}(p',p)u(p')dp'$$

$$=: (E - K_{l,s})u(p), \qquad (2.3.40)$$

where E is the operator of multiplication by $e(p) = \sqrt{p^2 + 1}$, with domain $L^2(\mathbb{R}_+; e^2(p)dp)$ and

$$K_{l,s}u(p) := \frac{\gamma}{\pi}\int_0^\infty k_{l,s}(p',p)u(p')dp'. \qquad (2.3.41)$$

Also, we have shown in Theorem 2.3.2, that for $u \in L^2(\mathbb{R}_+; e(p)dp)$,

$$|(K_{l,s}u, u)| \leq \frac{\gamma}{\gamma_c}\int_0^\infty e(p)|u(p)|^2dp.$$

Hence $P_{l,s} := E^{-1/2}K_{l,s}E^{-1/2}$ is a bounded symmetric operator on $L^2(\mathbb{R}_+)$ with norm satisfying

$$\|P_{l,s}\| \leq \gamma/\gamma_c.$$

We shall prove in this section that the operators $K_{l,s}$ with domain $L^2(\mathbb{R}_+; e^2(p)dp)$ are bounded relative to E and also determine the E-bound. This then yields important information about the self-adjoint extensions of $\mathfrak{b}_{l,s}$. We first consider the operators

$$\left(K_{l,s}^0 u\right)(p) := \frac{\gamma}{\pi}\int_0^\infty k_{l,s}^0(p',p)u(p')dp', \quad u \in L^2(\mathbb{R}_+; p^2dp), \qquad (2.3.42)$$

where

$$k_{l,s}^0(p',p) = \frac{1}{2}\left\{Q_l\left(\frac{1}{2}\left[\frac{p'}{p} + \frac{p}{p'}\right]\right) + Q_{l+2s}\left(\frac{1}{2}\left[\frac{p'}{p} + \frac{p}{p'}\right]\right)\right\}. \qquad (2.3.43)$$

Note that $k_{l,s}^0$ is the kernel corresponding to the case of zero mass m and $c = 1$ in the original notation of (2.1.1).

Lemma 2.3.9. *For $u \in L^2(\mathbb{R}_+; p^2dp)$,*

$$\|K_{l,s}^0 : L^2(\mathbb{R}_+; p^2dp) \to L^2(\mathbb{R}_+)\| = \frac{\gamma}{2\pi}\left\{I_{l,1/2} + I_{l+2s,1/2}\right\}, \qquad (2.3.44)$$

where

$$I_{l,1/2} = \frac{\pi}{\sqrt{2}}\int_{-1}^1 \frac{P_l(t)}{\sqrt{1-t}}dt.$$

In particular,

$$\|K_{l,1/2}^0 : L^2(\mathbb{R}_+; p^2dp) \to L^2(\mathbb{R}_+)\| \begin{cases} = \frac{4\gamma}{3}, & l = 0, \\ = \frac{7\gamma}{15}, & l = 1, \\ \leq \frac{7\gamma}{15}, & l > 1 \end{cases}$$

and

$$\|K^0_{l,-1/2} : L^2(\mathbb{R}_+; p^2 dp) \to L^2(\mathbb{R}_+)\| \begin{cases} = \frac{4\gamma}{3}, & l = 1, \\ = \frac{7\gamma}{15}, & l = 2, \\ \le \frac{7\gamma}{15}, & l > 2. \end{cases}$$

Proof. The proof is similar to that of Lemma 2.2.8. We may, and shall, assume, without loss of generality, that u is real. Let $v(p) = pu(p)$. Then $v \in L^2(\mathbb{R}_+)$ and

$$(\pi/\gamma)^2 \|K^0_{l,s} u\|^2_{L^2(\mathbb{R}_+)} = \int_0^\infty \int_0^\infty k^1_{l,s}(p', p'') u(p') u(p'') dp' dp''$$

$$= \int_0^\infty \int_0^\infty (p'p'')^{-1} k^1_{l,s}(p', p'') v(p') v(p'') dp' dp'',$$

where

$$k^1_{l,s}(p', p'') = \int_0^\infty k^0_{l,s}(p', p) k^0_{l,s}(p'', p) dp.$$

The kernel $(p'p'')^{-1} k^1_{l,s}(p', p'')$ is homogeneous of degree -1 and hence by Hilbert's inequality (1.7.10), we have

$$(\pi/\gamma)^2 \|K^0_{l,s} u\|^2_{L^2(\mathbb{R}_+)} \le C_{l,s} \int_0^\infty v^2(p) dp,$$

where the optimal value of the constant is

$$C_{l,s} = \int_0^\infty p'^{-3/2} k^1_{l,s}(p', 1) dp'$$

$$= \int_0^\infty p^{-1/2} k^0_{l,s}(1, p) dp \int_0^\infty x^{-3/2} k^0_{l,s}(x, 1) dx.$$

Let

$$I_{l,\beta} := \int_0^\infty Q_l \left(\frac{1}{2} [x + 1/x] \right) x^{-\beta} dx, \quad \beta = 1/2, 3/2.$$

Then $I_{l,1/2} = I_{l,3/2}$ and

$$C_{l,s} = \left\{ \frac{1}{2} [I_{l,1/2} + I_{l+2s,1/2}] \right\}^2. \tag{2.3.47}$$

Furthermore

$$I_{l,1/2} = \int_0^\infty \left(\frac{1}{2} \int_{-1}^1 \frac{P_l(t)}{\frac{1}{2}[x + 1/x] - t} dt \right) x^{-1/2} dx$$

$$= \int_{-1}^1 \left\{ \int_0^\infty \frac{\sqrt{x}}{(x^2 + 1 - 2tx)} dx \right\} P_l(t) dt. \tag{2.3.48}$$

The equality (2.3.44) follows since

$$\int_0^\infty \frac{\sqrt{x}}{(x^2 + 1 - 2tx)} dx = \frac{\pi}{\sqrt{2(1-t)}}.$$

On substituting $P_0(t) = 1, P_1(t) = t, P_2(t) = t^2 - 1/3$, it is readily shown that $I_{0,1/2} = 2\pi, I_{1,1/2} = 2\pi/3, I_{2,1/2} = 4\pi/15$ and from these we deduce the rest of the lemma.   □

Lemma 2.3.10. *Let* $\mathfrak{b}_{l,s}^0 := E^0 - K_{l,s}^0$, *where* $(E^0 u)(p) = pu(p)$. *The operator* $\mathfrak{b}_{l,s}$ *in (2.3.40) is defined on* $L^2(\mathbb{R}_+; e^2(p)dp)$ *and* $\mathfrak{b}_{l,s} - \mathfrak{b}_{l,s}^0$ *can be extended to a bounded operator on* $L^2(\mathbb{R}_+)$.

Proof. It is easily seen that $0 < k_{l,s}(p',p) \leq 2k_{l,s}^0(p',p)$. Therefore by Lemma 2.3.9, $K_{l,s}$ is defined on $L^2(\mathbb{R}_+, e^2(p)dp)$. To complete the proof, it is sufficient to prove that

$$\sup\{|(u, [\mathfrak{b}_{l,s} - \mathfrak{b}_{l,s}^0]u)| : u \in L^2(\mathbb{R}_+)\} < \infty.$$

Since $0 < e(p) - p < 1$, it follows that we need only consider the potential energy terms, namely

$$\int_0^\infty \int_0^\infty \overline{u}(p')\{k_{l,s}(p',p) - k_{l,s}^0(p',p)\}u(p)dpdp'.$$

We have

$$k_{l,s}(p',p) - k_{l,s}^0(p',p) = \frac{1}{2}\left[\sqrt{1 + \frac{1}{e(p')}}\sqrt{1 + \frac{1}{e(p)}} - 1\right]Q_l\left(\frac{1}{2}\left[\frac{p'}{p} + \frac{p}{p'}\right]\right)$$

$$+ \frac{1}{2}\left[\sqrt{1 - \frac{1}{e(p')}}\sqrt{1 - \frac{1}{e(p)}} - 1\right]Q_{l+2s}\left(\frac{1}{2}\left[\frac{p'}{p} + \frac{p}{p'}\right]\right).$$

For positive constants a, b, $2\sqrt{1+a}\sqrt{1+b} \leq 2 + a + b$ and so, in view of (2.2.10), it is enough to show that

$$I := \int_0^\infty \int_0^\infty |u(p')|q(p',p)|u(p)|dpdp' < \infty,$$

where

$$q(p',p) = Q_0\left(\frac{1}{2}\left[\frac{p'}{p} + \frac{p}{p'}\right]\right)\left(\frac{1}{e(p)} + \frac{1}{e(p')}\right).$$

By the Cauchy–Schwarz inequality, on using the symmetry of $q(p',p)$, we have for an arbitrary positive measurable function h on $(0, \infty)$,

$$I = \int_0^\infty \int_0^\infty |u(p')|\left(q(p',p)\frac{h(p')}{h(p)}\right)^{1/2}\left(q(p,p')\frac{h(p)}{h(p')}\right)^{1/2}|u(p)|dpdp'$$

$$\leq \int_0^\infty |u(p)|^2 dp \int_0^\infty q(p',p)\frac{h(p')}{h(p)}dp'.$$

Now we choose $h(p) = p^{-a}, 0 < a < 1$. We are then left with showing that for $b = a$, $1 + a$,

$$\int_0^\infty Q_0\left(\frac{1}{2}\left[\frac{p'}{p} + \frac{p}{p'}\right]\right)\frac{h(p')}{h(p)}\frac{dp'}{e(p')} \leq \int Q_0(\frac{1}{2}[x + 1/x])x^{-b}dx < \infty.$$

These are satisfied since

$$Q_0\left(\frac{1}{2}[x + 1/x]\right) = \begin{cases} O(x), & \text{as } x \to 0, \\ O(1/x), & \text{as } x \to \infty. \end{cases}$$

This concludes the proof. $\square$

On applying Lemmas 2.3.9, 2.3.10 and (2.2.10) to Theorems 1.1.1 and 1.1.2 we derive

Theorem 2.3.11. *There exists a positive constant C such that for all $u \in L^2(\mathbb{R}_+; e^2(p)dp)$,*

$$\|K_{l,s}u\| \leq (4\gamma/3)\|Eu\| + C\|u\| \tag{2.3.49}$$

for $(l, s) = (0, 1/2)$ or $(1, -1/2)$, and

$$\|K_{l,s}u\| \leq (7\gamma/15)\|Eu\| + C\|u\| \tag{2.3.50}$$

for other values of l, s. Hence,

- *for $(l, s) = (0, 1/2)$ or $(1, -1/2)$, $\mathfrak{b}_{l,s}$ is self-adjoint if $\gamma < 3/4$, and essentially self-adjoint if $\gamma = 3/4$;*
- *for other values of l, s, $\mathfrak{b}_{l,s}$ is self-adjoint if $\gamma < 15/7$ and hence for $\gamma \leq \gamma_c$.*

Remark 2.3.12. In [Tix (1997b)], Tix proves results for the Brown–Ravenhall operator which are analogous to those of [Yaouanc *et al.* (1997)] for the quasi-relativistic operator described in Section 2.2.3. He proves that for $3/4 < \gamma \leq \gamma_c$, $\mathfrak{b}_{0,1/2}$ and $\mathfrak{b}_{1,-1/2}$ are closed symmetric operators with deficiency indices $(1, 1)$ and have a one-parameter family of self-adjoint extensions indexed by a parameter $\beta \in [-1/2, 1/2)$ determined by the equation

$$\frac{2}{\gamma} = \frac{\tan(\pi\beta/2)}{\beta} + \frac{\beta}{(1 - \beta^2)\tan(\pi\beta/2)}. \tag{2.3.51}$$

These are the Friedrichs extensions for $-1/2 \leq \beta \leq 0$, and other self-adjoint extensions for $0 < \beta < 1/2$. Note that $\beta = 1/2$ corresponds to $\gamma = 3/4$ when $\mathfrak{b}_{0,1/2}$ and $\mathfrak{b}_{1,-1/2}$ are essentially self-adjoint, but not closed, by Theorem 2.3.11

The domain of the Friedrich extension $\mathfrak{b}_{l,s}^F$ of $\mathfrak{b}_{l,s}$ is proved to be

$$\mathcal{D}(\mathfrak{b}_{l,s}^F) = \{\phi = \psi + \xi : \psi \in L^2(\mathbb{R}_+; e^2(p)dp), \xi \in \text{span}[\theta(p-1)p^{-1-\beta}]\},$$

where θ is the Heaviside function. The Friedrichs extension of the Brown–Ravenhall operator $\mathfrak{b}$ is the direct sum $\mathfrak{b}^F = \bigoplus_{(l,s)} \mathfrak{b}_{l,s}^F$ with domain

$$\mathcal{D}(\mathfrak{b}^F) = \{u : u = \sum_{(l,m,s)\in\mathfrak{I}} p^{-1}a_{l,m,s}(p)\Omega_{l,m,s}(\omega), \ a_{l,m,s} \in \mathcal{D}(\mathfrak{b}_{l,s}^F),$$

$$\sum_{(l,m,s)\in\mathfrak{I}} \|a_{l,m,s}\|_{L^2(\mathbb{R}_+}^2 < \infty\}.$$

For β in $\{z \in \mathbb{C} : -1/2 \leq \beta < 1/2\}$, Tix proves that the family of operators is a self-adjoint holomorphic family in the sense of Kato, in analogy with the result of [Yaouanc *et al.* (1997)] described in Remark 2.2.16. We refer to [Tix (1997b)] for a full treatment of these and other related results.

2.4 A unique continuation property

For solutions of Dirac-type equations $D\psi = 0$, there exist a number of methods to prove the weak continuation property for solutions, i.e., that if ψ vanishes on a non-empty open set, then it vanishes everywhere ([Booß-Bavnbeck (2000)]). One way of proving it is to apply the classical Aronszajn theorem in [Aronszajn (1957)] for second-order elliptic operators to the square of the operator D.

In [Bär and Strohmaier (2001)] a generalisation of the weak continuation property for semi-bounded restrictions of Dirac-type operators has been established, which is particularly relevant to our needs. The general result is as follows: let M be a connected Riemannian manifold and let D be a Dirac-type operator acting on smooth compactly supported sections of a Hermitian vector bundle over M. Suppose D has a self-adjoint extension A in the Hilbert space of square-integrable sections. Suppose that an L^2-section ψ is contained in a closed A-invariant subspace on which the restriction of A is semi-bounded. Then if ψ vanishes on a non-empty open subset of M, it vanishes on all M.

In the context of this book, we are concerned with the following special cases, avoiding the complicated topological structures, which are not necessary for our purpose.

Example 1 Let $M = \mathbb{R}^3$ and in the terminology of Section 2.1.3, $D = \mathbb{D}'_0 + P$, the Dirac operator with a Hermitian matrix-valued function P. Electromagnetic potentials and electric and magnetic anomalous moments are therefore included.

Example 2 Since the underlying manifold M need not be complete, it is possible to deal with singular potentials. For example, we can choose $M = \mathbb{R}^3 \setminus \{0\}$, $D = \mathbb{D}'_0 + P$ with domain $C_0^\infty(\mathbb{R}^3 \setminus \{0\}, \mathbb{C}^4)$ and P having a Coulomb singularity, subject to the assumptions of Theorem 2.1.13 to make D essentially self-adjoint.

More examples can be generated from Examples 1 and 2 by the procedure in the next example.

Example 3 If D is a Dirac-type operator on M acting on sections of a Hermitian vector bundle E (spinor fields in Examples 1 and 2), then a Dirac-type operator on $M \times \mathbb{R}$ is defined by

$$\mathcal{D} := \begin{pmatrix} D & \frac{\partial}{\partial t} \\ -\frac{\partial}{\partial t} & -D \end{pmatrix}$$

acting on a section of $E \oplus E$.

Theorem 2.4.1. *Let P satisfy the conditions of Theorem 2.1.13 and let ψ belong to an invariant subspace of the self-adjoint extension $A = \mathbb{D}_P$ of $\mathbb{D}'_P$, on which A is bounded below, i.e., for some λ_0, $\psi \in \chi_{[\lambda_0, \infty)}(A)$, the projection onto the subspace of A corresponding to the spectral subset $[\lambda_0, \infty)$. Then if $\psi = 0$ on a non-empty open subset $\mathcal{O}$ of $\mathbb{R}^3$ it vanishes identically on $\mathbb{R}^3$.*

Note that Corollary 1.7 in [Thaller (1992)] gives the case $P = 0$ of this theorem. Physically this means that one cannot localise a single electron in a bounded region.

The same applies for an electron described by a Brown–Ravenhall operator with magnetic field; see Section 4.7.2.

Proof. We give a sketch only, and refer to [Bär and Strohmaier (2001)] and [Roe (1988)] for details and precise statements. The proof requires what is known as the "finite propagation speed" for a Dirac-type operator D: for compactly supported ψ the support of $\exp(itD)\psi$ grows at most with speed one. We wish to show that $\psi = 0$.

Without loss of generality we may assume that ψ belongs to an A-invariant subspace where A is strictly bounded below by 1, i.e., $\chi_{(-\infty,1]}(A)\psi = 0$. In other words, ψ lies in the image of the projection $\chi_{(1,\infty)}(A)$. The family of functions f_z,

$$f_z(\lambda) = \begin{cases} \lambda^{-1}e^{iz\lambda}, & \lambda \geq 1 \\ 0, & \lambda < 1, \end{cases}$$

is uniformly bounded by 1 for all $z \in \mathbb{C}_-$ and for each fixed λ it is continuous in z. Therefore the family of bounded operators $f_z(A)$ is continuous in $z \in \mathbb{C}_-$ in the strong operator topology, and the map $z \mapsto f_z(A)\psi$ is continuous from $\mathbb{C}_-$ into the set of $L^2(\mathbb{R}^3, \mathbb{C}^4)$-valued functions.

We now fix $z_0 \in \mathbb{C}_+$ and consider the family of functions g_z defined by

$$g_z(\lambda) = \begin{cases} \lambda^{-1}\frac{e^{iz\lambda}-e^{iz_0\lambda}}{z-z_0}, & z \neq z_0 \\ ie^{iz_0\lambda}, & z = z_0, \end{cases}$$

for $\lambda \geq 1$ and $g_z(\lambda) = 0$ for $\lambda < 1$. The family g_z is also uniformly bounded and continuous in z. This shows that

$$\lim_{z \to z_0} \frac{f_z(A)\psi - f_{z_0}(A)\psi}{z - z_0}$$

exists and hence $z \mapsto f_z(A)\psi$ is holomorphic on $\mathbb{C}_+$.

Fix a non-empty relatively compact open subset $\tilde{\mathcal{O}}$ of $\mathcal{O}$. If u is a smooth function with support inside $\tilde{\mathcal{O}}$, then by the finiteness of propagation speed, there exists $\epsilon > 0$ such that the support of $e^{itA}u$ is contained in $\mathcal{O}$ for $|t| < \epsilon$. Hence for all t with $|t| < \epsilon$ we have

$$0 = (\psi, e^{-itA}u) = (e^{itA}\psi, u)$$
$$= (Af_t(A)\psi, u)$$
$$= (f_t(A)\psi, Au).$$

The identity $e^{izA}\psi = Af_z(A)\psi$ follows from the assumption that ψ lies in the image of the projection $\chi_{(1,\infty)}(A)$. The function $z \mapsto (f_z(A)\psi, Au)$ is continuous on $\mathbb{C}_-$ and holomorphic on $\mathbb{C}_+$. Moreover, it vanishes on $[-\epsilon, \epsilon]$. From the Schwarz reflection principle we therefore have that

$$(f_z(A)\psi, Au) = 0$$

for all $z \in \mathbb{C}_-$. In particular for $z = it$, $t > 0$ we obtain

$$(e^{-tA}\psi, u) = (f_{it}(A)\psi, Au) = 0$$

for all u with support contained in $\tilde{O}$. Hence $e^{-tA}\psi$ vanishes on $\tilde{O}$ for all $t > 0$.

The last step of the proof involves the Dirac-type operators $\mathcal{D}$ defined on the half cylinder $X = \mathbb{R}^3 \setminus \{0\} \times \mathbb{R}_+$ and acting on $\mathbb{C}^4 \oplus \mathbb{C}^4$-valued functions:

$$\mathcal{D} := \begin{pmatrix} D & \frac{\partial}{\partial t} \\ -\frac{\partial}{\partial t} & -D \end{pmatrix}.$$

We define a distribution Ψ_1 in $\mathbb{C}^4$ over X by

$$\Psi_1(v) := \int_0^\infty (e^{-tA}\psi, v(\cdot, t))_{L^2(\mathbb{R}^3, \mathbb{C}^4)} \, dt$$

for all $v \in C_0^\infty(X, \mathbb{C}^4)$. The differential equation

$$\frac{d}{dt} e^{-tA}\psi = -A e^{-tA}\psi$$

shows that the distribution

$$\Psi := \begin{pmatrix} \Psi_1 \\ \Psi_1 \end{pmatrix}$$

in $\mathbb{C}^4 \oplus \mathbb{C}^4$ satisfies $\mathcal{D}\Psi = 0$ in the distributional sense. Then Ψ is regular by elliptic regularity theory. Since Ψ vanishes on the open subset $\tilde{O} \times (0, \infty)$ of X, the standard unique continuation property of $\mathcal{D}$ implies $\Psi = 0$. Hence $\Psi_1 = 0$ and $e^{-tA}\psi = 0$ for all $t > 0$. On taking the limit as $t \to +0$, we obtain $\psi = 0$. $\qquad\square$

Chapter 3

Spectra

3.1 The Dirac operator

3.1.1 *Preliminary lemmas*

The first lemma will be needed to determine the essential spectra of the Dirac, quasi-relativistic and Brown–Ravenhall operators. In it $\mathbb{H}_0 = \sqrt{-\Delta + 1}$ and $\mathbb{D}_0$ is the free Dirac operator.

Lemma 3.1.1. *Let $V(\mathbf{x}) = \gamma/|\mathbf{x}|$. Then the maps*

$$|V|^{1/2} : H^1(\mathbb{R}^3) \to L^2(\mathbb{R}^3), \quad |V| : H^2(\mathbb{R}^3) \to L^2(\mathbb{R}^3)$$

are compact. Equivalently $|V|^{1/2}\mathbb{H}_0^{-1}$ and $|V|\mathbb{H}_0^{-2}$ are compact on $L^2(\mathbb{R}^3)$, or, in the terminology of Section 1.4, $|V|^{1/2}$ is $\mathbb{H}_0$-compact and $|V|$ is $\mathbb{H}_0^2$-compact.

In view of (2.1.11) and (2.1.13), it follows that $|V|^{1/2}\mathbb{D}_0^{-1}$ and $|V|\mathbb{D}_0^{-2}$ are compact on $L^2(\mathbb{R}^3, \mathbb{C}^4)$.

Proof. Let $\theta \in C_0^\infty(\mathbb{R}^+)$ be such that $0 < \theta < 1$,

$$\theta(t) = \begin{cases} 1, & 0 < t < 1, \\ 0, & t > 2 \end{cases}$$

and, for $\varepsilon > 0$, $\theta_\varepsilon(t) = \theta(\varepsilon t)$. Then, for all $u \in H^1(\mathbb{R}^3)$,

$$\||V|^{1/2}u\|^2 \le \gamma \left\{ \int_{|\mathbf{x}|<1/\varepsilon} \frac{|\theta_\varepsilon u|^2}{|\mathbf{x}|} d\mathbf{x} + \varepsilon\|u\|^2 \right\}.$$

Also, by the Cauchy–Schwarz and Hardy inequalities,

$$\int_{|\mathbf{x}|<1/\varepsilon} \frac{|\theta_\varepsilon u|^2}{|\mathbf{x}|} d\mathbf{x} \le \left(\int_{|\mathbf{x}|<1/\varepsilon} |u|^2 d\mathbf{x} \right)^{1/2} \left(\int_{\mathbb{R}^3} \frac{|u|^2}{|\mathbf{x}|^2} d\mathbf{x} \right)^{1/2}$$

$$\le 2 \left(\int_{|\mathbf{x}|<1/\varepsilon} |u|^2 d\mathbf{x} \right)^{1/2} \|\nabla u\|. \tag{3.1.1}$$

It follows from this that $|V|^{1/2} : H^1(\mathbb{R}^3) \to L^2(\mathbb{R}^3)$ is compact. For, if u_m tends weakly to zero in $H^1(\mathbb{R}^3)$, it is bounded in $H^1(\mathbb{R}^3)$, and tends strongly to zero in

$L^2(B_{1/\varepsilon})$, where $B_{1/\varepsilon}$ is the ball with centre the origin and radius $1/\varepsilon$. Hence, from (3.1.1), for a positive constant K such that $\|u_m\| \leq K$,

$$\limsup_{m \to \infty} \||V|^{1/2}u_m\|^2 \leq \gamma K^2 \varepsilon,$$

whence $|V|^{1/2}u_m \to 0$ in $L^2(\mathbb{R}^3)$, since ε is arbitrary. Our assertion about $|V|^{1/2}$ follows.

Suppose next that $\{u_m\}$ is a sequence in $H^2(\mathbb{R}^3)$ that converges weakly to zero. Then, with θ_ε as before,

$$\|V\,u_m\|^2 \leq \int_{|\mathbf{x}|<2/\varepsilon} |V\theta_\varepsilon u_m|^2 d\mathbf{x} + \int_{|\mathbf{x}|>1/\varepsilon} |V\theta_\varepsilon u_m|^2 d\mathbf{x}$$

$$\leq 4\gamma^2 \int_{|\mathbf{x}|<2/\varepsilon} |\mathbf{\nabla}[\theta_\varepsilon u_m]|^2 d\mathbf{x} + \gamma^2 \varepsilon^2 \|u_m\|^2$$

$$\leq 4\gamma^2 \|u_m\|_{H^1(B_{2/\varepsilon})}^2 + \gamma^2 \varepsilon^2 \|u_m\|^2,$$

on using Hardy's inequality. Since the embedding $H^2(\mathbb{R}^3) \hookrightarrow H^1(B_{2/\varepsilon})$ is compact, it follows that

$$\limsup_{m \to \infty} \|Vu_m\|^2 \leq (\gamma K \varepsilon)^2,$$

which implies the compactness of $V : H^2(\mathbb{R}^3) \to L^2(\mathbb{R}^3)$.

The remaining statements in the lemma merely reflect the facts that $\mathbb{H}_0^{-1}$ and $\mathbb{D}_0^{-1}$ are homeomorphisms of $L^2(\mathbb{R}^3)$ onto $H^1(\mathbb{R}^3)$ and $L^2(\mathbb{R}^3; \mathbb{C}^4)$ onto $H^1(\mathbb{R}^3; \mathbb{C}^4)$, respectively. $\qquad\square$

The second lemma is an abstract virial theorem proved in [Balinsky and Evans (1998)]; it is the basis of the proof of Weidmann's virial theorem for Schrödinger operators in [Weidmann (1967)], and is modelled on Weidmann's proof.

Lemma 3.1.2. *Let $U(a), a \in \mathbb{R}^+$, be a one-parameter family of unitary operators on a Hilbert space $\mathcal{H}$, which converges strongly to the identity as $a \to 1$. Let T be a self-adjoint operator in $\mathcal{H}$ and $T_a := f(a)U(a)TU(a)^{-1}$, where $f(1) = 1$ and $f'(1)$ exists. If $\phi \in \mathcal{D}(T) \cap \mathcal{D}(T_a)$ is an eigenvector of T corresponding to an eigenvalue λ, then*

$$\lim_{a \to 1} \left(\phi_a, \left[\frac{T_a - T}{a - 1} \right] \phi \right) = \lambda f'(1) \|\phi\|^2, \tag{3.1.2}$$

where $\phi_a = U(a)\phi$ and $(\cdot), \| \cdot \|$ are, respectively, the inner product and norm of $\mathcal{H}$.

Proof. From $T\phi = \lambda\phi$ we have $T_a\phi_a = \lambda f(a)\phi_a$. Hence, $(\phi_a, T\phi) = \lambda(\phi_a, \phi)$ and $(T_a\phi_a, \phi) = \lambda f(a)(\phi_a, \phi)$. Consequently,

$$\left(\phi_a, \left[\frac{T_a - T}{a - 1} \right] \phi \right) = \lambda \left[\frac{f(a) - 1}{a - 1} \right] (\phi_a, \phi)$$

and the result follows on allowing $a \to 1$. $\qquad\square$

3.1.2 The essential spectrum of $\mathbb{D}$

We recall from Theorem 2.1.6 that for $|\gamma| \leq \sqrt{3}/2$, $\mathbb{D}_0 + \gamma/|\cdot|$ on $C_0^\infty(\mathbb{R}^3)$ is essentially self-adjoint, and so the associated Dirac operator $\mathbb{D}$ is its closure. Also, by Theorem 2.1.13, if $|\gamma| < \sqrt{3}/2$, $\mathcal{D}(\mathbb{D}) = \mathcal{D}(\mathbb{D}_0)$.

Theorem 3.1.3. *Suppose $V(\mathbf{x}) = \gamma/|\mathbf{x}|$, where $|\gamma| < \sqrt{3}/2$. Then $\mathbb{D}$ has essential spectrum $\mathbb{R} \setminus (-1, 1)$.*

Proof. We infer from $\mathcal{D}(\mathbb{D}) = \mathcal{D}(\mathbb{D}_0)$ that

$$(\mathbb{D}_0 + i)^{-1} + (\mathbb{D} + i)^{-1}\left((\mathbb{D}_0 + i) - (\mathbb{D} + i)\right)(\mathbb{D}_0 + i)^{-1} = (\mathbb{D} + i)^{-1}$$

and hence

$$(\mathbb{D} + i)^{-1} - (\mathbb{D}_0 + i)^{-1} = (\mathbb{D} + i)^{-1}V(\mathbb{D}_0 + i)^{-1}. \tag{3.1.3}$$

We showed in Lemma 3.1.1 that $A := |V|^{1/2}(\mathbb{D}_0 + i)^{-1}$ is compact on $L^2(\mathbb{R}^3; \mathbb{C}^4)$ and hence, so is

$$(\mathbb{D}_0 - i)^{-1}|V|^{1/2}|V|^{1/2}(\mathbb{D}_0 + i)^{-1} \subseteq A^*A.$$

This implies the compactness of

$$(\mathbb{D}_0 + i)^{-1}|V|^{1/2}|V|^{1/2}(\mathbb{D}_0 + i)^{-1}$$

since $(\mathbb{D}_0 + i)^{-1}(\mathbb{D}_0 - i)$ is bounded on $\mathcal{D}(\mathbb{D}_0) = H^1(\mathbb{R}^3; \mathbb{C}^4)$.

The fact that $\mathcal{D}(\mathbb{D}) = \mathcal{D}(\mathbb{D}_0)$, and the Closed Graph Theorem, ensure that the operator $(\mathbb{D} + i)^{-1}(\mathbb{D}_0 + i)$ is bounded on $\mathcal{D}(\mathbb{D}_0)$ furnished with the graph norm. Hence, by (3.1.3) and Section 1.4, $(\mathbb{D} + i)$ and $(\mathbb{D}_0 + i)$ have the same essential spectrum, namely $\mathbb{R} \setminus (-1, 1)$.

$\square$

In [Thaller (1992)], Theorem 4.7, it is shown that $\mathbb{D} = \mathbb{D}_0 + V$ has essential spectrum $\mathbb{R} \setminus (-1, 1)$ for any potential V which is such that $\mathbb{D}$ is self-adjoint, V is $\mathbb{D}_0$-bounded and $\lim_{R\to\infty} \|V(\mathbb{D}_0 - i)^{-1}\chi(|\mathbf{x}| > R)\| = 0$, where χ denotes the characteristic function. This covers the Coulomb potentials $V(\mathbf{x}) = \gamma/|\mathbf{x}|$ for $|\gamma| \leq \sqrt{3}/2$.

In order to prove that the essential spectrum of $\mathbb{D}$ does not contain any embedded eigenvalues, we apply Lemma 3.1.2 to establish a virial theorem for $\mathbb{D}$. The same method is used in [Thaller (1992)], Theorem 4.21, for more general potentials V.

Theorem 3.1.4. *Suppose $V(\mathbf{x}) = \gamma/|\mathbf{x}|, |\gamma| \leq \sqrt{3}/2$. Then $\mathbb{D} = \mathbb{D}_0 + V$ has no eigenvalues in $\mathbb{R}^3 \setminus (-1, 1)$. In fact the spectrum in $\mathbb{R}^3 \setminus (-1, 1)$ is absolutely continuous.*

Proof. To prove that there are no eigenvalues in $\mathbb{R}^3 \setminus [-1, 1]$, we apply Lemma 3.1.2 with $f(x) = x$ and $U(a)$ the dilation operator defined by $U(a)\phi(\mathbf{x}) = a^{3/2}\phi(a\mathbf{x})$ and $T = \mathbb{D} = \mathbb{D}_0 + V$. Then,

$$(T_a\phi)(\mathbf{x}) = a\left\{\frac{1}{a}[(-i\boldsymbol{\alpha}\cdot\boldsymbol{\nabla})\phi](\mathbf{x}) + \beta\phi(\mathbf{x}) + V(a\mathbf{x})\right\}$$

$$= (T\phi)(\mathbf{x}) + (a - 1)\beta\phi(\mathbf{x}).$$

Hence, $\mathcal{D}(T_a) = \mathcal{D}(T)$ and from Lemma 3.1.2, if $T\phi = \lambda\phi$, and $\phi_a := U(a)\phi$,

$$\lambda\|\phi\|^2 = \lim_{a \to 1}(\phi_a, \beta\phi)$$
$$= (\phi, \beta\phi).$$

From this we infer that $-1 \le \lambda \le 1$ and, hence, there are no eigenvalues in $\mathbb{R}^3 \setminus [-1, 1]$.

Suppose $T\phi = \phi$. Then, from above, $\|\phi\|^2 = (\phi, \beta\phi)$ and this implies that

$$\|\phi_3\|^2 + \|\phi_4\|^2 = 0,$$

where $\phi_j, j = 1, 2, 3, 4$ are the components of ϕ, and thus $\phi_3 = \phi_4 = 0$. Let u be the 2-spinor

$$u = \begin{pmatrix} \phi_1 \\ \phi_2 \end{pmatrix}.$$

Then, on substituting (2.1.4), it follows that $-i(\boldsymbol{\sigma} \cdot \boldsymbol{\nabla})u = 0$ which gives, on taking Fourier transforms,

$$(\mathbf{p} \cdot \boldsymbol{\sigma})\hat{u}(\mathbf{p}) = \sum_{j=1}^{3} p_j \sigma_j \hat{u}(\mathbf{p}) = 0$$

and hence $|\mathbf{p}|^2|\hat{u}(\mathbf{p})|^2 = 0$. This in turn implies that $\hat{u} = 0$ and hence $\phi_1 = \phi_2 = 0$. Therefore $\lambda = 1$ is not an eigenvalue. A similar argument proves that there is no eigenvalue at $\lambda = -1$.

The absolute continuity of the spectrum is established in [Weidmann (1982)].

$\square$

3.1.3 *Eigenvalues in* $(-1, 1)$

Theorem 3.1.5. *Suppose* $V(\mathbf{x}) = -\gamma/|\mathbf{x}|, 0 \le \gamma \le \sqrt{3}/2$. *Then the eigenvalues of the Dirac partial wave operators* $\mathbb{D}_{l,s} = \mathbb{D}_{0;l,s} + V$ *in the gap* $(-1, 1)$ *are given by*

$$\lambda_{n;l,s} = \left\{ 1 + \frac{\gamma^2}{(n + \sqrt{\kappa_{l,s}^2 - \gamma^2})^2} \right\}^{-1/2}, \quad n \in \mathbb{N}_0, \tag{3.1.4}$$

where $\kappa_{l,s} = l + s + 1/2$. *Hence, from Theorem 2.1.1, the set of eigenvalues of* $\mathbb{D} = \mathbb{D}_0 + V$ *is*

$$\bigcup_{n \in \mathbb{N}_0, (l,s) \in \mathcal{I}} \lambda_{n;l,s},$$

where $\mathcal{I}$ *is the index set defined in* (2.1.27). *Thus, there is an infinity of positive eigenvalues accumulating at* 1.

Proof. Set $\kappa = \kappa_{l,s}$ and consider the operator τ in (2.1.52) which typifies the partial wave operators, namely

$$\tau = -i\sigma_2 \frac{d}{dr} - \frac{\kappa}{r}\sigma_1 + \sigma_3 + V.$$

Note that $\kappa_{l,s} \geq 1$ since, in the notation $(l, s) \in \mathcal{J}$, $l \in \mathbb{N}_0$ for $s = 1/2$ and $l \in \mathbb{N}$ for $s = -1/2$. We follow the treatment in [Thaller (1992)], Section 7.4, but avoid the explicit use of supersymmetry and the abstract Foldy–Wouthuysen transformation made therein.

In

$$i\sigma_2(\tau - \lambda) = I_2\frac{d}{dr} + \begin{pmatrix} -\kappa & -\gamma \\ \gamma & \kappa \end{pmatrix}\frac{1}{r} + \begin{pmatrix} 0 & -\lambda - 1 \\ \lambda - 1 & 0 \end{pmatrix}$$

the second matrix on the right-hand side has eigenvalues $\pm t$, $t = \sqrt{\kappa^2 - \gamma^2}$, and is diagonalised by

$$A = \begin{pmatrix} \kappa - t & -\gamma \\ -\gamma & \kappa - t \end{pmatrix}.$$

It follows that

$$i\sigma_2 A^{-1} i\sigma_2(\tau - \lambda)f = \begin{pmatrix} -1 - \frac{\kappa\lambda}{t} & \tau_0^*(\lambda) \\ \tau_0(\lambda) & 1 - \frac{\kappa\lambda}{t} \end{pmatrix} u, \quad u = A^{-1}f, \tag{3.1.5}$$

where

$$\tau_0(\lambda) := -\frac{d}{dr} - \frac{t}{r} + \frac{\gamma\lambda}{t}$$

and $\tau_0^*(\lambda)$ is its Lagrangian adjoint $\frac{d}{dr} - \frac{t}{r} + \frac{\gamma\lambda}{t}$. The operators $\tau, \tau_0(\lambda)$ and $\tau_0^*(\lambda)$ have domains

$$\mathcal{D}(\tau) = \{u : u \in AC_{loc}(\mathbb{R}_+, \mathbb{C}^2), u, \tau u \in L^2(\mathbb{R}_+, \mathbb{C}^2)\},$$
$$\mathcal{D}(\tau_0(\lambda)) = \{v : v \in AC_{loc}(\mathbb{R}_+), v, \tau_0(\lambda)v \in L^2(\mathbb{R}_+)\},$$
$$\mathcal{D}(\tau_0^*(\lambda)) = \{v : v \in AC_{loc}(\mathbb{R}_+), v, \tau_0(\lambda)^*v \in L^2(\mathbb{R}_+)\}.$$

From (3.1.5) we infer that f is an eigenvector of τ corresponding to the eigenvalue λ if and only if $u = A^{-1}f = \begin{pmatrix} u_1 \\ u_2 \end{pmatrix}$ satisfies

$$\tau_0(\lambda)u_1 = (\frac{\kappa\lambda}{t} - 1)u_2$$

$$\tau_0^*(\lambda)u_2 = (\frac{\kappa\lambda}{t} + 1)u_1. \tag{3.1.6}$$

Note, in particular, that $u_1 \in \mathcal{D}(\tau_0(\lambda))$ and $u_2 \in \mathcal{D}(\tau_0^*(\lambda))$, and, by (3.1.6), $\tau_0(\lambda)u_1 \in \mathcal{D}(\tau_0^*(\lambda))$ and $\tau_0^*(\lambda)u_2 \in \mathcal{D}(\tau_0(\lambda))$. We therefore have

$$\tau_0^*(\lambda)\tau_0(\lambda)u_1 = (\frac{\kappa^2\lambda^2}{t^2} - 1)u_1 \tag{3.1.7}$$

$$\tau_0(\lambda)\tau_0^*(\lambda)u_2 = (\frac{\kappa^2\lambda^2}{t^2} - 1)u_2. \tag{3.1.8}$$

Since $\tau_0^*(\lambda)\tau_0(\lambda) \geq 0$, we must have $|\lambda| \geq t/\kappa$. The solutions of the equation $\tau_0^*(\lambda)v = 0$ are constant multiples of

$$v(r) = r^t e^{-(\gamma\lambda/t)r},$$

which lies in $L^2(\mathbb{R}_+)$ for $\lambda > 0$, whereas the solutions of $\tau_0(\lambda)v = 0$ are constant multiples of

$$v(r) = r^{-t} e^{(\gamma\lambda/t)r}$$

and, hence, only the trivial solution lies in $L^2(\mathbb{R}_+)$. Therefore $\tau_0(\lambda)$ has trivial null space, and so has $\tau_0^*(\lambda)\tau_0(\lambda)$ by (1.2.4). We conclude that the smallest eigenvalue of τ is

$$\lambda_0 = t/\kappa = \left\{1 + \frac{\gamma^2}{\kappa^2 - \gamma^2}\right\}^{-1/2} \tag{3.1.9}$$

with corresponding eigenvector

$$f_0 = Au^{(0)}, \quad u^{(0)} = \begin{pmatrix} 0 \\ r^t \exp\left(-\frac{\gamma\lambda_0}{t}r\right) \end{pmatrix}. \tag{3.1.10}$$

Let

$$\tau_1(\lambda) := -\frac{d}{dr} - \frac{t+1}{r} + \frac{\gamma\lambda}{t+1}.$$

Then

$$\tau_0^*(\lambda)\tau_0(\lambda) - \frac{\gamma^2\lambda^2}{t^2} = \tau_1(\lambda)\tau_1^*(\lambda) - \frac{\gamma^2\lambda^2}{(t+1)^2} \tag{3.1.11}$$

and hence (3.1.7) is identical with

$$\tau_1(\lambda)\tau_1^*(\lambda)u_1 = \left(\frac{\kappa_1^2\lambda^2}{(t+1)^2} - 1\right)u_1 \tag{3.1.12}$$

where $\kappa_1 = \sqrt{(t+1)^2 + \gamma^2}$. As before, the smallest value of λ for which this equation has an $L^2(\mathbb{R}_+)$ solution is

$$\lambda_1 = \frac{t+1}{\kappa_1} = \left\{1 + \frac{\gamma^2}{(t+1)^2}\right\}^{-1/2} \tag{}$$

corresponding to the solution

$$u_1^{(1)}(r) = r^{t+1} \exp\left(-\frac{\gamma\lambda_1}{(t+1)}r\right), \tag{3.1.13}$$

which lies in the null space of $\tau_1^*(\lambda_1)$. It follows from (3.1.11) that $u_1^{(1)}$ is an eigenvector of $\tau_0(\lambda_1)^*\tau_0(\lambda_1)$ corresponding to the positive eigenvalue $(\kappa^2\lambda_1^2/t^2) - 1$. Since $\tau_0(\lambda_1)$ has trivial null space, $\tau_0(\lambda_1)u_1^{(1)} \neq 0$ and, by the remark following (1.2.6), $\tau_0(\lambda_1)u_1^{(1)}$ is an eigenvector of $\tau_0(\lambda_1)\tau_0(\lambda_1)^*$ corresponding to the same eigenvalue $(\kappa^2\lambda_1^2/t^2) - 1$, and all the eigenvectors corresponding to this eigenvalue are constant multiples of $\tau_0(\lambda_1)u_1^{(1)}$. Also $\lambda_1 > \lambda_0$ and since the eigenvalues $\lambda > \lambda_0$ of τ are in one-one correspondence with the values of λ for which (3.1.12) has a non-trivial solution, it follows that there are no eigenvalues of τ between λ_0 and λ_1. Moreover, the eigenvector of τ corresponding to λ_1 is

$$f^{(1)} = Au^{(1)}, u^{(1)} = \begin{pmatrix} u_1^{(1)} \\ \tau_0(\lambda_1)u_1^{(1)} \end{pmatrix}. \tag{3.1.14}$$

An iterative procedure determines the remaining eigenvalues and eigenvectors of τ. For $\lambda > \lambda_1$ and $n \geq 2$, let

$$\tau_n(\lambda) := -\frac{d}{dr} - \frac{t+n}{r} + \frac{\gamma\lambda}{t+n}.$$

The smallest value of λ for which the equation

$$\tau_n(\lambda)\tau_n(\lambda)^*\phi = \left(\frac{\kappa_n^2\lambda^2}{(t+n)^2} - 1 \right)\phi, \quad \kappa_n = \sqrt{(t+n)^2 + \gamma^2} \tag{3.1.15}$$

has an $L^2(\mathbb{R}_+)$ solution is

$$\lambda_n = \frac{t+n}{\kappa_n} = \left\{ 1 + \frac{\gamma^2}{(n + \sqrt{\kappa^2 - \gamma^2})} \right\}^{-1/2} \tag{3.1.16}$$

with corresponding $L^2(\mathbb{R}_+)$ solution

$$\phi_n(r) = r^{t+n} \exp\left(-\frac{\gamma\lambda_n}{(t+n)}r \right), \tag{3.1.17}$$

which lies in the null space of $\tau_n^*(\lambda_n)$. A crucial property of the $\tau_n(\lambda)$ is

$$\tau_{n-1}(\lambda)^*\tau_{n-1}(\lambda) - \frac{\gamma^2\lambda^2}{(t+n-1)^2} = \tau_n(\lambda)\tau_n(\lambda)^* - \frac{\gamma^2\lambda^2}{(t+n)^2},$$

which has the following consequences. Firstly, $\tau_{n-1}(\lambda)^*\tau_{n-1}(\lambda) > 0$, so that $\tau_{n-1}(\lambda)$ has trivial null space, and secondly

$$\tau_{n-1}(\lambda_n)^*\tau_{n-1}(\lambda_n)\phi_n = \left(\frac{\kappa_{n-1}^2\lambda_n^2}{(t+n-1)^2} - 1 \right)\phi_n, \tag{3.1.18}$$

where $\lambda_n > (t+n-1)/\kappa_{n-1} = \lambda_{n-1}$. Furthermore $\psi := \tau_{n-1}(\lambda_n)\phi_n$ is a non-trivial $L^2(\mathbb{R}_+)$ solution of

$$\tau_{n-1}(\lambda_n)\tau_{n-1}(\lambda_n)^*\psi = \left(\frac{\kappa_{n-1}^2\lambda_n^2}{(t+n-1)^2} - 1 \right)\psi. \tag{3.1.19}$$

It follows that

$$\psi = \tau_1(\lambda_2)\tau_2(\lambda_3)\cdots\tau_{n-1}(\lambda_n)\phi_n$$

satisfies

$$\tau_0^*(\lambda_1)\tau_0(\lambda_1)\psi = \left(\frac{\kappa^2\lambda_1^2}{t^2} - 1 \right)\psi$$

and the non-trivial $L^2(\mathbb{R}_+)$ solutions of

$$\tau_0(\lambda_1)\tau_0^*(\lambda_1)\phi = \left(\frac{\kappa^2\lambda_1^2}{t^2} - 1 \right)\phi$$

are constant multiples of $\tau_0(\lambda_1)\psi$. Consequently, eigenvectors of τ corresponding to the eigenvalue λ_n in (3.1.16) for $n \geq 2$ are constant multiples of

$$f^{(n)} = Au^{(n)}, \quad u^{(n)} = \begin{pmatrix} \tau_1(\lambda_2)\tau_2(\lambda_3)\cdots\tau_{n-1}(\lambda_n)\phi_n \\ \tau_0(\lambda_1)\tau_1(\lambda_2)\tau_2(\lambda_3)\cdots\tau_{n-1}(\lambda_n)\phi_n \end{pmatrix}, \tag{3.1.20}$$

where ϕ_n is given by (3.1.17). $\qquad\square$

Remark 3.1.6. If $\sqrt{3}/2 < \gamma < 1$, Theorem 3.1.5 continues to be true for the distinguished self-adjoint extension of $\mathbb{D}_0' - \gamma/|\cdot|$ mentioned in Remark 2.1.9. But this requires the imposition of a boundary condition like

$$\lim_{r \to 0} u(r) = 0.$$

The proof above then remains valid since $L^2(\mathbb{R}_+)$ solutions of $\tau_0(\lambda)v = 0$ are still excluded. This is tantamount to choosing a distinguished self-adjoint extension of the operator T_0' generated by τ on $C_0^\infty(\mathbb{R}_+, \mathbb{C}^2)$ (see Section 2.1.2), for those values of κ for which τ is not in the limit-point case at the origin.

3.1.4   *Min-max characterisation*

In [Dolbeault *et al.* (2000b)] and [Dolbeault *et al.* (2000a)], Dolbeault, Esteban and Séré establish a min-max characterisation of the eigenvalues in a gap within the essential spectrum of an unbounded self-adjoint operator, which is applicable to the eigenvalues in $(-1, 1)$ of Dirac operators with Coulomb-type potentials. In their result for $\mathbb{D} = \mathbb{D}_0 + V, V(\mathbf{x}) = \gamma/|\mathbf{x}|$, quoted in the following theorem, Λ_+ and $\Lambda_- = 1 - \Lambda_+$ are the projections onto the positive and negative spectral subspaces respectively of $\mathbb{D}_0$, $\mathcal{H} = L^2(\mathbb{R}^3, \mathbb{C}^4)$, $\mathcal{H}_\pm = \Lambda_\pm \mathcal{H}$ and $F_\pm := \Lambda_\pm C_0^\infty(\mathbb{R}^3, \mathbb{C}^4)$.

Theorem 3.1.7. *Let*

$$\lambda_k = \inf_{M_k} \quad \sup_{\substack{\psi \in M_k \oplus F_- \\ \|\psi\| = 1}} \quad (\mathbb{D}\psi, \psi), \quad k \geq 1, \tag{3.1.21}$$

where the infimum is over all k-dimensional subspaces M_k of F_+. Then, $\lambda_k = \mu_k$, the kth positive eigenvalue of $\mathbb{D}$, counting multiplicity and $\lim_{k \to \infty} \lambda_k = 1$.

In [Dolbeault *et al.* (2000b)], it is assumed that $\gamma < \gamma_c = 4\pi/(\pi^2 + 4) \approx 0.9$, the critical value for the positivity of the Brown–Ravenhall operator as established in Theorem 2.3.7. The range of γ is extended in [Dolbeault *et al.* (2000a)] to $0 < \lambda < 1$ and so the eigenvalues of the distinguished self-adjoint operator $\tilde{\mathbb{D}}$ referred to in Remark 2.1.9 and Remark 2.1.14 are covered. Results on a min-max principle for abstract self-adjoint operators with a spectral gap may also be found in [Dolbeault *et al.* (2000a)], [Griesemer and Siedentop (1999)], [Griesemer *et al.* (1999)].

3.2   The quasi-relativistic operator

3.2.1   *The essential spectrum*

Theorem 3.2.1. *Suppose $0 \leq \gamma \leq 2/\pi$. Then the essential spectrum of $\mathbb{H} = \mathbb{H}_0 + V$, $V(\mathbf{x}) = -\gamma/|\mathbf{x}|$, coincides with that of $\mathbb{H}_0 = \sqrt{-\Delta + 1}$, namely, $[1, \infty)$. In $(0, 1)$,* *the spectrum of $\mathbb{H}$ consists of an infinite number of isolated eigenvalues of finite multiplicity, which accumulate only at 1. If $\gamma < 2/\pi$, the smallest eigenvalue λ_0 satisfies $\lambda_0 \geq (1 - [\pi\gamma/2]^2)^{1/2}$.*

Proof. From Lemma 3.1.1, $A := |V|^{1/2}(\mathbb{H}_0+1)^{-1}$ is compact on $L^2(\mathbb{R}^3)$ and hence so is $(\mathbb{H}_0 + 1)^{-1}|V|(\mathbb{H}_0 + 1)^{-1} = A^*A$. Therefore, in the terminology of Section 1.4, V is form $\mathbb{H}_0^2$-compact. Also, by Theorem 2.2.6, $\mathbb{H}_0 + V \geq 0$ and this implies that $\mathbb{Q}(|V|) \supset \mathbb{Q}(\mathbb{H}_0)$. The fact that $\mathbb{H}$ has essential spectrum $[1, \infty)$, then follows from Theorem 1.4.2 in the case $\gamma < 2/\pi$ when $\mathbb{H}$ is the form sum of $\mathbb{H}_0$ and V. The proof in the critical case $\gamma = 2/\pi$, when $\mathbb{H}$ is the Friedrichs extension of $\mathbb{H}_0$ and V, is included in [Yaouanc *et al.* (1997)], Lemma 4.

Since $\mathbb{H}_0 = \sqrt{-\Delta + 1} \leq 1 - \Delta$, and $-\Delta - \gamma/|\cdot|$ has an infinite number of negative eigenvalues (see [Edmunds and Evans (1987)], Theorem XI.1.5), it follows that $\mathbb{H}$ has infinitely many eigenvalues smaller than 1. The lower bound for the smallest eigenvalue λ_0 is established in Theorem 2.2.7 when $\gamma < 2/\pi$. When $\gamma = 2/\pi$, λ_0 is still positive, as noted in the following Remark. $\square$

Remark 3.2.2. When $\gamma = 2/\pi$, the lower bound for λ_0 in Theorem 3.2.1 vanishes. However, even in this critical case, there are numerical indications of Hardekopf and Sucher in [Hardekopf and Sucher (1985)] that the lower bound is positive. This was confirmed by Martin, Raynal, Roy, Singh and Stubbe in [Raynal *et al.* (1994)] who proved that, for $0 < \gamma \leq 2/\pi$,

$$\lambda_0 \geq \inf_{\mathbf{p} \in \mathbb{R}^3} \frac{(\mathbb{H}\phi)(\mathbf{p})}{\phi(\mathbf{p})}$$

for any positive trial function ϕ, and used this to yield

$$0.4825 < \lambda_0 < 0.4843$$

when $\gamma = 2/\pi$.

We now use Lemma 3.1.2 to prove a virial theorem for $\mathbb{H}$ which is similar to Herbst's virial theorem in [Herbst (1977)], Theorem 2.4.

Theorem 3.2.3. *Let $\gamma < 2/\pi$ and $\mathbb{H}\phi = \lambda\phi$. Then*

$$(\phi, \mathbb{H}_0^{-1}\phi) = \lambda\|\phi\|^2 \tag{3.2.1}$$

and

$$(V\phi, \phi) = \|\mathbb{H}_0^{-1/2}\phi\|^2 - \|\mathbb{H}_0^{1/2}\phi\|^2. \tag{3.2.2}$$

It follows that $\mathbb{H}$ has no embedded eigenvalues in $(1, \infty)$. The spectrum in $[1, \infty)$ is absolutely continuous.

Proof. Let $\mathbb{H}_{0,m} := (-\Delta + m^2)^{1/2}$ where $m > 0$. Theorem 2.2.6 applies also to the operator $\mathbb{H}_m = \mathbb{H}_{0,m} + V$: in particular, it is an operator sum for $\gamma < 1/2$ and a form sum for $1/2 \leq \gamma < 2/\pi$. Furthermore, if $m_1, m_2 > 0$, then

$$\mathbb{H}_{0,m_1} - \mathbb{H}_{0,m_2} = (m_1^2 - m_2^2)(\mathbb{H}_{0,m_1} + \mathbb{H}_{0,m_2})^{-1} =: B_{m_1,m_2}$$

is a bounded operator on $L^2(\mathbb{R}^3)$. Thus, if $\gamma < 1/2$, it follows that

$$\mathbb{H}_{m_1} = \mathbb{H}_{m_2} + B_{m_1,m_2} \tag{3.2.3}$$

and $\mathcal{D}(\mathbb{H}_{m_1}) = \mathcal{D}(\mathbb{H}_{m_2})$. The identity (3.1.3) is still true in the range $1/2 \leq \gamma < 2/\pi$, but since $\mathbb{H}_{m_1}$ and $\mathbb{H}_{m_2}$ are form sums, the derivation is not as immediate. We first assert that their domains coincide. For, on applying the Parseval formula, with $u, v \in \mathcal{Q}(\mathbb{H}_{0,m_1}) = \mathcal{Q}(\mathbb{H}_{0,m_2}) = H^{1/2}(\mathbb{R}^3)$, we have

$$(\mathbb{H}_{0,m_1}^{1/2} u, \mathbb{H}_{0,m_1}^{1/2} v) = (\mathbb{H}_{0,m_2}^{1/2} u, \mathbb{H}_{0,m_2}^{1/2} v)$$
$$+ \int_{\mathbb{R}^3} \left\{ \sqrt{p^2 + m_1^2} - \sqrt{p^2 + m_2^2} \right\} \hat{u}(\mathbf{p}) \overline{\hat{v}(\mathbf{p})} d\mathbf{p}$$
$$= (\mathbb{H}_{0,m_2}^{1/2} u, \mathbb{H}_{0,m_2}^{1/2} v) + (B_{m_1,m_2} u, v)$$

and so, for all $u, v \in \mathcal{Q}(\mathbb{H}_{m_1}) = \mathcal{Q}(\mathbb{H}_{m_2})$,

$$h_{m_1}[u, v] = h_{m_2}[u, v] + (B_{m_1,m_2} u, v),$$

where $h_{m_j}[\cdot, \cdot]$ is the quadratic form associated with $\mathbb{H}_{m_j}$, $j = 1, 2$. Hence, if $u \in \mathcal{D}(\mathbb{H}_{m_1})$ and $v \in \mathcal{D}(\mathbb{H}_{m_2})$ it follows from the Second Representation Theorem that

$$(\mathbb{H}_{m_1} u, v) = (u, \mathbb{H}_{m_2} v) + (B_{m_1,m_2} u, v).$$

Since B_{m_1,m_2} is bounded and $\mathbb{H}_{m_1}$ and $\mathbb{H}_{m_2}$ are self-adjoint, it follows that $\mathcal{D}(\mathbb{H}_{m_1}) = \mathcal{D}(\mathbb{H}_{m_2})$ and (3.1.3) holds.

We now apply Lemma 3.1.2 with $T = \mathbb{H} \equiv \mathbb{H}_1$, $f(t) = t$, and $U(a)$ the unitary operators defined by $U(a)\phi(\mathbf{x}) = a^{3/2}\phi(a\mathbf{x})$. Then it is readily verified that $T_a = aU(a)\mathbb{H}_1 U(a)^{-1} = \mathbb{H}_a$ and so

$$\left(\phi_a, \left[\frac{\mathbb{H}_a - \mathbb{H}}{a - 1}\right]\phi\right) = (a + 1)\left(\phi_a, [\mathbb{H}_{0,a} + \mathbb{H}_0]^{-1}\phi\right)$$
$$\rightarrow (\phi, \mathbb{H}_0^{-1}\phi)$$

as $a \rightarrow 1$, whence (3.1.1).

Furthermore,

$$(\phi, \mathbb{H}_0^{-1}\phi) = \lambda\|\phi\|^2$$
$$= (\phi, \mathbb{H}\phi)$$
$$= \|\mathbb{H}_0^{1/2}\phi\|^2 + (V\phi, \phi) \tag{3.2.4}$$

whence (3.2.2).

Suppose $\mathbb{H}\phi = \lambda\phi$. Then from (3.2.4),

$$(V\phi, \phi) = (\mathbb{H}_0^{-1}\phi, \phi) - (\mathbb{H}_0\phi, \phi)$$

and so

$$\lambda\|\phi\|^2 = (\mathbb{H}_0^{-1}\phi, \phi)$$
$$\leq \|\phi\|^2.$$

Hence, there are no eigenvalues in $(1, \infty)$.

The absence of singular continuous spectrum, and, hence, the absolute continuity of the spectrum in $[1, \infty)$, is proved in [Herbst (1977)], Theorem 2.3.   □

3.3 The Brown–Ravenhall operator

3.3.1 *The essential spectrum*

The analogue of Theorem 3.2.1 is the following

Theorem 3.3.1. Let $V(\mathbf{x}) = -\gamma/|\mathbf{x}|$, $0 \le \gamma \le \gamma_c = 2\left(\frac{\pi}{2} + \frac{2}{\pi}\right)^{-1}$. Then the Brown–Ravenhall operator $\mathbb{B} = \Lambda_+(\mathbb{D}_0 + V)\Lambda_+$ in (2.3.10) has the same essential spectrum as $\Lambda_+\mathbb{D}_0\Lambda_+$, namely $[1, \infty)$. In $[0, 1)$, the spectrum of $\mathbb{B}$ consists of an infinite number of isolated eigenvalues of finite multiplicity which accumulate at 1. The smallest eigenvalue λ_0 satisfies $\lambda_0 \ge 1 - \gamma_c > 0.09$.

Proof. We shall prove the result only in the case $\gamma < \gamma_c$, when, in view of Remark 2.3.8(v), $\mathbb{B}$ is the form sum of $\Lambda_+\mathbb{D}_0\Lambda_+ = \mathbb{D}_0\Lambda_+$ and $\Lambda_+V\Lambda_+$, and the form domain $\mathbb{Q}(\mathbb{B})$ coincides with that of $\Lambda_+\mathbb{D}_0\Lambda_+ = \mathbb{D}_0\Lambda_+$. Set $\mathbb{B}_0 := \Lambda_+\mathbb{D}_0\Lambda_+ = \mathbb{D}_0\Lambda_+$ and $C = \Lambda_+W\Lambda_+, W = -V$. Note that the operators in these forms act in $L^2(\mathbb{R}^3, \mathbb{C}^4)$; see (2.3.10) and (2.3.11). We have

$$(\mathbb{B}\psi, \phi) = (\mathbb{B}_0\psi, \phi) - (C\psi, \phi), \quad \psi, \phi \in \mathbb{Q}(\mathbb{B}),$$

in the form sense. On taking, for arbitrary $f, g \in L^2(\mathbb{R}^3, \mathbb{C}^4), \psi = \mathbb{B}^{-1}f, \phi = \mathbb{B}_0^{-1}g$, so that $\psi \in \mathcal{D}(\mathbb{B})$ and $\phi \in \mathcal{D}(\mathbb{B}_0) \subset \mathbb{Q}(\mathbb{B}_0) = \mathbb{Q}(\mathbb{B})$, it follows from the First Representation Theorem that

$$(f, \mathbb{B}_0^{-1}g) = (\mathbb{B}^{-1}f, g) - (\mathbb{B}_0^{-1}C\mathbb{B}^{-1}f, g)$$

and, hence, in view of the self-adjointness of $\mathbb{B}^{-1}$,

$$\begin{aligned}
\mathbb{B}^{-1} - \mathbb{B}_0^{-1} &= \mathbb{B}_0^{-1}C\mathbb{B}^{-1} \\
&= [\mathbb{B}_0^{-1}\Lambda_+W^{1/2}][W^{1/2}\Lambda_+\mathbb{B}^{-1}];
\end{aligned} \tag{3.3.1}$$

recall from Theorem 2.3.7(i) that $\mathbb{B}$ is strictly positive and is thus invertible. We shall prove that $\mathbb{B}^{-1} - \mathbb{B}_0^{-1}$ is compact. By (2.3.5), $\mathbb{B}_0 = \mathbb{H}_0\Lambda_+$, where $\mathbb{H}_0 = \sqrt{-\Delta + 1}$, and so the first factor on the right-hand side of (3.3.1) can be written

$$\mathbb{B}_0^{-1}\Lambda_+W^{1/2} = \Lambda_+\mathbb{H}_0^{-1}W^{1/2}.$$

We know from Lemma 3.1.1 that $W^{1/2} : H^1(\mathbb{R}^3; \mathbb{C}^4) \to L^2(\mathbb{R}^3, \mathbb{C}^4)$ is compact. Consequently, $\mathbb{H}_0^{-1}W^{1/2} : H^1(\mathbb{R}^3; \mathbb{C}^4) \to H^1(\mathbb{R}^3; \mathbb{C}^4)$ is compact and $\mathbb{B}_0^{-1}\Lambda_+W^{1/2}$ is compact on $H^1(\mathbb{R}^3; \mathbb{C}^4)$. To show that the second factor in (3.3.1) is bounded, we write it as

$$W^{1/2}\Lambda_+\mathbb{B}^{-1} = [W^{1/2}\Lambda_+\mathbb{B}_0^{-1/2}][\mathbb{B}_0^{1/2}\mathbb{B}^{-1}]. \tag{3.3.2}$$

The range of $\mathbb{B}^{-1}$, namely $\mathcal{D}(\mathbb{B})$, lies in $\mathbb{Q}(\mathbb{B}) = \mathbb{Q}(\mathbb{B}_0) = \mathcal{D}(\mathbb{B}_0^{1/2})$, and hence $\mathbb{B}_0^{1/2}\mathbb{B}^{-1}$ is defined on $L^2(\mathbb{R}^3, \mathbb{C}^4)$ and is a closed operator. Hence by the Closed Graph Theorem, $\mathbb{B}_0^{1/2}\mathbb{B}^{-1}$ is bounded on $L^2(\mathbb{R}^3, \mathbb{C}^4)$. Next, we write the first factor on the right-hand side of (3.3.2) as

$$W^{1/2}\Lambda_+\mathbb{B}_0^{-1/2} = W^{1/2}\mathbb{H}_0^{-1/2}\Lambda_+.$$

By Kato's inequality, we have

$$\|W^{1/2}f\|^2 \leq (\gamma\pi/2)\int_{\mathbb{R}^3} |\mathbf{p}||\hat{f}(\mathbf{p})|^2 d\mathbf{p}$$

$$\leq (\gamma\pi/2)\int_{\mathbb{R}^3} \sqrt{|\mathbf{p}|^2+1}|\hat{f}(\mathbf{p})|^2 d\mathbf{p} = (\gamma\pi/2)\|\mathbb{H}_0^{1/2}f\|^2$$

and, hence, for all $g \in L^2(\mathbb{R}^3, \mathbb{C}^4)$,

$$\|W^{1/2}\mathbb{H}_0^{-1/2}\Lambda_+ g\| \leq (\gamma\pi/2)^{1/2}\|\Lambda_+ g\|.$$

Therefore, $W^{1/2}\mathbb{H}_0^{-1/2}\Lambda_+$ is bounded on $L^2(\mathbb{R}^3, \mathbb{C}^4)$, which implies the boundedness of $W^{1/2}\Lambda_+\mathbb{B}^{-1}$ in (3.3.2) and completes the proof of the claim that $\mathbb{B}^{-1} - \mathbb{B}_0^{-1}$ is compact on $L^2(\mathbb{R}^3, \mathbb{C}^4)$, thus establishing that $\mathbb{B}$ and $\mathbb{B}_0$ have the same essential spectrum, namely $[1, \infty)$.

The case $\gamma = \gamma_c$ is proved by Tix in [Tix (1998)], Theorem 8, by subtly adapting the above proof and using results from [Herbst (1977)] and [Yaouanc *et al.* (1997)].

By Theorem 3.1.7, the kth eigenvalues of $\mathbb{D}$ and $\mathbb{B}$ satisfy the inequality

$$\lambda_k(\mathbb{D}) \geq \inf_{M_k} \sup_{\psi \in M_k, \|\psi\|=1} (\mathbb{D}\psi, \psi) \geq \lambda_k(\mathbb{B}),$$

where M_k is a k-dimensional subspace of $\Lambda_+ C_0^\infty(\mathbb{R}^3, \mathbb{C}^4)$. Since $\mathbb{D}$ has an infinity of positive eigenvalues in $(0, 1)$, it follows that for all $k \in \mathbb{N}$, there exists a k-dimensional subspace M_k on which $(\mathbb{B}\psi, \psi) \leq 1$. Consequently, $\mathbb{B}$ has an infinity of isolated eigenvalues of finite multiplicity in $[0, 1)$, which accumulate at 1. Note that if $\gamma \leq 2/\pi$, since $\mathbb{B}$ is then the restriction of $\mathbb{H}I_4$ to $\mathcal{H} = \Lambda_+ L^2(\mathbb{R}^3, \mathbb{C}^4)$, it follows that $\lambda_k(\mathbb{H}) \leq \lambda_k(\mathbb{B}) \leq \lambda_k(\mathbb{D})$.

The final assertion in the theorem that $\lambda_0 \geq 1 - \gamma_c > 0.09$ is due to Tix [Tix (1998)]; see also [Tix (1997a)].

$$\square$$

3.3.2 *The virial theorem*

Our main results are the following theorem and corollary, which result from the application of Lemma 3.1.2 to the Brown–Ravenhall operator $\mathfrak{b}$ of (2.3.18). In Remark 2.3.8(iv) we noted that for $\gamma < \gamma_c$, $\mathfrak{b}$ is the form sum of E and $-K$ with form domain $L^2(\mathbb{R}^3; \sqrt{p^2+1}dp) \otimes \mathbb{C}^2$, where E is the operator of multiplication by $e(p) = \sqrt{p^2+1}$,

$$(Ku)(\mathbf{p}) = \frac{\gamma}{2\pi^2}\int_{\mathbb{R}^3} K(\mathbf{p}', \mathbf{p})u(\mathbf{p}')d\mathbf{p}'$$

with kernel

$$K(\mathbf{p}', \mathbf{p}) = \frac{(e(p')+1)(e(p)+1)I_2 + (\mathbf{p}'\cdot\sigma)(\mathbf{p}\cdot\sigma)}{n(p')|\mathbf{p}'-\mathbf{p}|^2 n(p)},$$

and $n(p) = \sqrt{2e(p)[e(p)+1]}$.

Theorem 3.3.2. *Let $\gamma < \gamma_c$. If λ is an eigenvalue of $\mathfrak{b}$ with eigenvector ϕ, then*

$$\lambda\|\phi\|^2 = \int_{\mathbb{R}^3} \frac{1}{e(p)} |\phi(\mathbf{p})|^2 d\mathbf{p}$$

$$- \frac{\gamma}{4\pi^2} \iint_{\mathbb{R}^3 \times \mathbb{R}^3} \phi^*(\mathbf{p}')K^1(\mathbf{p}',\mathbf{p})\phi(\mathbf{p})\left[\frac{1}{e(p)} - \frac{1}{e(p)^2} + \frac{1}{e(p')} - \frac{1}{e(p')^2}\right] d\mathbf{p}'d\mathbf{p}$$

$$+ \frac{\gamma}{4\pi^2} \iint_{\mathbb{R}^3 \times \mathbb{R}^3} \phi^*(\mathbf{p}')K^2(\mathbf{p}',\mathbf{p})\phi(\mathbf{p})\left[\frac{1}{e(p)} + \frac{1}{e(p)^2} + \frac{1}{e(p')} + \frac{1}{e(p')^2}\right] d\mathbf{p}'d\mathbf{p}, \quad (3.3.3)$$

where

$$K^1(\mathbf{p}',\mathbf{p}) = \frac{[e(p')+1][e(p)+1]I_2}{n(p')|\mathbf{p}-\mathbf{p}'|^2 n(p)}, \qquad (3.3.4)$$

$$K^2(\mathbf{p}',\mathbf{p}) = \frac{(\mathbf{p}'\cdot\boldsymbol{\sigma})(\mathbf{p}\cdot\boldsymbol{\sigma})}{n(p')|\mathbf{p}-\mathbf{p}'|^2 n(p)}. \qquad (3.3.5)$$

Corollary 3.3.3. *Let $\gamma < \gamma_c$ and $\mathfrak{b}\phi = \lambda\phi$. Then*

$$(\lambda-1)\int_{\mathbb{R}^3} |\phi(\mathbf{p})|^2 \left\{1 - \frac{1}{e(p)} + \frac{1}{e(p)^2}\right\} d\mathbf{p}$$

$$= \frac{\gamma}{2\pi^2} \iint_{\mathbb{R}^3 \times \mathbb{R}^3} \phi^*(\mathbf{p}')K^2(\mathbf{p}',\mathbf{p})\phi(\mathbf{p})\left[\frac{1}{e(p')} + \frac{1}{e(p)}\right] d\mathbf{p}'d\mathbf{p}$$

$$- \int_{\mathbb{R}^3} |\phi(\mathbf{p})|^2 \frac{[e(p)-1][2e(p)-1]}{e(p)^2} d\mathbf{p}. \qquad (3.3.6)$$

Before proving Theorem 3.3.2 and Corollary 3.3.3, we need the following lemma, which is similar to Lemma 2.3.10. In it $\mathfrak{b}^m$ denotes the operator with $e(p)$ replaced by $e_m(p) := \sqrt{p^2 + m^2}$ in $\mathfrak{b}$ and associated changes to make $\mathfrak{b}^m$ the Brown–Ravenhall operator when the mass $m \neq 1$. To be specific, $\mathfrak{b}^m = E_m - K^m$, where E_m is multiplication by $e_m(p)$ and K^m is the integral operator with kernel

$$K^m(\mathbf{p}',\mathbf{p}) = \frac{[e_m(p')+m][e_m(p)+m]I_2 + (\mathbf{p}'\cdot\boldsymbol{\sigma})(\mathbf{p}\cdot\boldsymbol{\sigma})}{n_m(p')|\mathbf{p}-\mathbf{p}'|^2 n_m(p)},$$

where $n_m(p) = \sqrt{2e_m(p)[e_m(p)+m]}$. In particular $\mathfrak{b}^1 \equiv \mathfrak{b}$ and

$$(\mathfrak{b}^0 u)(\mathbf{p}) = pu(\mathbf{p}) - \frac{\gamma}{2\pi^2}(K^0 u)(\mathbf{p})$$

where K^0 has kernel

$$K^0(\mathbf{p}',\mathbf{p}) = \frac{1}{2}\left\{\frac{pp'I_2 + (\mathbf{p}'\cdot\boldsymbol{\sigma})(\mathbf{p}\cdot\boldsymbol{\sigma})}{p'p|\mathbf{p}-\mathbf{p}'|^2}\right\}.$$

Lemma 3.3.4. *For any $m \in \mathbb{R}^+$, the operators $\mathfrak{b}^m$ and $\mathfrak{b}^0$ have the same domain and $\mathfrak{b}^m - \mathfrak{b}^0$ can be extended to a bounded operator on $L^2(\mathbb{R}^3, \mathbb{C}^2)$.*

Proof. We have

$$0 \le e_m(p) - p \le m,$$

and, on using the inequality $2\sqrt{1+a}\sqrt{1+b} \le 2 + |a| + |b|$ for $a, b \in (-1, 1)$, we can write

$$|K^m(\mathbf{p}', \mathbf{p}) - K^0(\mathbf{p}', \mathbf{p})| = \frac{1}{2|\mathbf{p} - \mathbf{p}'|^2} \left\{ \left[\sqrt{1 + \frac{m}{e_m(p')}} \sqrt{1 + \frac{m}{e_m(p)}} - 1 \right] \right.$$
$$+ \left. \left[\sqrt{1 - \frac{m}{e_m(p')}} \sqrt{1 - \frac{m}{e_m(p)}} - 1 \right] \frac{(\mathbf{p}' \cdot \boldsymbol{\sigma})(\mathbf{p} \cdot \boldsymbol{\sigma})}{p'p} \right\}$$
$$\le \frac{m}{2|\mathbf{p} - \mathbf{p}'|^2} \left\{ \frac{1}{e_m(p)} + \frac{1}{e_m(p')} \right\}.$$

Hence, for ϕ in the form domain of $\mathfrak{b}^m - \mathfrak{b}^0$, we have

$$|(\phi, \mathfrak{b}^m \phi) - (\phi, \mathfrak{b}^0)| \le$$

$$\le m\|\phi\|^2 + \frac{m\gamma}{4\pi^2} \iint\limits_{\mathbb{R}^3 \times \mathbb{R}^3} \frac{1}{|\mathbf{p} - \mathbf{p}'|^2} \left\{ \frac{1}{e_m(p)} + \frac{1}{e_m(p')} \right\} |\phi(\mathbf{p})||\phi(\mathbf{p}')| d\mathbf{p}' d\mathbf{p}.$$

By the Cauchy–Schwarz inequality, the last integral is no greater than

$$\iint\limits_{\mathbb{R}^3 \times \mathbb{R}^3} |\phi(\mathbf{p})|^2 \frac{1}{|\mathbf{p} - \mathbf{p}'|^2} \left\{ \frac{1}{e_m(p)} + \frac{1}{e_m(p')} \right\} \frac{h(\mathbf{p}')}{h(\mathbf{p})} d\mathbf{p}' d\mathbf{p},$$

where we choose $h(\mathbf{p}) = |\mathbf{p}|^{-\alpha}$, $1 < \alpha < 2$. Then (see [Lieb and Loss (1997)], (3) in the proof of Corollary 5.10)

$$\int_{\mathbb{R}^3} \frac{1}{|\mathbf{p} - \mathbf{p}'|^2} \frac{h(\mathbf{p}')}{h(\mathbf{p})e_m(p)} d\mathbf{p}' = \frac{|\mathbf{p}|^\alpha}{e_m(p)} \int_{\mathbb{R}^3} \frac{1}{|\mathbf{p} - \mathbf{p}'|^2} \frac{1}{|\mathbf{p}'|^\alpha} d\mathbf{p}'$$
$$= O(|\mathbf{p}|^{\alpha-1}|\mathbf{p}|^{1-\alpha}) = O(1)$$

and

$$\int_{\mathbb{R}^3} \frac{1}{|\mathbf{p} - \mathbf{p}'|^2} \frac{h(\mathbf{p}')}{h(\mathbf{p})e_m(p')} d\mathbf{p}' \le |\mathbf{p}|^\alpha \int_{\mathbb{R}^3} \frac{1}{|\mathbf{p} - \mathbf{p}'|^2} \frac{1}{|\mathbf{p}'|^{\alpha+1}} d\mathbf{p}' = O(1).$$

Hence, for some positive constant C,

$$|(\phi, \mathfrak{b}^m \phi) - (\phi, \mathfrak{b}^0)| \le Cm\|\phi\|^2$$

and $(\cdot, \mathfrak{b}^m \cdot) - (\cdot, \mathfrak{b}^0 \cdot)$ can be extended to a bounded quadratic form on $\left\{ L^2(\mathbb{R}^3, \mathbb{C}^2) \right\} \times \left\{ L^2(\mathbb{R}^3, \mathbb{C}^2) \right\}$. It follows that $\mathfrak{b}^m$ and $\mathfrak{b}^0$ have the same domain and this in turn implies the rest of the lemma. $\square$

Proof of Theorem 3.3.2. We apply Lemma 3.1.2 to $T = \mathfrak{b}$, $f(a) = a$ and $U(a)$ defined by $U(a)\phi(\mathbf{p}) = a^{-3/2}\phi(\mathbf{p}/a) =: \phi_a(\mathbf{p})$. Then $U(a) \to I$ strongly in $L^2(\mathbb{R}^3, \mathbb{C}^2)$ and

$$T_a := aU(a)\mathfrak{b}U(a)^{-1} = \mathfrak{b}^a.$$

By Lemma 3.3.4, an eigenvector ϕ of $\mathfrak{b}$, lies in $\mathcal{D}(\mathfrak{b}^a)$ for any a. Thus, all that remains is to evaluate the limit on the left-hand side of (3.1.2). We have

$$\left(\phi_a, \left[\frac{\mathfrak{b}^a - \mathfrak{b}}{a - 1}\right]\phi\right) = \int_{\mathbb{R}^3}\left[\frac{e_a(p) - e(p)}{a - 1}\right]\phi_a(\mathbf{p})\phi(\mathbf{p})^* d\mathbf{p}$$

$$- \frac{\gamma}{2\pi^2}\iint_{\mathbb{R}^3 \times \mathbb{R}^3} \phi_a(\mathbf{p}')^*\left[\frac{K^a(\mathbf{p}', \mathbf{p}) - K(\mathbf{p}', \mathbf{p})}{a - 1}\right]\phi(\mathbf{p})d\mathbf{p}'d\mathbf{p}$$

$$=: I_1 - \frac{\gamma}{2\pi^2}I_2, \qquad (3.3.7)$$

and (3.3.3) will follow if we can justify taking the limit as $a \to 1$ under the integral signs on the right-hand side of (3.3.7).

In I_1, $0 < \{e_a(p) - e(p)\}/(a - 1) \leq 1$ and, hence, as $a \to 1$, since $\|\phi_a - \phi\| \to 0$, we have

$$\lim_{a \to 1} I_1 = \lim_{a \to 1}\int_{\mathbb{R}^3}\left[\frac{e_a(p) - e(p)}{a - 1}\right]|\phi(\mathbf{p})|^2 d\mathbf{p} = \int_{\mathbb{R}^3}\frac{1}{e(p)}|\phi(\mathbf{p})|^2 d\mathbf{p},$$

on applying the Dominated Convergence Theorem. We write I_2 as

$$\iint_{\mathbb{R}^3 \times \mathbb{R}^3} [\phi_a(\mathbf{p}') - \phi(\mathbf{p}')]^*\left[\frac{K^a(\mathbf{p}', \mathbf{p}) - K(\mathbf{p}', \mathbf{p})}{a - 1}\right]\phi(\mathbf{p})d\mathbf{p}'d\mathbf{p}$$

$$+ \iint_{\mathbb{R}^3 \times \mathbb{R}^3} \phi(\mathbf{p}')^*\left[\frac{K^a(\mathbf{p}', \mathbf{p}) - K(\mathbf{p}', \mathbf{p})}{a - 1}\right]\phi(\mathbf{p})d\mathbf{p}'d\mathbf{p} =: I_3 + I_4,$$

and make use of the readily verified estimate

$$\left|\frac{K^a(\mathbf{p}', \mathbf{p}) - K(\mathbf{p}', \mathbf{p})}{a - 1}\right| \leq k_a/|\mathbf{p} - \mathbf{p}'|^2,$$

where $k_a \to 1$ as $a \to 1$. Thus I_3 is bounded by

$$\iint_{\mathbb{R}^3 \times \mathbb{R}^3} |\phi_a(\mathbf{p}') - \phi(\mathbf{p}')||\phi(\mathbf{p})|\frac{d\mathbf{p}'d\mathbf{p}}{|\mathbf{p} - \mathbf{p}'|^2}$$

and, on using the Parseval identity and $\mathbb{F}\left(\frac{1}{|\cdot|}\right) = \sqrt{\frac{2}{\pi}}\frac{1}{|\cdot|^2}$, this is equal to

$$2\pi^2\int_{\mathbb{R}^3} u(\mathbf{x})v(\mathbf{x})\frac{d\mathbf{x}}{|\mathbf{x}|},$$

where $\widehat{u}(\mathbf{p}) = |\phi_a(\mathbf{p}) - \phi(\mathbf{p})|$ and $\widehat{v}(\mathbf{p}) = |\phi(\mathbf{p})|$. This in turn is bounded by a constant multiple of

$$\left(\int_{\mathbb{R}^3} |\phi_a(\mathbf{p}) - \phi(\mathbf{p})|^2|\mathbf{p}|d\mathbf{p}\right)^{1/2}\left(\int_{\mathbb{R}^3} |\phi(\mathbf{p})|^2|\mathbf{p}|d\mathbf{p}\right)^{1/2}$$

on using the Cauchy–Schwarz and Kato inequalities. We claim that the first of these integrals tends to 0 as $a \to 1$. This is obviously so if $\phi = \psi \in C_0(\mathbb{R}^3, \mathbb{C}^2)$. Also

$$\int_{\mathbb{R}^3} |\phi_a(\mathbf{p}) - \psi_a(\mathbf{p})|^2|\mathbf{p}|d\mathbf{p} = a\int_{\mathbb{R}^3} |\phi(\mathbf{p}) - \psi(\mathbf{p})|^2|\mathbf{p}|d\mathbf{p}.$$

Since $C_0(\mathbb{R}^3, \mathbb{C}^2)$ is dense in $L^2(\mathbb{R}^3, \mathbb{C}^2)$, our claim is justified and we have that $I_3 \to 0$ as $a \to 1$. In I_4 the integrand is majorised by the function $|\mathbf{p} - \mathbf{p}'|^{-2}|\phi(\mathbf{p}')||\phi(\mathbf{p})|$, which is integrable by Kato's inequality. Hence the Dominated Convergence Theorem applies and it is easily verified that (3.3.3) follows. $\qquad\square$

Proof of Corollary 3.3.3. From (3.3.3) we have

$$
\lambda|\phi|^2 = \int_{\mathbb{R}^3} |\phi(\mathbf{p})|^2 \frac{d\mathbf{p}}{e(p)}
$$
$$
- \frac{\gamma}{2\pi^2} \mathrm{Re}\left[\iint_{\mathbb{R}^3 \times \mathbb{R}^3} \phi^*(\mathbf{p}')K(\mathbf{p}',\mathbf{p})\phi(\mathbf{p})\left\{\frac{1}{e(p)} - \frac{1}{e(p)^2}\right\}d\mathbf{p}'d\mathbf{p}\right]
$$
$$
+ \frac{\gamma}{2\pi^2} \iint_{\mathbb{R}^3 \times \mathbb{R}^3} \phi^*(\mathbf{p}')K^2(\mathbf{p}',\mathbf{p})\phi(\mathbf{p})\left\{\frac{1}{e(p)} + \frac{1}{e(p')}\right\}d\mathbf{p}'d\mathbf{p}. \quad (3.3.8)
$$

Also, for all $\psi \in \mathbb{Q}(\mathfrak{b})$,

$$
\int_{\mathbb{R}^3} e(p)\psi^*(\mathbf{p})\phi(\mathbf{p})d\mathbf{p} - \frac{\gamma}{2\pi^2} \iint_{\mathbb{R}^3 \times \mathbb{R}^3} \psi^*(\mathbf{p}')K(\mathbf{p}',\mathbf{p})\phi(\mathbf{p})d\mathbf{p}'d\mathbf{p}
$$
$$
= \lambda \int_{\mathbb{R}^3} \psi^*(\mathbf{p})\phi(\mathbf{p})d\mathbf{p}. \quad (3.3.9)
$$

We choose $\psi(\mathbf{p}) = [e(p)^{-1} - e(p)^{-2}]\phi(\mathbf{p})$ in (3.3.9): clearly $\psi \in \mathbb{Q}(\mathfrak{b})$. Then

$$
\frac{\gamma}{2\pi^2} \iint_{\mathbb{R}^3 \times \mathbb{R}^3} \phi^*(\mathbf{p}')K(\mathbf{p}',\mathbf{p})\phi(\mathbf{p})\left\{\frac{1}{e(p)} - \frac{1}{e(p)^2}\right\}d\mathbf{p}'d\mathbf{p}
$$
$$
= \int_{\mathbb{R}^3} [e(p) - \lambda]\left\{\frac{1}{e(p)} - \frac{1}{e(p)^2}\right\}|\phi(\mathbf{p})|^2 d\mathbf{p}. \quad (3.3.10)
$$

On substituting (3.3.10) in (3.3.9),

$$
\lambda|\phi|^2 = \int_{\mathbb{R}^3} |\phi(\mathbf{p})|^2 \frac{d\mathbf{p}}{e(p)} - \int_{\mathbb{R}^3} [e(p) - \lambda]\left\{\frac{1}{e(p)} - \frac{1}{e(p)^2}\right\}|\phi(\mathbf{p})|^2 d\mathbf{p}
$$
$$
+ \frac{\gamma}{2\pi^2} \iint_{\mathbb{R}^3 \times \mathbb{R}^3} \phi^*(\mathbf{p}')K^2(\mathbf{p}',\mathbf{p})\phi(\mathbf{p})\left\{\frac{1}{e(p)} + \frac{1}{e(p')}\right\}d\mathbf{p}, d\mathbf{p}'
$$

whence

$$
\lambda \int_{\mathbb{R}^3} |\phi(\mathbf{p})|^2 \left\{1 - \frac{1}{e(p)} + \frac{1}{e(p)^2}\right\}d\mathbf{p} = \int_{\mathbb{R}^3} |\phi(\mathbf{p})|^2 \left\{\frac{1}{e(p)} - 1 + \frac{1}{e(p)}\right\}d\mathbf{p}
$$
$$
+ \frac{\gamma}{2\pi^2} \iint_{\mathbb{R}^3 \times \mathbb{R}^3} \phi^*(\mathbf{p}')K^2(\mathbf{p}',\mathbf{p})\phi(\mathbf{p})\left[\frac{1}{e(p')} + \frac{1}{e(p)}\right]d\mathbf{p}'d\mathbf{p},
$$

and

$$(\lambda - 1) \int_{\mathbb{R}^3} |\phi(\mathbf{p})|^2 \left\{ 1 - \frac{1}{e(p)} + \frac{1}{e(p)^2} \right\} dp$$

$$= \int_{\mathbb{R}^3} |\phi(\mathbf{p})|^2 \frac{[e(p) - 1][1 - 2e(p)]}{e(p)^2} dp$$

$$+ \frac{\gamma}{2\pi^2} \iint_{\mathbb{R}^3 \times \mathbb{R}^3} \phi^*(\mathbf{p}') K^2(\mathbf{p}', \mathbf{p}) \phi(\mathbf{p}) \left[\frac{1}{e(p')} + \frac{1}{e(p)} \right] dp' dp. \quad (3.3.11)$$

The corollary is therefore proved. $\qquad\square$

3.4 The absence of embedded eigenvalues

Theorem 3.4.1. *If $\gamma < \gamma_c$, the operator $\mathfrak{h}$ has no eigenvalues in $[1, \infty)$. The spectrum in $[1, \infty)$ is absolutely continuous.*

Proof. From (3.3.6)

$$\left(\lambda - 1 \right) \int_{\mathbb{R}^3} |\phi(\mathbf{p})|^2 \left\{ 1 - \frac{1}{e(p)} + \frac{1}{e(p)^2} \right\} dp$$

$$\leq \frac{\gamma}{2\pi^2} \iint_{\mathbb{R}^3 \times \mathbb{R}^3} \phi^*(\mathbf{p}') H(\mathbf{p}', \mathbf{p}) |\phi(\mathbf{p})| dp' dp$$

$$- \int_{\mathbb{R}^3} |\phi(\mathbf{p})|^2 \frac{[e(p) - 1][2e(p) - 1]}{e(p)^2} dp, \quad (3.4.1)$$

where

$$H(\mathbf{p}', \mathbf{p}) = \frac{pp'}{n(p')n(p)|\mathbf{p} - \mathbf{p}'|^2} \left[\frac{1}{e(p')} + \frac{1}{e(p)} \right].$$

Set

$$J(\mathbf{p}', \mathbf{p}) = \frac{1}{|\mathbf{p} - \mathbf{p}'|^2} \left[\frac{1}{e(p')} + \frac{1}{e(p)} \right]$$

and let h be an arbitrary positive function, which will be chosen later. Then,

$$\iint_{\mathbb{R}^3 \times \mathbb{R}^3} H(\mathbf{p}', \mathbf{p}) |\phi(\mathbf{p}')| |\phi(\mathbf{p})| dp' dp =$$

$$= \iint_{\mathbb{R}^3 \times \mathbb{R}^3} \left(\frac{p}{n(p)} |\phi(\mathbf{p})| J^{1/2}(\mathbf{p}', \mathbf{p}) \left[\frac{h(\mathbf{p}')}{h(\mathbf{p})} \right]^{1/2} \right) \times$$

$$\times \left(\frac{p'}{n(p')} |\phi(\mathbf{p}')| J^{1/2}(\mathbf{p}', \mathbf{p}) \left[\frac{h(\mathbf{p})}{h(\mathbf{p}')} \right]^{1/2} \right) dp dp'$$

$$\leq \int_{\mathbb{R}^3} \frac{p^2}{n^2(p)} |\phi(\mathbf{p})|^2 dp \int_{\mathbb{R}^3} \frac{h(\mathbf{p}')}{h(\mathbf{p})} J(\mathbf{p}', \mathbf{p}) dp', \quad (3.4.2)$$

by the Cauchy–Schwarz inequality. We make the choice $h(\mathbf{p}) = p^{-3/2}$ and so

$$\int_{\mathbb{R}^3} \frac{h(\mathbf{p}')}{h(\mathbf{p})} J(\mathbf{p}', \mathbf{p}) d\mathbf{p}'$$

$$= \int_{\mathbb{R}^3} \frac{1}{|\mathbf{p}' - \mathbf{p}|^2} \frac{1}{\sqrt{p'^2 + 1}} \frac{p^{3/2}}{p'^{3/2}} d\mathbf{p}' + \int_{\mathbb{R}^3} \frac{1}{|\mathbf{p}' - \mathbf{p}|^2} \frac{1}{\sqrt{p^2 + 1}} \frac{p^{3/2}}{p'^{3/2}} d\mathbf{p}'$$

$$=: I_1 + I_2 \quad (3.4.3)$$

say. We now follow [Jakabussa-Amundsen (2004)], Section I.5, to obtain more precise estimates for I_1 and I_2 than those in the proof of Theorem 3.1 in [Balinsky and Evans (1998)]. These are based on the simple estimate

$$\frac{1}{\sqrt{q^2 + 1}} \leq \begin{cases} 1, & \text{if } q \leq 1, \\ \frac{1}{q}, & \text{if } q > 1. \end{cases}$$

We first note that

$$\int_{\mathbb{S}^2} \frac{1}{|\mathbf{p}' - \mathbf{p}|^2} d\omega' = \frac{2\pi}{pp'} \ln \left| \frac{p + p'}{p - p'} \right|.$$

Hence,

$$I_1 \leq 2\pi p^{1/2} \left\{ \int_0^1 p'^{-1/2} \ln \left| \frac{p + p'}{p - p'} \right| dp' + \int_1^\infty p'^{-3/2} \ln \left| \frac{p + p'}{p - p'} \right| dp' \right\}$$

$$= 2\pi p \int_0^{1/p} t^{-1/2} \ln \left| \frac{1 + t}{1 - t} \right| dt + 2\pi \int_{1/p}^\infty t^{-3/2} \ln \left| \frac{1 + t}{1 - t} \right| dp'. \quad (3.4.4)$$

Also

$$I_2 = 2\pi \frac{p}{\sqrt{p^2 + 1}} \int_0^\infty t^{-1/2} \ln \left| \frac{1 + t}{1 - t} \right| dt.$$

The next step is to substitute the following integrals evaluated in [Jakabussa-Amundsen (2004)], Appendix A; see also [Gradshteyn and Ryzhik (1965)], p. 205–206.

$$\int_0^a t^{-1/2} \ln \left| \frac{1 + t}{1 - t} \right| dt = 2\sqrt{a} \ln \left| \frac{1 + a}{1 - a} \right| + 4 \arctan \sqrt{a} - 2 \ln \left| \frac{\sqrt{a} + 1}{\sqrt{a} - 1} \right|,$$

$$\int_0^\infty t^{-1/2} \ln \left| \frac{1 + t}{1 - t} \right| dt = 2\pi, \quad (3.4.5)$$

$$\int_a^\infty t^{-3/2} \ln \left| \frac{1 + t}{1 - t} \right| dt = 2\pi - 2 \ln \left| \frac{\sqrt{a} + 1}{\sqrt{a} - 1} \right| - 4 \arctan \sqrt{a} + \frac{2}{\sqrt{a}} \ln \left| \frac{1 + a}{1 - a} \right|.$$

It follows that

$$I_1 \leq 4\pi^2 + 4\pi \left\{ 2\sqrt{p} \ln \left| \frac{1 + p}{1 - p} \right| + 2(p - 1) \arctan(1/\sqrt{p}) \right.$$

$$\left. -(p + 1) \ln \left| \frac{1 + \sqrt{p}}{1 - \sqrt{p}} \right| \right\} =: 4\pi^2 \alpha(p) \quad (3.4.6)$$

say, and

$$I_2 = \frac{4\pi^2 p}{\sqrt{p^2 + 1}}.$$

On substituting these in (3.4.2), and then in (3.4.1), the resulting inequality can be rewritten in the following form:

$$0 \leq \int_{\mathbb{R}^3} |\phi(\mathbf{p})|^2 \left\{ 1 - \frac{1}{e(p)} + \frac{1}{e(p)^2} \right\} \{ 1 - \lambda + \Phi(p) \} \, d\mathbf{p}, \qquad (3.4.7)$$

where

$$\Phi(p) \left\{ 1 - \frac{1}{e(p)} + \frac{1}{e(p)^2} \right\} = \frac{2\gamma p^2}{n^2(p)} \left(\alpha(p) + \frac{p}{\sqrt{p^2 + 1}} \right) - \frac{[e(p) - 1][2e(p) - 1]}{e(p)^2}.$$

This gives

$$\Phi(p) = \frac{p^2 g_1(p)}{p^2 + 2 - \sqrt{p^2 + 1}} \left\{ \gamma - \frac{g_0(p)}{g_1(p)} \right\},$$

where

$$g_0(p) = \frac{2\sqrt{p^2 + 1} - 1}{\sqrt{p^2 + 1} + 1}, \qquad g_1(p) = \frac{p + \alpha(p)\sqrt{p^2 + 1}}{\sqrt{p^2 + 1} + 1}.$$

A numerical calculation gives $\min_{p \in \mathbb{R}_+} g_0(p)/g_1(p) = 0.973$. Since $\gamma_c = 0.906$, it follows that Φ is negative for $\gamma \leq \gamma_c$. Hence, from (3.4.7), we conclude that $\lambda < 1$, thus proving that there are no eigenvalues embedded in the essential spectrum $[1, \infty)$. The absence of a singular continuous spectrum is proved in [Evans *et al.* (1996)], Theorem 2. The proof is therefore complete. $\qquad \Box$

It is also of interest to analyse the implications of the virial theorem for the operators $b_{l,s}$.

In the partial wave decomposition of $\mathbb{B}$ and $\mathfrak{b}$ in Section 2.3.1, spinors $\phi \in L^2(\mathbb{R}^3, \mathbb{C}^2)$ are expanded in terms of the spherical spinors $\Omega_{l,n,s}$,

$$\phi(\mathbf{p}) = \sum_{(l,m,s) \in \mathfrak{I}} p^{-1} \phi_{l,m,s}(p) \Omega_{l,m,s}(\omega) \qquad (3.4.8)$$

over the index set $\mathfrak{I} = \{ l \in \mathbb{N}_0, m = -l - 1/2, \cdots, l + 1/2, s = \pm 1/2, \Omega_{l,m,s} \neq 0 \}$ and

$$\int_{\mathbb{R}^3} |\phi(\mathbf{p})|^2 \, d\mathbf{p} = \sum_{(l,m,s) \in \mathfrak{I}} \int_0^\infty |\phi_{l,m,s}(p)|^2 \, dp. \qquad (3.4.9)$$

This yields the representation (2.3.21), namely,

$$(\phi, \mathfrak{b}\psi) = \sum_{(l,m,s) \in \mathfrak{I}} (\phi_{l,m,s}, \mathfrak{b}_{l,s}\psi_{l,m,s}), \qquad (3.4.10)$$

where

$$(\phi_{l,m,s}, \mathfrak{b}_{l,s}\psi_{l,m,s}) = \int_0^\infty e(p)\phi_{l,m,s}(p)\overline{\psi}_{l,m,s}(p)dp$$

$$-\frac{\gamma}{\pi}\int_0^\infty \int_0^\infty \phi_{l,m,s}(p')k_{l,s}(p',p)\overline{\psi}_{l,m,s}(p)dp'dp \quad (3.4.11)$$

and

$$k_{l,s}(p',p) = \frac{[e(p')+1]Q_l(\frac{1}{2}[\frac{p'}{p}+\frac{p}{p'}])[e(p)+1]}{n(p')n(p)} + \frac{p'Q_{l+2s}(\frac{1}{2}[\frac{p'}{p}+\frac{p}{p'}])p}{n(p')n(p)}$$

$$=: k_{l,m}^1(p',p) + k_{l,m}^2(p',p) \quad (3.4.12)$$

say. The nature of the associated self-adjoint operators $\mathfrak{b}_{l,s}$ is described in Theorem 2.3.11 and Remark 2.3.12.

Theorem 3.4.2. *For all values of (l,s), the operators $b_{l,s}$ have no eigenvalues in $[1,\infty)$ if $\gamma \leq \gamma_c$.*

Proof. The analogue of (3.3.6) is

$$(\lambda-1)\int_0^\infty |\phi_{l,m,s}(p)|^2\left\{1 - \frac{1}{e(p)} + \frac{1}{e(p)^2}\right\}dp$$

$$= \frac{\gamma}{\pi}\int_0^\infty \int_0^\infty \phi_{l,m,s}(p')k_{l,s}^2(p',p)\overline{\phi}(p)\left[\frac{1}{e(p')}+\frac{1}{e(p)}\right]dp'dp$$

$$-\int_0^\infty |\phi_{l,m,s}(p)|^2\frac{[e(p)-1][2e(p)-1]}{e(p)^2}dp. \quad (3.4.13)$$

Since $Q_0(t) \geq \cdots \geq Q_l(t) \geq 0$ for all $t > 1$, it follows that for all $(l,s) \neq (1,-1/2)$

$$k_{l,s}^2(p',p) \leq \frac{p'p}{n(p')n(p)}Q_1\left(\frac{1}{2}\left[\frac{p'}{p}+\frac{p}{p'}\right]\right)$$

and, hence, on writing ϕ for $\phi_{l,m,s}$ in (3.4.13), we get

$$(\lambda-1)\int_0^\infty |\phi(p)|^2\left\{1 - \frac{1}{e(p)} + \frac{1}{e(p)^2}\right\}dp$$

$$\leq \frac{\gamma}{\pi}\int_0^\infty |\phi(p)|^2\frac{p^2}{n^2(p)h(p)}dp\int_0^\infty h(p')Q_1\left(\frac{1}{2}\left[\frac{p'}{p}+\frac{p}{p'}\right]\right)\left(\frac{1}{e(p')}+\frac{1}{e(p)}\right)dp'$$

$$-\int_0^\infty |\phi(p)|^2\frac{[e(p)-1][2e(p)-1]}{e(p)^2}dp. \quad (3.4.14)$$

for any positive function $h(\cdot)$, on applying the Cauchy–Schwarz inequality. We choose $h(t) = 1/t$, so that in the first term on the right-hand side of (3.4.14) we have

$$\frac{\gamma}{\pi} \int_0^\infty |\phi(p)|^2 \frac{p^2}{n^2(p)} \left\{ \frac{p}{e(p)} \int_0^\infty \frac{1}{p'} Q_1\left(\frac{1}{2}\left[\frac{p'}{p} + \frac{p}{p'}\right]\right) dp' \right.$$

$$\left. + p \int_0^\infty \frac{1}{p'e(p')} Q_1\left(\frac{1}{2}\left[\frac{p'}{p} + \frac{p}{p'}\right]\right) dp' \right\} dp$$

$$= \frac{\gamma}{\pi} \int_0^\infty |\phi(p)|^2 \frac{p^2}{n^2(p)} \left\{ I_1 + I_2 \right\} dp$$

say. Let $g_1(u) = Q_1(\frac{1}{2}[u + \frac{1}{u}])$. Then, by (2.2.8)

$$I_1 \le \int_0^\infty g_1(u) \frac{du}{u} = 2,$$

and since $g_1(u) = g_1(1/u)$,

$$I_2 \le \int_0^\infty g_1(u) \frac{du}{u^2} = \int_0^\infty g_1(u) du.$$

Furthermore,

$$\int_0^1 g_1(u) du = \frac{1}{2} \int_0^1 u \ln\left(\frac{u+1}{1-u}\right) du + \frac{1}{2} \int_0^1 \frac{1}{u} \ln\left(\frac{u+1}{1-u}\right) du - \int_0^1 du$$

$$= \frac{1}{2} + \frac{1}{2} \frac{\pi^2}{4} - 1 = \frac{\pi^2}{8} - \frac{1}{2}$$

and

$$\int_1^\infty g_1(u) du = \frac{1}{2} \int_1^\infty \frac{1}{u} \ln\left(\frac{u+1}{u-1}\right) du + \int_1^\infty \left\{ \frac{u}{2} \ln\left(\frac{u+1}{u-1}\right) - 1 \right\} du$$

$$= \frac{\pi^2}{8} + \left[\frac{1}{4}(u^2 - 1) \ln\left(\frac{u+1}{u-1}\right) - \frac{u}{2} \right]_1^\infty = \frac{\pi^2}{8} + \frac{1}{2}.$$

Hence, $I_2 \le \frac{\pi^2}{4}$, and from (3.4.14)

$$\left(\lambda - 1\right) \int_0^\infty |\phi(p)|^2 \left\{ 1 - \frac{1}{e(p)} + \frac{1}{e(p)^2} \right\} dp$$

$$\le \frac{\gamma}{\pi} \int_0^\infty |\phi(p)|^2 \frac{p^2}{n^2(p)} \left(\frac{\pi^2}{4} + 2\right) dp - \int_0^\infty |\phi(p)|^2 \frac{[e(p) - 1][2e(p) - 1]}{e(p)^2} dp.$$

On simplifying, we obtain

$$0 < \int_0^\infty |\phi(\mathbf{p})|^2 \left(1 - \frac{1}{\sqrt{p^2+1}} + \frac{1}{p^2+1}\right)\left(1 - \lambda + \Psi(p)\right)dp,$$

where

$$\Psi(p) = \frac{p^2(\sqrt{p^2+1})}{(\sqrt{p^2+1}+1)(p^2+2-\sqrt{p^2+1})}\left\{\pi\left(\frac{\gamma}{2\pi^2}\right)\left(\frac{\pi^2}{4}+2\right) - \frac{2\sqrt{p^2+1}-1}{\sqrt{p^2+1}}\right\}$$

$$\leq \frac{p^2\sqrt{p^2+1}}{(\sqrt{p^2+1}+1)(p^2+2-\sqrt{p^2+1})}\left\{\frac{2}{\pi^2+4}\left(\frac{\pi^2}{4}+2\right)-1\right\} < 0$$

for $\gamma \leq \gamma_c$. Hence, $\lambda < 1$.

For the case $(l,s) = (1,-1/2)$, the Legendre function Q_1 in (3.4.14) has to be replaced by Q_0, and we have to consider

$$\frac{1}{h(p)}\int_0^\infty h(p')Q_0\left(\frac{1}{2}\left[\frac{p'}{p}+\frac{p}{p'}\right]\right)\left[\frac{1}{e(p')}+\frac{1}{e(p)}\right]dp'. \qquad (3.4.15)$$

The more precise estimates used in the proof of Theorem 3.4.1 are necessary in this case. We now make the choice $h(u) = 1/\sqrt{u}$, and, with $g_0(u) = Q_0(\frac{1}{2}[u+\frac{1}{u}]) = \ln\left|\frac{u+1}{u-1}\right|$, (3.4.15) becomes $J_1 + J_2$ say, where on using (3.4.5),

$$J_1 = p\int_0^\infty \frac{1}{\sqrt{u}e(pu)}g_0(u)du$$

$$\leq p\int_0^{1/p}\frac{1}{\sqrt{u}}g_0(u)du + \int_{1/p}^\infty \frac{1}{u^{3/2}}g_0(u)du$$

$$= 2p^{1/2}\ln\left|\frac{1+1/p}{1-1/p}\right| + 4p\arctan\sqrt{1/p} - 2p\ln\left|\frac{1+\sqrt{1/p}}{1-\sqrt{1/p}}\right|$$

$$+ 2\pi - 2\ln\left|\frac{1+\sqrt{1/p}}{1-\sqrt{1/p}}\right| - 4\arctan\sqrt{1/p} + 2\sqrt{p}\ln\left|\frac{1+1/p}{1-1/p}\right|$$

$$= 2\pi + 4(p-1)\arctan\sqrt{1/p} + 4\sqrt{p}\ln\left|\frac{1+1/p}{1-1/p}\right| - 2(p+1)\ln\left|\frac{1+\sqrt{1/p}}{1-\sqrt{1/p}}\right|$$

$$= 2\pi\alpha(p)$$

and

$$J_2 = \frac{p}{e(p)}\int_0^\infty \frac{1}{\sqrt{u}}g_0(u)du = 2\pi\frac{p}{\sqrt{p^2+1}}.$$

On substituting in (3.4.14) (with Q_1 replaced by Q_0) we deduce that

$$0 \leq \int_0^\infty |\phi(p)|^2 \left(1 - \frac{1}{\sqrt{p^2+1}} + \frac{1}{p^2+1}\right)\{1 - \lambda + \Psi(p)\}dp,$$

where

$$\left(1 - \frac{1}{\sqrt{p^2+1}} + \frac{1}{p^2+1}\right)\Psi(p) = \frac{\gamma p^2[\alpha(p)\sqrt{p^2+1}+p]}{(p^2+1)[\sqrt{p^2+1}+1]} - \frac{p^2[2\sqrt{p^2+1}-1]}{(p^2+1)(\sqrt{p^2+1}+1}$$

and, hence,

$$\Psi(p) = \frac{p^2(\alpha(p)\sqrt{p^2+1}+p)}{[p^2+2-\sqrt{p^2+1}][\sqrt{p^2+1}+1]}\left\{\gamma - \frac{2\sqrt{p^2+1}-1}{\alpha(p)\sqrt{p^2+1}+p}\right\}.$$

Thus Ψ is negative if

$$\gamma < \min_{p\in\mathbb{R}_+} \frac{2\sqrt{p^2+1}-1}{\alpha(p)\sqrt{p^2+1}+p} =: \gamma_0.$$

As we observed at the end of the proof of Theorem 3.4.1, $\gamma_0 = 0.973 > 0.906 = \gamma_c$. It follows that $\lambda < 1$ and the proof is complete. $\qquad\square$

Chapter 4

Miscellany

4.1 Stability of matter

A system of N electrons and K nuclei is known to be stable when governed by non-relativistic quantum mechanics. This means that the ground state energy E_0 associated with the Hamiltonian, which, in appropriate units, is of the form

$$H_{N,K} = \sum_{j=1}^{N} (-\Delta_j) + \alpha V_C \tag{4.1.1}$$

satisfies

$$E_0 \geq \text{const.} \ (N + K). \tag{4.1.2}$$

In (4.1.1), α is Sommerfeld's fine-structure constant, whose physical value is approximately $1/137$, and V_C is the Coulomb potential of the electrons and nuclei, namely,

$$V_C(\mathbf{x}_1, \cdots, \mathbf{x}_N) = -Z \sum_{j=1}^{N} \sum_{k=1}^{K} \frac{1}{|\mathbf{x}_j - R_k|} + \sum_{1 \leq j < l \leq N} \frac{1}{|\mathbf{x}_j - \mathbf{x}_l|}$$

$$+ Z^2 \sum_{1 \leq j < k \leq K} \frac{1}{|R_j - R_k|}, \tag{4.1.3}$$

where the $\mathbf{x}_j, R_k$ denote the co-ordinates of the jth electron and kth nucleus, respectively, and we have taken the nuclear charges to be the same, namely Z, for simplicity: in fact for the problem of stability, this does not involve any loss of generality, for, as proved in [Daubechies and Lieb (1983)], if there are separate charges Z_j, the energy is concave in each Z_j and stability holds in $\{0 \leq Z_j \leq Z : j = 1, \cdots, K\}$ if it holds when all the $Z_j = Z$.

The result (4.1.2) was first established by Dyson and Lenard in their pioneering work [Dyson and Lenard (1967)] and [Dyson and Lenard (1968)]. Subsequently, a different, and considerable easier proof was given by Lieb and Thirring in [Lieb and Thirring (1975)], who also derived a much better constant. Stability became a favourite topic of research and a number of interesting generalisations have been achieved. We now give a brief summary of some of the highlights. A significant advance of relevance to this book was made by Conlon in [Conlon (1984)] who investigated the quasi-relativistic Hamiltonian

$$\mathbb{H}_{N,K} = \sum_{j=1}^{N} \{(-\Delta_j + 1)^{1/2} - 1\} + \alpha V_C \tag{4.1.4}$$

and proved stability as long as $\alpha Z < 10^{-200}$; Fefferman and de la Llave improved on Conlon's bound for αZ in [Fefferman and de la Llave (1986)] to $\alpha Z \leq \pi/5$, with the improvement $\alpha Z \leq \pi/2.06$ on using a computer assisted proof. Lieb and Yau then proved in [Lieb and Yau (1988)] for $\alpha Z \leq 2/\pi$, the optimal range in view of Theorem 2.2.6, that (4.1.4) is stable if and only if α is less than some critical value $\alpha_c > 1/94$. Lieb, Loss and Siedentop introduced a magnetic field, $\mathbf{B}(\mathbf{x}) = \text{curl } \mathbf{A}(\mathbf{x})$ in [Lieb *et al.* (1996)] and considered

$$\mathbb{H}_{N,K}(\mathbf{A}) = \sum_{j=1}^{N} | -i\boldsymbol{\nabla}_j + \mathbf{A}(\mathbf{x}_j)| + \alpha V_C. \tag{4.1.5}$$

They proved that for arbitrary magnetic potential $\mathbf{A}$, (4.1.5) is stable if $\alpha < \tilde{\alpha}$, where

$$1/\tilde{\alpha} = (\pi/2)Z + 2.80Z^{2/3} + 1.30. \tag{4.1.6}$$

For $\alpha = 1/137$, this allows $Z \leq 59$. The presence of the magnetic field, for arbitrary $\mathbf{A}$, actually eases the stability problem for the magnetic Schrödinger operator

$$H_{N,K}(\mathbf{A}) = \sum_{j=1}^{N} (-i\boldsymbol{\nabla}_j + \mathbf{A}(\mathbf{x}_j))^2 + \alpha V_C \tag{4.1.7}$$

on account of the *diamagnetic inequality* in Theorem 4.5.1 below. However, difficulties are encountered on replacing the one-electron kinetic energy by the non-relativistic Pauli operator,

$$\mathbb{P}_{\mathbf{A}} = [\boldsymbol{\sigma} \cdot (-i\boldsymbol{\nabla} + \mathbf{A})]^2 = \sum_{j=1}^{3} \{\sigma_j (-i\partial_j + A_j)\}^2$$

$$= (-i\boldsymbol{\nabla} + \mathbf{A})^2 I_2 + \boldsymbol{\sigma} \cdot \mathbf{B},$$

where $\boldsymbol{\sigma} = (\sigma_1, \sigma_2, \sigma_3)$ is the triple of Pauli matrices defined in (2.1.5), $\mathbf{A} = (A_1, A_2, A_3)$ and I_2 is the 2×2 identity matrix. The presence of the *Zeeman* term $\boldsymbol{\sigma} \cdot \mathbf{B}$ allows for the possibility that arbitrarily large magnetic fields $\mathbf{B}$ can cause even the hydrogen Hamiltonian

$$\mathbb{P}_{\mathbf{A}} + \alpha V_C, \quad V_C(\mathbf{x}) = -Z/|\mathbf{x}|,$$

to have arbitrarily large negative energies, despite the fact that $\mathbb{P}_{\mathbf{A}}$ is non-negative for any magnetic field. It is the existence of zero modes of the Pauli operator which makes this possible: a *zero mode* of $\mathbb{P}_{\mathbf{A}}$ is an eigenvector φ corresponding to an eigenvalue at 0, thus $\mathbb{P}_{\mathbf{A}}\varphi = 0$. To see this, let φ be a zero mode, and for $\kappa > 0$ define $\varphi_\kappa(\mathbf{x}) = \kappa^{3/2}\varphi(\kappa\mathbf{x})$, so that $\|\varphi_\kappa\| = \|\varphi\|$. Then

$$\mathbb{P}_{\mathbf{A}_\kappa}\varphi_\kappa = 0,$$

where $\mathbf{A}_\kappa(\mathbf{x}) := \kappa\mathbf{A}(\kappa\mathbf{x})$ and hence $\mathbf{B}_\kappa(\mathbf{x}) = [\text{curl } \mathbf{A}_\kappa](\mathbf{x}) = \kappa\mathbf{B}(\kappa\mathbf{x})$. Furthermore,

$$([\mathbb{P}_{\mathbf{A}_\kappa} + \alpha V_C]\varphi_\kappa, \varphi_\kappa) = -\alpha Z \int_{\mathbb{R}^3} \frac{1}{|\mathbf{x}|^2}|\varphi_\kappa(\mathbf{x})|^2 dx$$

$$= -\alpha Z\kappa \int_{\mathbb{R}^3} \frac{1}{|\mathbf{x}|^2}|\varphi(\mathbf{x})|^2 dx.$$

Therefore, on allowing $\kappa \to \infty$, we see that $\mathbb{P}_{\mathbf{A}} + \alpha V_C$ can have arbitrarily large negative energy for arbitrarily large magnetic fields. To remedy this problem for arbitrary large fields, the *field energy*

$$H_F = (8\pi\alpha^2)^{-1} \int_{\mathbb{R}^3} |\mathbf{B}(\mathbf{x})|^2 dx$$

is added to give, in general, the Hamiltonian

$$\mathbb{P}_{N,K}(\mathbf{A}) = \sum_{j=1}^{N} \mathbb{P}_{\mathbf{A}}(\mathbf{x}_j) + \alpha V_C + H_F. \tag{4.1.8}$$

In this case, stability was first announced in an unpublished note by Fefferman for sufficiently small α and Z, and a proof, with physically realistic constants, obtained by Lieb, Loss and Solovej in [Lieb *et al.* (1995)]. A special case of the latter is $\alpha = 1/137, Z \leq 1050$. As in the quasi-relativistic example above, a feature of the result is that the system is unstable if α is greater than some critical value, no matter how small Z is.

A Hamiltonian, which is both relativistic and also includes magnetic fields, is the following modelled on the Brown–Ravenhall operator,

$$\mathbb{B}_{N,K}(\mathbf{A}) = \Lambda_+ \left(\sum_{j=1}^{N} \mathbb{D}_0^{(j)}(\mathbf{A}) + \alpha V_C + H_F \right) \Lambda_+, \tag{4.1.9}$$

where

$$\mathbb{D}_0^{(j)}(\mathbf{A}) = \boldsymbol{\alpha} \cdot (-i\boldsymbol{\nabla} + \mathbf{A}(\mathbf{x}_j)) + \beta.$$

In (4.1.9) there are two obvious candidates for Λ_+, namely the projection $\Lambda_{+,N}$ onto

$$\mathcal{H}_N = \bigwedge_{j=1}^{N} \mathcal{H}_+^{(j)},$$

where $\mathcal{H}_+^{(j)}$ is the positive spectral subspace of the free Dirac operator $\mathbb{D}_0^{(j)}$ acting in the space of the jth electron, and the projection $\Lambda_{+,N}(\mathbf{A})$ onto

$$\mathcal{H}_N(\mathbf{A}) = \bigwedge_{j=1}^{N} \mathcal{H}_+^{(j)}(\mathbf{A})$$

where $\mathcal{H}_+^{(j)}(\mathbf{A})$ is the positive spectral subspace of $\mathbb{D}_0^{(j)}(\mathbf{A})$. However, it is proved in [Lieb *et al.* (1997)] that the first choice yields instability for any α, while the second yields stability for small enough α and Z. In particular, with the choice of $\Lambda_{+,N}(\mathbf{A})$, there is stability for $Z \leq 56$ and $\alpha = 1/137$.

In the case of no magnetic field and with $\Lambda_+ = \Lambda_{+,N}$, it is established in [Balinsky and Evans (1999)] that the one-electron operator $\mathbb{B}_{1,K}(0)$ is stable if $\alpha Z \leq 2(2/\pi + \pi/2)^{-1}$ and $\alpha \leq 2\pi\{(\pi^2 + 4)(2 + \sqrt{1 + \pi/2}\}^{-1}$. This is the analogue of the result proved by Daubechies and Lieb in [Daubechies and Lieb (1983)] for the quasi-relativistic operator $\mathbb{H}_{1,K}$. Hoever and Siedentop [Hoever and Siedentop (1999)] consider arbitrary values of N and K, but with the *massless* case of the Dirac operator, that is, $\mathbb{D}_0 = -i\boldsymbol{\alpha} \cdot \boldsymbol{\nabla}$. They prove that for $\gamma < 2(2/\pi + \pi/2)^{-1}$, there exists an α_0 such that there is stability (in fact $\mathbb{B}_{N,K}(0) \geq 0$) for all $\alpha Z < \gamma$ and $\alpha < \alpha_0$. With $\alpha = 1/137$, this allows $Z \leq 88$.

Our main concern in this chapter will be with two features of the problem of stability which have a pivotal role in the problem and many of the works quoted above. The first is the existence, or otherwise, of zero modes of the kinetic energy part of the Hamiltonian. We have already hinted above that the Pauli operator can have zero modes. This was not established until Loss and Yau gave the first example in [Loss and Yau (1986)]. This and other examples of potentials that give rise to zero modes for the Pauli operator will be discussed in Section 4.2. Note that $-\Delta$ and $\mathbb{H}_0$ have no zero modes and so zero modes are not relevant to (4.1.1) and (4.1.4). If $\mathbb{P}_{\mathbf{A}}\psi = 0$, then the kinetic energy term in (4.1.8) vanishes for $\Psi(\mathbf{x}) = \prod_{j=1}^N \psi(\mathbf{x}_j)$, and

$$(\mathbb{P}_{N,K}(\mathbf{A})\Psi, \Psi) = ([\alpha V_C + H_F]\Psi, \Psi),$$

which is not bounded below for large α and Z leading to instability. The existence of zero modes of $\mathbb{P}_{\mathbf{A}}$ has other important physical consequences; see [Adam *et al.* (1999)], [Frölich *et al.* (1986)], [Lieb and Loss (1986)] and [Loss and Yau (1986)]. An important mathematical implication is that, even if the scalar potential V in the one-body operator $\mathbb{P}_{\mathbf{A}} + V$, is such that the operator is bounded below, there can't be an analogue of the celebrated Cwikel, Lieb, Rosenblum inequality for the negative eigenvalues of $\mathbb{P}_{\mathbf{A}} + V$, in terms of some $L^p(\mathbb{R}^3)$ norm of V, since any small negative perturbation V would produce negative eigenvalues, contrary to such an inequality if V is sufficiently small.

The second feature referred to above concerns the Lieb–Thirring inequalities for the sum of the negative eigenvalues of an operator. The need for such an estimate appears naturally in some of the proofs of stability (e.g. in [Lieb *et al.* (1995)], [Lieb *et al.* (1997)]) in which the operator is bounded below by the sum of N identical copies of a one-body operator $\mathfrak{h}$, say. Since electrons, being fermions, satisfy the *Exclusion Principle*, a lower bound for the many-body operator is then given by the sum of the negative eigenvalues of $\mathfrak{h}$. The results in [Lieb *et al.* (1995)], [Lieb *et al.* (1997)] will be discussed in Section 4.6 below, but the original papers should be consulted for full details.

4.2 Eigenvalues of the operators $\mathbb{D}_\mathbf{A}$ and $\mathbb{P}_\mathbf{A}$

We shall see that there are distinct advantages in approaching the problem for the Pauli operator $\mathbb{P}_\mathbf{A}$ through the *Weyl–Dirac* or *massless Dirac* operator

$$\mathbb{D}_\mathbf{A} = \boldsymbol{\sigma} \cdot (-i\boldsymbol{\nabla} + \mathbf{A}). \tag{4.2.1}$$

Note that unlike the Dirac operator, this acts on 2-spinors. We then have

$$\mathbb{P}_\mathbf{A} = \mathbb{D}_\mathbf{A}^2 = (-i\boldsymbol{\nabla} + \mathbf{A})^2 + \boldsymbol{\sigma} \cdot \mathbf{B}. \tag{4.2.2}$$

If $|\mathbf{B}(\mathbf{x})| \to 0$ as $|\mathbf{x}| \to \infty$, it is known (see [Thaller (1992)], Theorem 7.7) that the Dirac operator $\mathbb{D}_\mathbf{A} + \beta$ has essential spectrum $\mathbb{R} \setminus (-1, 1)$ and so $\mathbb{D}_\mathbf{A}$ has essential spectrum $\mathbb{R}$. It is shown in [Helffer *et al.* (1989)] that this remains true under conditions that are much weaker on $\mathbf{B}$, without even the requirement that $|\mathbf{B}(\mathbf{x})| \to 0$ as $|\mathbf{x}| \to \infty$. If $\mathbf{A}$ has continuously differentiable components and the associated magnetic field $\mathbf{B}$ satisfies the conditions

$$\lim_{|\mathbf{x}|\to\infty} |\mathbf{B}(\mathbf{x})| = 0, \quad \lim_{|\mathbf{x}|\to\infty} |\mathbf{x} \times \mathbf{B}(\mathbf{x})| = 0, \tag{4.2.3}$$

it is proved in [Berthier and Georgescu (1987)] that $\mathbb{D}_\mathbf{A}$ has no eigenvalues embedded in $\mathbb{R} \setminus \{0\}$. An eigenfunction corresponding to an assumed eigenvalue in $\mathbb{R} \setminus \{0\}$ is proved to have compact support and is therefore identically zero on account of the unique continuation principle in Theorem 2.4.1.

4.2.1 *Examples of zero modes*

Loss–Yau examples

The first example of a magnetic potential $\mathbf{A}$ and magnetic field $\mathbf{B} = \boldsymbol{\nabla} \times \mathbf{A}$ for which $\mathbb{P}_\mathbf{A}$ has zero modes is the following constructed in [Loss and Yau (1986)]:

$$\mathbf{A}(\mathbf{x}) = \frac{3}{(1+r^2)^2} \begin{pmatrix} 2x_1x_3 - 2x_2 \\ 2x_2x_3 + 2x_1 \\ 1 - x_1^2 - x_2^2 + x_3^2 \end{pmatrix} \tag{4.2.4}$$

and

$$\mathbf{B}(\mathbf{x}) = \frac{12}{(1+r^2)^3} \begin{pmatrix} 2x_1x_3 - 2x_2 \\ 2x_2x_3 + 2x_1 \\ 1 - x_1^2 - x_2^2 + x_3^2 \end{pmatrix} \tag{4.2.5}$$

where $\mathbf{x} = (x_1, x_2, x_3)$ and $r = |\mathbf{x}|$. Their example is based on finding a 2-spinor ψ and a real-valued function λ such that

$$[(\boldsymbol{\sigma} \cdot \mathbf{p})\, \psi](\mathbf{x}) = \lambda(\mathbf{x})\psi(\mathbf{x}), \tag{4.2.6}$$

the magnetic potential $\mathbf{A} = (A_1, A_2, A_3)$ that they seek then being given by

$$A_i(\mathbf{x}) = \lambda(\mathbf{x})\frac{\langle \psi(\mathbf{x}), \sigma_i\psi(\mathbf{x})\rangle}{|\psi(\mathbf{x})|^2}. \tag{4.2.7}$$

The above example is determined by the solution

$$\psi(\mathbf{x}) = \frac{1}{(1+r^2)^{3/2}}(1+i\mathbf{x}\cdot\boldsymbol{\sigma})\begin{pmatrix}1\\0\end{pmatrix} \qquad (4.2.8)$$

of the Loss–Yau equation (4.2.6) with $\lambda(\mathbf{x}) = 3/(1+r^2)$.

It is observed in [Loss and Yau (1986)] that in addition to their simplest solution (4.2.8) of (4.2.6), there are similar solutions with higher angular momenta. Using instead of the constant spinor

$$\begin{pmatrix}1\\0\end{pmatrix},$$

the spinor

$$\Phi_{l,m} = \begin{pmatrix}\sqrt{l+m+1/2}\ \ Y_{l,m-1/2}\\ -\sqrt{l-m+1/2}\ \ Y_{l,m+1/2}\end{pmatrix},$$

where $m \in [-l-1/2, l+1/2]$ and Y are spherical harmonics, they found the solution

$$\psi_{l,m}(\mathbf{x}) = \frac{r^l}{(1+r^2)^{l+3/2}}(1+i\mathbf{x}\cdot\boldsymbol{\sigma})\Phi_{l,m} \qquad (4.2.9)$$

of (4.2.6) with

$$\lambda(\mathbf{x}) = \frac{2l+3}{1+r^2}.$$

For the special case of maximal number $m = l+1/2$, these solutions simplify to

$$\psi_l(\mathbf{x}) = \psi_{l,l+1/2} = \frac{Y_{l,l}r^l}{(1+r^2)^{l+3/2}}(1+i\mathbf{x}\cdot\boldsymbol{\sigma})\begin{pmatrix}1\\0\end{pmatrix} \qquad (4.2.10)$$

and the corresponding magnetic potential and magnetic field are

$$\mathbf{A}(l)(\mathbf{x}) = \frac{3+2l}{(1+r^2)^2}\begin{pmatrix}2x_1x_3 - 2x_2\\ 2x_2x_3 + 2x_1\\ 1 - x_1^2 - x_2^2 + x_3^2\end{pmatrix} \qquad (4.2.11)$$

and

$$\mathbf{B}(l)(\mathbf{x}) = \frac{4(3+2l)}{(1+r^2)^3}\begin{pmatrix}2x_1x_3 - 2x_2\\ 2x_2x_3 + 2x_1\\ 1 - x_1^2 - x_2^2 + x_3^2\end{pmatrix}. \qquad (4.2.12)$$

The following two features of the Loss–Yau examples are of particular relevance to what follows in Section 4.3. These are

$$|\mathbf{B}| \in L^p(\mathbb{R}^3), \quad \text{for any}\quad p > 3/4, \qquad (4.2.13)$$

$$\tfrac{1}{3}(3+2l)\mathbf{B}(\mathbf{x}), \ \ l \in \mathbb{N}, \quad \text{also yields zero modes.} \qquad (4.2.14)$$

In [Adam *et al.* (2000a)], it is proved that with the above $\mathbf{A}(l)$, the multiplicity of the zero modes is at least $l + 1$. This was the first result to demonstrate that degeneracy of zero modes can occur in three dimensions. However, multiple zero modes were also constructed in [Erdős and Solovej (2000)] and [Erdős and Solovej (2001)].

In an attempt to explain the origin of zero modes, Erdős and Solovej in [Erdős and Solovej (2000)] and [Erdős and Solovej (2001)] consider a more geometric view-point. Using the known behaviour of the Dirac operator under conformal trans-formations, and the knowledge that $\mathbb{R}^3$ is conformally equivalent to the punctured 3-dimensional sphere $\mathbb{S}^3$, they establish their zero modes as pull-backs of zero modes of $\mathbb{S}^2$ under the Hopf map $\mathbb{S}^3 \to \mathbb{S}^2$. In many cases they are able to compute the multiplicity of zero modes precisely, with examples of arbitrary degeneracy being exhibited. In $\mathbb{R}^2$, the celebrated Aharonov–Casher Theorem gives a precise value to the dimension of the space of zero modes: see (4.3.2) below.

Elton's examples
In [Elton (2000)], Elton constructed smooth magnetic potentials $\mathbf{A}^{(1)}$ and $\mathbf{A}^{(2)}$ yielding zero modes of $\mathbb{P}_{\mathbf{A}}$, which have the following properties:
1) $\mathbf{A}^{(1)}$ has compact support and the associated zero mode is contained in $L^2(\mathbb{R}^3)$;
2) $\mathbf{A}^{(2)} = (-x_2, x_1, 0) + \widetilde{\mathbf{A}}^{(2)}$, where $\widetilde{\mathbf{A}}^{(2)}$ is bounded and supported on $\{|x_3| \le 1\}$, and the associated zero mode is in the Schwartz class.

The construction of Elton's examples follows from the "if" part of the next lemma, which is Proposition 1 in [Elton (2000)] and is a summary of observations made in the first part of Section 3 in [Loss and Yau (1986)]).

Lemma 4.2.1. *Suppose that ψ is a smooth spinor field on $\mathbb{R}^3$ with $|\psi(\mathbf{x})| > 0$ for all $\mathbf{x} \in \mathbb{R}^3$, and $\mathbf{A}$ is a real magnetic potential. Then ψ and A satisfy*

$$\boldsymbol{\sigma} \cdot (\mathbf{p} + \mathbf{A})\psi = 0, \quad \mathbf{p} = -i\boldsymbol{\nabla},$$

if and only if they satisfy the pair of equations

$$\mathbf{A} = \frac{1}{|\psi|^2}\left(\frac{1}{2}\mathrm{curl} < \psi, \boldsymbol{\sigma}\psi > + \mathrm{Im} < \psi, \boldsymbol{\nabla}\psi >\right) \tag{4.2.15}$$

and

$$\mathrm{div} < \psi, \boldsymbol{\sigma}\psi > = 0, \tag{4.2.16}$$

where $< \psi, \boldsymbol{\sigma}\psi >$ is the vector-valued function with components $< \psi, \sigma_i\psi >$, $i = 1, 2, 3$.

The construction of both of Elton's examples use a common auxiliary function g with the following properties:
    (A1) $g : \mathbb{R} \to \mathbb{R}$ is smooth, compactly supported and non-negative;
    (A2) $g(t) = (4 - t^2)^{1/2}$ for $t \in [-1/2, 1/2]$;
    (A3) $\mathrm{supp}(g) \subseteq [-1, 1]$;
    (A4) $\pm g'(t) \le 0$ for $\pm t \ge 0$.
Then $\mathcal{R}(g) = [0, 2]$ and the radially symmetric function g is smooth on $\mathbb{R}^3$. It is

helpful to note that for any spinor $\psi = \begin{pmatrix} \psi_1 \\ \psi_2 \end{pmatrix}$, we have

$$< \psi, \boldsymbol{\sigma}\psi > = \begin{pmatrix} 2\text{Re}(\overline{\psi_1}\psi_2) \\ 2\text{Im}(\overline{\psi_1}\psi_2) \\ |\psi_1|^2 - |\psi_2|^2 \end{pmatrix}.$$

Elton's first step in the construction of his example $\mathbf{A}^{(1)}$ is to define the real-valued function

$$f(r) = r^{-3}\left(-\int_0^r t^4 (g^2)^{'}(t)dt\right)^{1/2}$$

for all $r \geq 0$. Then, $f(r)$ is a smooth, strictly positive, radially symmetric function on $\mathbb{R}^3$ and $f(r) = 1/\sqrt{3}$ for $r \in (0, 1/2]$. Also the property (A3) of g implies that $f(r) = C_1 r^{-3}$ for all $r \geq 1$, where C_1 is the strictly positive constant defined by

$$C_1^2 = -\int_0^1 t^4 (g^2)^{'}(t)dt.$$

Elton then defines a smooth spinor field $\psi^{(1)}$ on $\mathbb{R}^3$ by

$$\psi^{(1)}(\mathbf{x}) = f(r)\begin{pmatrix} x_3 \\ x_1 + ix_2 \end{pmatrix} + \begin{pmatrix} ig(r) \\ 0 \end{pmatrix}.$$

Thus

$$|\psi^{(1)}(\mathbf{x})|^2 = r^2 f^2 + g^2.$$

Since $f(r) > 0$ for all $r \in \mathbb{R}_+$ and $g(0) = 2$, it follows that $|\psi^{(1)}(\mathbf{x})|$ never vanishes. Furthermore,

$$< \psi^{(1)}, \boldsymbol{\sigma}\psi^{(1)} > = \begin{pmatrix} 2x_1 x_3 f^2 + 2x_2 fg \\ 2x_2 x_3 f^2 - 2x_1 fg \\ (x_3^2 - x_1^2 - x_2^2)f^2 + g^2 \end{pmatrix}.$$

A straightforward calculation gives that $\text{div} < \psi^{(1)}, \boldsymbol{\sigma}\psi^{(1)} > = 0$.

The magnetic potential $\mathbf{A}^{(1)}$ is that given by Lemma 4.2.1 with $\psi = \psi^{(1)}$. For $r \geq 1$

$$\psi^{(1)} = C_1 r^{-3}\begin{pmatrix} x_3 \\ x_1 + ix_2 \end{pmatrix}$$

and

$$< \psi^{(1)}(\mathbf{x}), \boldsymbol{\sigma}\psi^{(1)}(\mathbf{x}) > = C_1^2 r^{-6}\begin{pmatrix} 2x_1 x_3 \\ 2x_2 x_3 \\ (x_3^2 - x_1^2 - x_2^2) \end{pmatrix}.$$

Further calculations then give

$$\text{curl} < \psi^{(1)}(\mathbf{x}), \boldsymbol{\sigma}\psi^{(1)}(\mathbf{x}) > = -2C_1^2 r^{-6}\begin{pmatrix} -x_2 \\ x_1 \\ 0 \end{pmatrix}$$

and

$$< \psi^{(1)}(\mathbf{x}), \nabla \psi^{(1)}(\mathbf{x}) > = C_1^2 r^{-6} \begin{pmatrix} -2x_1 - ix_2 \\ ix_1 - 2x_2 \\ -2x_3 \end{pmatrix}.$$

Combining these expressions we thus have from Lemma 4.2.1 that $\mathbf{A}^{(1)} = 0$ for $r \geq 1$. Therefore $\mathbf{A}^{(1)}$ is supported in the unit ball $\{\mathbf{x} : |\mathbf{x}| \leq 1\}$. Moreover, $\psi^{(1)}$ is smooth and $|\psi^{(1)}(\mathbf{x})|^2 = C_1^2 r^{-4}$ when $r \geq 1$. Thus, $\psi^{(1)} \in L^2(\mathbb{R}^3)$.

Since $\mathcal{R}(g) = [0, 2]$ we have $4 - g^2 \geq 0$. Thus the following function is well defined:

$$h(t) = \begin{cases} -(4 - g^2(t))^{1/2} & \text{for } t \leq 0, \\ (4 - g^2(t))^{1/2} & \text{for } t \geq 0. \end{cases}$$

Using h, define another smooth function k on $\mathbb{R}$ by $k(t) = \int_0^t h(s)ds$ and consider g, h and k as functions on $\mathbb{R}^3$ depending on x_3 and constant in x_1 and x_2. Define a smooth function u and spinor field $\psi^{(2)}$ on $\mathbb{R}^3$ by

$$u = e^{-\rho^2/2 - k} \quad \text{and} \quad \psi^{(2)} = u \begin{pmatrix} (\rho^2 - 1)(-h + ig) \\ 2(x_1 + ix_2) \end{pmatrix},$$

where $\rho = (x_1^2 + x_2^2)^{1/2}$. A calculation gives that div $< \psi^{(2)}(\mathbf{x}), \boldsymbol{\sigma}\psi^{(2)}(\mathbf{x}) > = 0$. Elton's second example $\mathbf{A}^{(2)}$ is the real magnetic potential given by Lemma 4.2.1 with $\psi = \psi^{(2)}$. A tedious, but straightforward, calculation leads to

$$\mathbf{A}^{(2)} = (-x_2, x_1, 0) + \tilde{\mathbf{A}}^{(2)},$$

where $\tilde{\mathbf{A}}^{(2)}$ is bounded and supported on $\{|x_3| \leq 1\}$, and the associated zero mode is in the Schwartz class. We refer to [Elton (2000)] for more details.

Adam, Muratori and Nash examples

Elton's magnetic potential $\mathbf{A}^{(1)}$ was the first example of a compactly supported magnetic potential giving rise to a zero mode, but this example belongs to the type studied in [Adam et al. (1999)]. In this section we present results from [Adam et al. (2000d)] where a whole class of zero modes of Pauli operators with compactly supported magnetic potentials have been constructed. This demonstrates that the possibility of having zero modes in magnetic fields of finite range is not just a curiosity, but a rather general feature of the Pauli operator. This is particularly interesting from a physical perspective since magnetic fields with finite range are the types of magnetic fields that may be realised experimentally.

Following [Adam et al. (2000d)], our objective is to study specific solutions $\Psi \in L^2(\mathbb{R}^3)$ of the equation

$$\boldsymbol{\sigma} \cdot (\mathbf{p} + \mathbf{A})\Psi = 0$$

where the magnetic potential $\mathbf{A} = (A_1, A_2, A_3)$ and magnetic field $\mathbf{B} = \text{curl } \mathbf{A}$ are non-singular everywhere in $\mathbb{R}^3$ and are different from zero only in a finite region of space. Further we require Ψ, A_i and B_i to be smooth everywhere.

We first observe that the spinor field

$$\Psi^0(\mathbf{x}) = \frac{i}{r^3} \begin{pmatrix} x_3 \\ x_1 + ix_2 \end{pmatrix}$$

solves the equation

$$(\boldsymbol{\sigma} \cdot \mathbf{p})\Psi^0 = 0$$

in $\mathbb{R}^3 \setminus \{0\}$. The spinor field Ψ^0 is singular at $r = 0$ but it is well behaved for large r. So, it is natural to ask whether there exist spinor fields that are equal to Ψ^0 outside a ball of radius $r = R$, and inside the ball they solve the Dirac equation for some non-zero A_i and are such that they are non-singular and smooth everywhere. In [Adam *et al.* (2000d)], the authors construct a whole class of such zero modes. Their construction is based on the following ansatz for zero modes from [Adam *et al.* (1999)]:

$$\Psi = g(r)e^{\left(if(r)\frac{\mathbf{x}}{r}\boldsymbol{\sigma}\right)} \begin{pmatrix} 1 \\ 0 \end{pmatrix}$$

$$= g(r) \left[\cos(f(r)) + i\sin(f(r))\frac{\mathbf{x}}{r}\boldsymbol{\sigma}\right] \begin{pmatrix} 1 \\ 0 \end{pmatrix} \tag{4.2.17}$$

for some real function f on $\mathbb{R}_+$ to be determined. The spinor field (4.2.17) leads to a zero mode for the magnetic potential $\mathbf{A}$ with components

$$A_i(\mathbf{x}) = h(r)\frac{\langle \Psi(\mathbf{x}), \sigma_i \Psi(\mathbf{x}) \rangle}{|\Psi(\mathbf{x})|^2} \tag{4.2.18}$$

provided that $g(r)$ and $h(r)$ are given in terms of the independent function $f(r)$ by

$$\frac{dg}{dr} = -\frac{2}{r}\frac{t^2}{1+t^2}g$$

$$h = \frac{1}{1+t^2}\left(\frac{dt}{dr} + \frac{2}{r}t\right),$$

where

$$t(r) := \tan(f(r)).$$

A sufficient condition on $t(r)$ to yield smooth, non-singular spinor fields in $L^2(\mathbb{R}^3, \mathbb{C}^2)$, and smooth, non-singular magnetic potentials with finite magnetic field energy $\int_{\mathbb{R}^3} |\mathbf{B}|^2 d\mathbf{x}$ is

$$\begin{aligned} &t(0) = 0, \\ &t(r) \sim c_1 r + o(r^2) \quad \text{for} \ \ r \to 0, \\ &t(\infty) = \infty, \end{aligned} \tag{4.2.19}$$

which we shall assume hereafter. If $t \to \infty$, then Ψ becomes Ψ^0 and $\mathbf{A}$ vanishes. So, the main idea in [Adam *et al.* (2000d)] was to find some function t that becomes infinite at some finite $r = R$ in a "smooth way" and stays infinite for $r > R$. To get a more manageable condition, we rewrite these functions in terms of the function

$$c(r) := \cos(f(r)) = \frac{1}{(1 + t(r)^2)^{1/2}}.$$

Then,

$$\frac{dg}{dr} = -\frac{2}{r}\left(1 - c^2\right)g$$

and

$$h = c\left(-\frac{1}{(1 - c^2)^{1/2}}\frac{dc}{dr} + \frac{2}{r}\left(1 - c^2\right)^{1/2}\right).$$

We need c to behave like

$$c(r) \sim 1 - c_2 r^2 + \dots \quad \text{for} \quad r \to 0.$$

Let us assume that c approaches zero in a smooth way at $r = R$ and stays zero for $r \geq R$. Also, suppose that $dc/dr = 0$ at $r = R$, and that the second derivative d^2c/dr^2 is finite at $r = R$. These assumptions then imply that

$$\mathbf{A} = 0 \quad \text{for} \quad r \geq R,$$

and

$$g(r) = k \cdot r^2 \quad \text{for} \quad r \geq R, \quad \text{with} \quad k = \exp\left(-2\int_0^R \frac{1 - c(r)^2}{r}\,dr\right).$$

The last identity leads to $\Psi = \text{const} \cdot \Psi^0$ for $r > R$.

By choosing different functions c many example of zero modes with finite range magnetic potentials can be produced. For example, the choice $R = 1$ and

$$c(r) = \left(1 - r^2\right)^2 \quad \text{for} \quad r < 1, \quad c(r) = 0 \quad \text{for} \quad r \geq 1$$

gives

$$g(r) = \exp\left(-4r^2 + 3r^4 - \frac{4}{3}r^6 + \frac{1}{4}r^8\right) \quad \text{for} \quad r < 1,$$

$$g(r) = \exp\left(-\frac{25}{12}\right)r^{-2} \quad \text{for} \quad r \geq 1.$$

and

$$h(r) = \frac{2\left(1 - r^2\right)^2\left(2 - 4r^2 + 4r^4 - r^6\right)}{(4 - 6r^2 + 4r^4 - r^6)^{1/2}} \quad \text{for} \quad r < 1$$

$$h(r) = 0 \quad \text{for} \quad r \geq 1.$$

Another example is

$$c(r) = \exp\left(\frac{r^2}{r^2 - 1}\right) \quad \text{for} \quad r < 1, \quad c(r) = 0 \quad \text{for} \quad r \geq 1.$$

4.2.2 *Chern–Simons actions for magnetic fields with zero modes*

The Chern–Simons action for a magnetic potential $\mathbf{A}$ is the integral

$$CS(\mathbf{A}) = \frac{1}{16\pi^2} \int_{\mathbb{R}^3} \mathbf{A} \cdot \mathbf{B} \, dx,$$

where $\mathbf{B}$ is a magnetic field with magnetic potential $\mathbf{A}$. A conjecture that stood for many years was that the Pauli operator $\mathbb{P}_{\mathbf{A}}$ has a zero mode if and only if there exists a certain functional of the magnetic potential which admits only discrete values. The simplest possible functional that could be imagined was the Chern–Simons action. It is a topological invariant of the magnetic field, i.e., independent of the metric, and therefore, if the existence and degeneracy of zero modes is related to some topological feature, the Chern–Simons action would be an obvious candidate. The conjecture fitted in well with the results in [Balinsky and Evans (2001)] and [Balinsky and Evans (2002b)] (see also Section 4.3 below), that the existence of zero modes is a rare event. It was also supported by examples of a whole class of magnetic potentials with an arbitrary number of zero modes from [Adam *et al.* (2000b)]. For all these magnetic potentials, which are characterised by an arbitrary function and an integer l (the number of zero modes for a given gauge potential), the Chern–Simon action indeed admits only the quantised values

$$\frac{1}{16\pi^2} \int_{\mathbb{R}^3} \mathbf{A} \cdot \mathbf{B} \, dx = \frac{1}{4} \left(l + \frac{1}{2} \right)^2.$$

The Chern–Simons actions for all the magnetic potentials from [Adam *et al.* (2000b)] were calculated explicitly in [Adam *et al.* (2000c)].

However, in [Adam *et al.* (2003)] the conjecture was contradicted with the counterexample which we now present. We show that the Chern–Simons action may admit a continuous range of values for magnetic potentials with zero modes, so it is not quantised in general. To see this let us look again into ansatz (4.2.17) with conditions (4.2.19). For the corresponding magnetic potential (4.2.18), the Chern–Simons action may be expressed (after performing the angular integration) as

$$\int_{\mathbb{R}^3} \mathbf{A} \cdot \mathbf{B} dx = 4\pi(2\pi - 4) \int_0^\infty \frac{rth^2}{1+t^2} dr + 4\pi(8 - 2\pi) \int_0^\infty \frac{r^2 t' h^2}{1+t^2} dr$$

$$= 4\pi \int_0^\infty (rf' + \sin(2f))[(8 - 2\pi)rf' + (\pi - 2)\sin(2f)] \frac{dr}{r},$$

where $t' = dt/dr$ and $f' = df/dr$.

Consider now the one-parameter family of functions

$$t_a(r) = r(a + r^2),$$

where a is an arbitrary real number. The functions t_a satisfy the sufficient condition (4.2.19) for all values of a, and therefore each t_a generates a magnetic potential with zero mode. From

$$t'_a = a + 3r^2, \quad h_a = \frac{3a + 5r^2}{1 + r^2(a + r^2)^2}$$

it follows that

$$\int A \cdot B d\mathbf{x} = 16\pi \int_0^\infty r^2 [a + (5 - \pi)r^2] \frac{(3a + 5r^2)^2}{(1 + r^2(a + r^2)^2)^3} dr. \qquad (4.2.20)$$

The last integral can easily be evaluated numerically and it is not difficult to see that the Chern–Simons action (4.2.20) is a continuous decreasing function of a taking all values greater than $9\pi^2$.

4.2.3 *Quaternionic description of zero modes*

We now give a formal treatment of zero modes in terms of *quaternions*, which clarifies the construction of examples of magnetic fields that give rise to zero modes. It is based on the discussion in [Balinsky and Evans (2003)] and is motivated by the constructions in [Loss and Yau (1986)], [Elton (2000)] and [Adam *et al.* (1999)], which were described above.

Let $\mathbb{H}$ denote the 4-dimensional real vector space of quaternions, i.e., $\mathbf{x} = (x_1, x_2, x_3, x_4) \in \mathbb{H}$ is represented in the form $\mathbf{x} = x_1 + x_2 i + x_3 j + x_4 k$, where the elements i, j, k anticommute,

$$ij = k, \ jk = i, \ ki = j, \quad i^2 = j^2 = k^2 = -1,$$

and x_1, x_2, x_3, x_4 are real. Multiplication in $\mathbb{H}$ is associative, but obviously not commutative. The *conjugate* $\bar{\mathbf{x}}$ of $\mathbf{x}$ is the element $\bar{\mathbf{x}} = x_1 - x_2 i - x_3 j - x_4 k$, $|\mathbf{x}|^2 := \mathbf{x}\bar{\mathbf{x}} = x_1^2 + x_2^2 + x_3^2 + x_4^2$ and $\mathbf{x}^{-1} := \bar{\mathbf{x}}/|\mathbf{x}|^2$. It follows that for all $\mathbf{a}, \mathbf{b} \in \mathbb{H}, \overline{\mathbf{ab}} = \bar{\mathbf{b}}\bar{\mathbf{a}}$. Quaternions of the form $\mathbf{x} = x_2 i + x_3 j + x_4 k$ are said to be *pure* and we denote the set of pure quaternions by Im $\mathbb{H}$. Note that, if $\mathbf{a} \in$ Im $\mathbb{H}$, then for all $\psi \in \mathbb{H}$,

$$\psi \mathbf{a} \bar{\psi} \in \text{Im } \mathbb{H}, \quad \psi \mathbf{a} \psi^{-1} \in \text{Im } \mathbb{H} \quad (\psi \neq 0). \qquad (4.2.21)$$

The maps

$$L_i : \mathbf{x} \mapsto i\mathbf{x}, \quad L_j : \mathbf{x} \mapsto j\mathbf{x}, \quad L_k : \mathbf{x} \mapsto k\mathbf{x},$$

of $\mathbb{H}$ into itself are real linear maps. A complex structure is introduced on $\mathbb{H}$ by the map $I : \mathbf{x} \mapsto \mathbf{x}i$, with respect to which L_i, L_j, L_k become complex linear maps on what is now a 2-dimensional complex vector space, for, since $\mathbb{H}$ is associative,

$$L_i(I(\mathbf{x})) = i(\mathbf{x}i) = (i\mathbf{x})i = I(L_i(\mathbf{x})),$$

and similarly, L_j and L_k commute with I. Furthermore, the L_i, L_j, L_k, anticommute and satisfy

$$L_i^2 = L_j^2 = L_k^2 = -1, \quad L_i L_j = L_k.$$

We can therefore make the identification (cf. [Rastall (1964)])

$$L_i = -i\sigma_1, \quad L_j = -i\sigma_2, \quad L_k = -i\sigma_3,$$

in terms of the Pauli matrices $\sigma_1, \sigma_2, \sigma_3$.

Consider now

$$\mathbb{D}_\mathbf{A}\psi = [\boldsymbol{\sigma}\cdot(-i\boldsymbol{\nabla}+A)]\psi = \sum_{j=1}^{3}(-i\sigma_j)\partial_j\psi + \sum_{j=1}^{3}(-i\sigma_j)(iA_j)\psi,$$

and regard ψ as having values in $\mathbb{H}$. Then, in particular, $iA_j\psi$ is now $I(A_j\psi)$, and hence

$$[\boldsymbol{\sigma}\cdot(-i\boldsymbol{\nabla}+\mathbf{A})]\psi = L_i\frac{\partial\psi}{\partial x} + L_j\frac{\partial\psi}{\partial y} + L_k\frac{\partial\psi}{\partial z} + L_i(IA_1\psi) + L_j(IA_2\psi) + L_k(IA_3\psi)$$
$$= \mathcal{D}\psi + \tilde{a}\psi i, \tag{4.2.22}$$

where

$$\mathcal{D} \;=\; i\frac{\partial}{\partial x} + j\frac{\partial}{\partial y} + k\frac{\partial}{\partial z} \tag{4.2.23}$$

and

$$\tilde{a} \;=\; iA_1 + jA_2 + kA_3 \;=\; A_1i + A_2j + A_3k \tag{4.2.24}$$

since A_1, A_2, A_3 are real-valued.

For ψ to be a zero mode of $\mathbb{D}_\mathbf{A}$, it must satisfy $\mathcal{D}\psi + \tilde{a}\psi i = 0$ for some $\tilde{a}$ of the form (4.2.24), i.e. $\tilde{a} \in \mathrm{Im}\,\mathbb{H}$. Of course, to be a zero mode, ψ must also lie in $L^2(\mathbb{R}^3, \mathbb{C}^2)$; see Remark 4.2.6 below.

Lemma 4.2.2. *Suppose that $\mathcal{D}\psi = \psi f$. Then $\mathcal{D}\psi + \tilde{a}\psi i = 0$ if and only if $\tilde{a} = \psi(fi)\psi^{-1}$. Also, $fi \in \mathrm{Im}\,\mathbb{H}$ if and only if $\tilde{a} \in \mathrm{Im}\,\mathbb{H}$.*

Proof.   We have that

$$\mathcal{D}\psi + \tilde{a}\psi i = \psi f + \tilde{a}\psi i = 0$$

if and only if $\tilde{a}\psi i = -\psi f$ and hence $\tilde{a}\psi = \psi(fi)$, i.e., $\tilde{a} = \psi(fi)\psi^{-1}$ and $fi = \psi^{-1}\tilde{a}\psi$. It follows from (4.2.21) that $fi \in \mathrm{Im}\,\mathbb{H}$ if and only if $\tilde{a} \in \mathrm{Im}\,\mathbb{H}$. $\square$

Remark 4.2.3. The special case of Lemma 4.2.2, when f is real is the basis of the technique used by Loss and Yau in [Loss and Yau (1986)], Section 2 (see Section 4.2.1 above), for the construction of the first example of a magnetic potential yielding zero modes.

If $a \in \mathrm{Im}\,\mathbb{H}$, then $a = a_1i + a_2j + a_3k$ and the correspondence $a \leftrightarrow (a_1, a_2, a_3)$ identifies $\mathrm{Im}\,\mathbb{H}$ with $\mathbb{R}^3$. From (4.2.21), if $\psi : \mathbb{R}^3 \to \mathbb{H}$ and $a : \mathbb{R}^3 \to \mathrm{Im}\,\mathbb{H}$, then $\bar\psi(\mathbf{x})a(\mathbf{x})\psi(\mathbf{x}) \in \mathrm{Im}\,\mathbb{H}$, and if $\bar\psi(\mathbf{x})a(\mathbf{x})\psi(\mathbf{x}) = A(\mathbf{x})i + B(\mathbf{x})j + C(\mathbf{x})k$ we can identify $\bar\psi a\psi$ with the vector field $(A, B, C) : \mathbb{R}^3 \to \mathbb{R}^3$. We now prove that this fact can be used to construct a zero mode of $\mathbb{D}_a$, for some magnetic potentials a, derived from appropriate vector fields as was done in [Loss and Yau (1986)].

Theorem 4.2.4. *Let* $\psi \in C^1(\mathbb{R}^3, \mathbb{H})$ *and denote the vector fields associated with* $\bar{\psi}i\psi$, $\bar{\psi}j\psi$, $\bar{\psi}k\psi$ *by* $\mathbf{A} = (A_1, A_2, A_3)$, $\mathbf{B} = (B_1, B_2, B_3)$, $\mathbf{C} = (C_1, C_2, C_3)$, *respectively. Then* $\mathbf{A}(\mathbf{x})$, $\mathbf{B}(\mathbf{x})$, $\mathbf{C}(\mathbf{x})$ *are orthogonal in* $\mathbb{R}^3$ *with equal norms* $|\psi(\mathbf{x})|^2$ *for all* $\mathbf{x} \in \mathbb{R}^3$, *and if* ψ *satisfies*

$$\mathcal{D}\psi + \tilde{a}\psi i = 0, \quad \tilde{a} \in \operatorname{Im} \mathbb{H}, \tag{4.2.25}$$

then the vector field (A_1, B_1, C_1) *has zero divergence.*

Conversely, given a non-vanishing vector field in $C^1(\mathbb{R}^3, \mathbb{R}^3)$ *with zero divergence, a function* $\psi \in C^1(\mathbb{R}^3, \mathbb{H})$ *with* $\psi(\mathbf{x}) \neq 0$ *can be constructed to satisfy (4.2.25) for* $\tilde{a} = (\mathcal{D}\psi)i\psi^{-1}$.

Proof. If

$$\mathcal{D}\psi + \tilde{a}\psi i = 0 \quad \text{with} \quad \tilde{a} \in \operatorname{Im} \mathbb{H},$$

then

$$\bar{\psi}(\mathcal{D}\psi) = -\bar{\psi}\tilde{a}\psi i$$

and from (4.2.21) it follows that $\bar{\psi}(\mathcal{D}\psi)$ has no i-component in $\mathbb{H}$. Equivalently

$$\bar{\psi}(\mathcal{D}\psi) - \overline{(\bar{\psi}(\mathcal{D}\psi))} = \bar{\psi}(\mathcal{D}\psi) - \overline{(\mathcal{D}\psi)}\psi$$

has no i-component in its range. On substituting

$$\mathcal{D}\psi = i\frac{\partial \psi}{\partial x} + j\frac{\partial \psi}{\partial y} + k\frac{\partial \psi}{\partial z}, \quad \overline{\mathcal{D}\psi} = -\frac{\partial \bar{\psi}}{\partial x}i - \frac{\partial \bar{\psi}}{\partial y}j - \frac{\partial \bar{\psi}}{\partial z}k$$

we get that

$$\bar{\psi}(\mathcal{D}\psi) - \overline{(\mathcal{D}\psi)}\psi = \frac{\partial \mathbf{A}}{\partial x} + \frac{\partial \mathbf{B}}{\partial y} + \frac{\partial \mathbf{C}}{\partial z}$$

has no i-component in its range, and hence

$$\operatorname{div}(A_1, B_1, C_1) = 0.$$

Also

$$\mathbf{A} \cdot \mathbf{B} = \sum_{l=1}^{3} A_l B_l = -\operatorname{Re}\big\{(\bar{\psi}i\psi)(\bar{\psi}j\psi)\big\}$$

$$= -|\psi|^2 \operatorname{Re}(\bar{\psi}k\psi) = 0$$

since $\psi\bar{\psi} = |\psi|^2$ and $\bar{\psi}k\psi \in \operatorname{Im} \mathbb{H}$. Similarly $\mathbf{A} \cdot \mathbf{C} = \mathbf{B} \cdot \mathbf{C} = 0$. Furthermore

$$|\mathbf{A}|^2 = |\psi|^4 = |\mathbf{B}|^2 = |\mathbf{C}|^2.$$

Conversely, let $\mathbf{X} = (A_1, B_1, C_1)$ be a non-vanishing vector field with div $\mathbf{X} = 0$. The 2-dimensional planes which are orthogonal to $\mathbf{X}$ at each $\mathbf{x} \in \mathbb{R}^3$, constitute a 2-dimensional real vector bundle over $\mathbb{R}^3$. Since each fibre has an induced scalar product and orientation we have an $SO(2)$-vector bundle. It is well-known that as $\mathbb{R}^3$ is contractible, this $SO(2)$-vector bundle is trivial, and hence there exist orthogonal vector fields $\mathbf{Y} = (A_2, B_2, C_2)$, $\mathbf{Z} = (A_3, B_3, C_3)$, which are orthogonal to $\mathbf{X}$, have the same norm and $\mathbf{X}, \mathbf{Y}, \mathbf{Z}$ has positive orientation. The columns of the matrix

$$Q = \begin{pmatrix} A_1 & A_2 & A_3 \\ B_1 & B_2 & B_3 \\ C_1 & C_2 & C_3 \end{pmatrix}$$

are therefore orthogonal with the same norm and $\det(Q) > 0$. Hence the rows $\mathbf{A} = (A_1, A_2, A_3)$, $\mathbf{B} = (B_1, B_2, B_3)$ and $\mathbf{C} = (C_1, C_2, C_3)$ are also orthogonal and have the same norms. On using (4.2.21), we see that any $\phi \in \mathbb{H}_1 = \{h \in \mathbb{H} : |h| = 1\}$ defines the orthogonal map $x \mapsto \bar{\phi}x\phi : \text{Im } \mathbb{H} \to \text{Im } \mathbb{H}$, and so, on identifying $\text{Im } \mathbb{H}$ with $\mathbb{R}^3$, a map $\Phi : \mathbb{H}_1 \to SO(3)$ is defined. Both ϕ and $-\phi$ are pre-images of the same member of $SO(3)$, and in fact Φ is a double covering of $SO(3)$. The orthonormal basis $\left(\frac{\mathbf{A}}{|\mathbf{A}|}, \frac{\mathbf{B}}{|\mathbf{B}|}, \frac{\mathbf{C}}{|\mathbf{C}|}\right)$ defines a function $\Psi : \mathbb{R}^3 \to SO(3)$. Since $\mathbb{R}^3$ is simply connected, it is well-known that Ψ can be lifted to $\tilde{\Psi} : \mathbb{R}^3 \to \mathbb{H}_1$, such that $\Psi = \tilde{\Psi} \circ \Phi$. The zero mode ψ that we seek is given by

$$\psi(\mathbf{x}) = \sqrt{|\mathbf{X}(\mathbf{x})|} \cdot \tilde{\Psi}(\mathbf{x}).$$

For, we now have

$$\bar{\psi}i\psi = \mathbf{A}, \quad \bar{\psi}j\psi = \mathbf{B}, \quad \bar{\psi}k\psi = \mathbf{C}.$$

Hence, as in the first part, $\bar{\psi}\mathcal{D}\psi$ has no i-component in $\mathbb{H}$ if $\text{div}(A_1, B_1, C_1) = 0$. Since this is given the theorem is proved. $\qquad \square$

The next theorem connects the geometrical approach of [Erdős and Solovej (2000, 2001)] and the analytical approach of [Adam *et al.* (2000a)] to construct multiple zero modes.

Theorem 4.2.5. *Let* $\psi_0 \in C^1(U, \mathbb{H})$, $U \subset \mathbb{R}^3$, *satisfy*

$$\mathcal{D}\psi_0 + \tilde{a}\psi_0 i = 0, \quad \tilde{a} : U \to \text{Im } \mathbb{H} \tag{4.2.26}$$

and let $g = u + iv \in C^1(U, \mathbb{C})$.

(a) If $\psi_0 g$ *is a solution of (4.2.26) in the domain* U, *so is* $\psi_0 g^n$ *for all* $n \in \mathbb{Z}$.

(b) Let $\mathbf{A}$, $\mathbf{B}$, $\mathbf{C}$ *be the vector fields in Theorem 4.2.4. Then* $\psi_0 g$ *satisfies (4.2.26) in the domain* U *if and only if, at every* $\mathbf{x} \in U$, (A_2, B_2, C_2) *and* (A_3, B_3, C_3) *lie in the plane of* grad u *and* grad v, *which is perpendicular to* (A_1, B_1, C_1), *and* g *is a conformal immersion of* U *into* $\mathbb{R}^2$, *i.e. the images of* (A_2, B_2, C_2) *and* (A_3, B_3, C_3) *under* g *are orthogonal in* $\mathbb{R}^2$ *and have the same length.*

Proof. Since the range of g is in $\mathbb{C}$, we have $gi = ig$. Hence

$$0 = (\mathcal{D}\psi_0)g + \tilde{a}\psi_0 ig$$
$$= (\mathcal{D}\psi_0)g + \tilde{a}(\psi_0 g)i$$
$$= \mathcal{D}(\psi_0 g) + \tilde{a}(\psi_0 g)i$$

if and only if

$$i\psi_0 \frac{\partial g}{\partial x} + j\psi_0 \frac{\partial g}{\partial y} + k\psi_0 \frac{\partial g}{\partial y} = 0. \tag{4.2.27}$$

This implies that if $\psi_0 g$ is a solution of (4.2.22), so is $\psi_0 g^n$ for all $n \in \mathbb{Z}$. From (4.2.27)

$$(\overline{\psi}_0 i\psi_0)\frac{\partial g}{\partial x} + (\overline{\psi}_0 j\psi_0)\frac{\partial g}{\partial y} + (\overline{\psi}_0 k\psi_0)\frac{\partial g}{\partial y} = 0$$

and so

$$(A_1 i + A_2 j + A_3 k)\left(\frac{\partial u}{\partial x} + i\frac{\partial v}{\partial x}\right) + (B_1 i + B_2 j + B_3 k)\left(\frac{\partial u}{\partial y} + i\frac{\partial v}{\partial y}\right) +$$

$$+ (C_1 i + C_2 j + C_3 k)\left(\frac{\partial u}{\partial z} + i\frac{\partial v}{\partial z}\right) = 0. \tag{4.2.28}$$

This yields in particular

$$(A_1, B_1, C_1) \cdot \nabla u = (A_1, B_1, C_1) \cdot \nabla v = 0 \tag{4.2.29}$$

so that the vector (A_1, B_1, C_1) is perpendicular to grad u and grad v. Since vectors (A_2, B_2, C_2) and (A_3, B_3, C_3) have been shown to be perpendicular to (A_1, B_1, C_1) in the proof of Theorem 4.2.4, the first part of (b) follows.

Let $\mathbf{x}(t) = (x(t), y(t), z(t))$ be a tangent curve to (A_2, B_2, C_2) in $\mathbb{R}^3$ with $(\dot{x}(0), \dot{y}(0), \dot{z}(0)) = (A_2, B_2, C_2)$. The image of the vector (A_2, B_2, C_2) under g is given by

$$\frac{d}{dt}\left(u(x(t), y(t), z(t)), v(x(t), y(t), z(t))\right)\big|_{t=0}$$

$$= \left(\frac{\partial u}{\partial x}A_2 + \frac{\partial u}{\partial y}B_2 + \frac{\partial u}{\partial z}C_2, \frac{\partial v}{\partial x}A_2 + \frac{\partial v}{\partial y}B_2 + \frac{\partial v}{\partial z}C_2\right). \tag{4.2.30}$$

Similarly the image of (A_3, B_3, C_3) under g is

$$\left(\frac{\partial u}{\partial x}A_3 + \frac{\partial u}{\partial y}B_3 + \frac{\partial u}{\partial z}C_3, \frac{\partial v}{\partial x}A_3 + \frac{\partial v}{\partial y}B_3 + \frac{\partial v}{\partial z}C_3\right). \tag{4.2.31}$$

But from (4.2.28)

$$\frac{\partial u}{\partial x}A_2 + \frac{\partial u}{\partial y}B_2 + \frac{\partial u}{\partial z}C_2 = -\left(\frac{\partial v}{\partial x}A_3 + \frac{\partial v}{\partial y}B_3 + \frac{\partial v}{\partial z}C_3\right)$$

and

$$\frac{\partial u}{\partial x}A_3 + \frac{\partial u}{\partial y}B_3 + \frac{\partial u}{\partial z}C_3 = \frac{\partial v}{\partial x}A_2 + \frac{\partial v}{\partial y}B_2 + \frac{\partial v}{\partial z}C_2.$$

It follows that (4.2.30) and (4.2.31) are orthogonal in $\mathbb{R}^2$ with the same length, as asserted.

□

Remark 4.2.6. As noted in the paragraph before Lemma 4.2.2, for ψ to be a zero mode of $\mathbb{D}_a$, it must lie in $L^2(\mathbb{R}^3, \mathbb{C}^2)$ as well as satisfy (4.2.25). Thus, while the above theorems provide useful indicators, and indeed, as we have observed, special cases were important steps in establishing the existence of zero modes in [Loss and Yau (1986)], difficult analysis remains in the problem of constructing $\tilde{a}$ for which a zero mode exists

4.3   Zero modes of $\mathbb{D}_A$ and $\mathbb{P}_A$ in $\mathbb{R}^n$, $n \geq 2$.

We shall consider a self-adjoint realisation of the Weyl–Dirac operator,

$$\mathbb{D}_A = \boldsymbol{\sigma} \cdot (\mathbf{p} + \mathbf{A}), \quad \mathbf{p} = -i\boldsymbol{\nabla}, \tag{4.3.1}$$

in $L^2(\mathbb{R}^n, \mathbb{C}^2)$, for $n \geq 2$, but our primary concern is with $n = 2$ and $n = 3$, and, indeed, the operators in preceding chapters have been defined with these values in mind. We shall be assuming conditions on $\mathbf{A}$ which ensure that $\mathbb{D}_A$ is defined as the operator sum of $\boldsymbol{\sigma} \cdot \mathbf{p}$ and $\boldsymbol{\sigma} \cdot \mathbf{A}$, and $\mathbb{P}_A = \mathbb{D}_A^2$. Hence nul $\mathbb{P}_A = $ nul $\mathbb{D}_A$.

The case $n = 2$ is of special interest in view of the Aharonov–Casher Theorem in $\mathbb{R}^2$, which gives a precise value for the nullity of $\mathbb{P}_A$ in terms of the magnetic flux, namely

$$\text{nul } \mathbb{P}_A = \left\{ \frac{1}{2\pi} \left| \int_{\mathbb{R}^2} B(\mathbf{x}) d\mathbf{x} \right| \right\}, \tag{4.3.2}$$

where $\{y\}$ denotes the largest integer strictly less than y and $\{0\} = 0$; see [Cycon *et al.* (1987)], Section 6.4. Note that in this 2-dimensional case,

$$\mathbb{P}_A = \sum_{j=1}^{2} \left\{ \sigma_j \left(\frac{1}{i} \partial_j + A_j \right) \right\}^2,$$

$\mathbf{A} = (A_1, A_2)$ and the magnetic field is identified with the scalar $B = \partial_1 A_2 - \partial_2 A_1$. In an unpublished work (see [Cycon *et al.* (1987)], Section 6.4) Avron and Tamares considered the analogous problem on the sphere $\mathbb{S}^2$. There are difficulties in defining the Dirac operator with a magnetic field on $\mathbb{S}^2$. The correct definition should be as an operator acting on sections of a $U(1)$-bundle over $\mathbb{S}^2$. The magnetic field is then the curvature 2-form of such a bundle and the nullity of $\mathbb{P}_A$ is given by $|(1/2\pi) \int_{\mathbb{S}^2} B| d\boldsymbol{\omega}$, which is the absolute value of the first Chern invariant C_1 of this bundle.

We follow standard practice and use $\mathbf{p}$ to denote the momentum operator, taking care to avoid the possibility of confusion with points in momentum space: thus $-i\boldsymbol{\nabla}$ is denoted by $\mathbf{p}$, rather than by $\mathbb{F}^{-1}\mathbf{p}\mathbb{F}$ in terms of the Fourier transform $\mathbb{F}$. It has operator core $C_0^\infty(\mathbb{R}^n, \mathbb{C}^2)$, that is, it is the closure of its restriction to $C_0^\infty(\mathbb{R}^n, \mathbb{C}^2)$. To simplify notation, we denote $L^2(\mathbb{R}^n, \mathbb{C}^2)$ by $\mathcal{H}_n$. The operators $|\mathbf{p}|$, $\sqrt{|\mathbf{p}|}$ and $1/\sqrt{|\mathbf{p}|}$ are non-negative self-adjoint operators in $\mathcal{H}_n$; they are also injective and have dense ranges $\mathcal{R}$ in $\mathcal{H}_n$. For $s = 1, 1/2, -1/2$, we define D_n^s to be the completion of the domain $\mathcal{D}(|\mathbf{p}|^s)$ of $|\mathbf{p}|^s$ with respect to the norm

$$\|u\|_{D_n^s} := \||\mathbf{p}|^s u\|_{\mathcal{H}_n}. \tag{4.3.3}$$

Note that $C_0^\infty(\mathbb{R}^n, \mathbb{C}^2)$ is dense in D_n^s for $s = 1, 1/2$.

For $\phi \in \mathcal{R}(\sqrt{|\mathbf{p}|}) = \mathcal{D}(1/\sqrt{|\mathbf{p}|})$, we have

$$\left\| \frac{1}{\sqrt{|\mathbf{p}|}} \phi \right\|_{D_n^{1/2}} = \|\phi\|_{\mathcal{H}_n}$$

and so $1/\sqrt{|\mathbf{p}|} : \mathcal{R}(\sqrt{|\mathbf{p}|}) \subset \mathcal{H}_n \to D_n^{1/2}$ is an isometry. Since $\mathcal{D}(\sqrt{|\mathbf{p}|})$ and $\mathcal{R}(\sqrt{|\mathbf{p}|})$ are dense in $D_n^{1/2}$ and $\mathcal{H}_n$, respectively, it follows that $1/\sqrt{|\mathbf{p}|}$ can be extended by continuity to a unitary map from $\mathcal{H}_n$ onto $D_n^{1/2}$. Furthermore, $1/\sqrt{|\mathbf{p}|}$ extends to a unitary map from $D_n^{-1/2}$ onto $\mathcal{H}_n$. Thus $1/\sqrt{|\mathbf{p}|}$ has three roles: a self-adjoint operator in $\mathcal{H}_n$, and unitary operators $\mathcal{H}_n \to D_n^{1/2}$ and $D_n^{-1/2} \to \mathcal{H}_n$. We shall continue to use the same notation for all three as the precise meaning will be clear from the context.

The following facts will be needed from Sections 1.6 and 1.7.

(i) There are natural embeddings

$$D_n^{\frac{1}{2}} \hookrightarrow L^{\frac{2n}{n-1}}(\mathbb{R}^n, \mathbb{C}^2), \quad n \ge 2, \tag{4.3.4}$$

$$D_n^1 \hookrightarrow L^{2n/(n-2)}(\mathbb{R}^n, \mathbb{C}^2), \tag{4.3.5}$$

$$H^1(\mathbb{R}^2, \mathbb{C}^2) \hookrightarrow L^q(\mathbb{R}^2, \mathbb{C}^2), \quad \forall q \in [2, \infty). \tag{4.3.6}$$

(ii) The dual $\left(D_n^{\frac{1}{2}}\right)^*$ of $D_n^{\frac{1}{2}}$ is $D_n^{-\frac{1}{2}}$.

(iii) Kato's inequality holds for all $n \ge 2$: for all $u \in D_n^{\frac{1}{2}}$,

$$\int_{\mathbb{R}^n} \frac{1}{|\mathbf{x}|} |u(\mathbf{x})|^2 \, d\mathbf{x} \le c_n \int_{\mathbb{R}^n} |\mathbf{p}| |\hat{u}(\mathbf{p})|^2 \, d\mathbf{p} = c_n \|u\|^2_{D_n^{\frac{1}{2}}} \tag{4.3.7}$$

where the best possible constant c_n is given by the case $p = 2$, $\alpha = 1/2$, of Theorem 1.7.1.

We shall write $\|\mathbf{A}\|_{L^n(\mathbb{R}^n)}$ to mean $\| |\mathbf{A}| \|_{L^n(\mathbb{R}^n)}$, where $|\mathbf{A}| = (\sum_{j=1}^n |A_j|^2)^{1/2}$.

Lemma 4.3.1. *Let* $|\mathbf{A}| \in L^n(\mathbb{R}^n)$, $n \ge 2$. *Then*

$$\frac{1}{\sqrt{|\mathbf{p}|}} (\boldsymbol{\sigma} \cdot \mathbf{A}) \frac{1}{\sqrt{|\mathbf{p}|}} : \mathcal{H}_n \longrightarrow \mathcal{H}_n \tag{4.3.8}$$

is compact. Moreover, for all $\phi \in D_n^{\frac{1}{2}}$,

$$\|(\boldsymbol{\sigma} \cdot \mathbf{A})\phi\|_{D_n^{-\frac{1}{2}}} \le \gamma_n^2 \|\mathbf{A}\|_{L^n(\mathbb{R}^n)} \|\phi\|_{D_n^{\frac{1}{2}}}, \tag{4.3.9}$$

where γ_n *is the norm of the embedding (4.3.4).*

Proof. We have seen that $1/\sqrt{|\mathbf{p}|}$ is a unitary map from $\mathcal{H}_n$ onto $D_n^{1/2}$ and from $D_n^{-1/2}$ onto $\mathcal{H}_n$. It is therefore sufficient to prove that the multiplication operator $\sigma \cdot \mathbf{A}$ is a compact map of $D_n^{1/2}$ into $D_n^{-1/2}$. Let $\phi_m \to 0$ weakly in $D_n^{1/2}$. Then $\|\phi_m\|_{D_n^{1/2}} \le k$ say. Given $\epsilon > 0$, set $|\mathbf{A}| = a_1 + a_2$, where $a_1 \in C_0^\infty(\Omega_\epsilon)$ for some ball Ω_ϵ and $|a_1(\mathbf{x})| \leq k_\epsilon$ say, and $\|a_2\|_{L^n(\mathbb{R}^n)} < \epsilon$. Then, for $\psi \in D_n^{1/2}$, we have the inequality

$$|([\boldsymbol{\sigma} \cdot \mathbf{A}]\phi_m, \psi)| \leq \int_{\mathbb{R}^n} (a_1 + a_2)|\phi_m||\psi| \, d\mathbf{x} := I_1 + I_2,$$

where, by Kato's inequality (4.3.7),

$$I_1 \leq k_\epsilon \int_{\Omega_\epsilon} \left[\sqrt{|\mathbf{x}|}|\phi_m(\mathbf{x})|\right]\left[\frac{1}{\sqrt{|\mathbf{x}|}}|\psi(\mathbf{x})|\right] d\mathbf{x}$$

$$\leq k_\epsilon' \left(\int_{\Omega_\epsilon} |\phi_m(\mathbf{x})|^2 \, d\mathbf{x}\right)^{\frac{1}{2}} \left(\int_{\mathbb{R}^n} \frac{1}{|\mathbf{x}|}|\psi(\mathbf{x})|^2 \, d\mathbf{x}\right)^{\frac{1}{2}}$$

$$\leq k_\epsilon'' \left(\int_{\Omega_\epsilon} |\phi_m(\mathbf{x})|^2 \, d\mathbf{x}\right)^{\frac{1}{2}} \|\psi\|_{D_n^{1/2}},$$

and

$$I_2 \leq \|a_2\|_{L^n(\mathbb{R}^n)} \|\phi_m\|_{L^{\frac{2n}{n-1}}(\mathbb{R}^n)} \|\psi\|_{L^{\frac{2n}{n-1}}(\mathbb{R}^n)}$$

$$\leq \epsilon \gamma_n^2 \|\phi_m\|_{D_n^{1/2}} \|\psi\|_{D_n^{1/2}} \leq \epsilon \gamma_n^2 k \|\psi\|_{D_n^{1/2}}.$$

Hence,

$$\|[\boldsymbol{\sigma} \cdot \mathbf{A}]\phi_m\|_{D_n^{-1/2}} \leq k_\epsilon'' \|\phi_m\|_{L^2(\Omega_\epsilon)} + \epsilon \gamma_n^2 k.$$

Since $D_n^{1/2} \hookrightarrow L^2(\Omega_\epsilon)$ is compact, we have

$$\limsup_{m \to \infty} \|(\boldsymbol{\sigma} \cdot \mathbf{A})\phi_m\|_{D_n^{-1/2}} \leq \epsilon \gamma_n^2 k,$$

whence the compactness of (4.3.8) as ϵ is arbitrary. The proof of (4.3.9) follows from that for I_2 above. □

Lemma 4.3.2. *Suppose that the following conditions are satisfied:*

(i) when $n \geq 3$, $|\mathbf{A}| \in L^n(\mathbb{R}^n)$,
(ii) when $n = 2$, $|\mathbf{A}| \in L^r(\mathbb{R}^2)$ for some $r > 2$.

Then, given any $\epsilon > 0$ there exists $k_\epsilon > 0$ such that for all $\phi \in D_n^1 \cap \mathcal{H}_n$

$$\|(\boldsymbol{\sigma} \cdot \mathbf{A})\phi\|^2 \leq \epsilon^2 \|\phi\|_{D_n^1}^2 + k_\epsilon \|\phi\|^2. \tag{4.3.10}$$

Proof. Choose $|\mathbf{A}| = a_1 + a_2$, where $|a_1(\mathbf{x})| \leq k_\epsilon$ and $\|a_2\|_{L^n(\mathbb{R}^n)} < \epsilon$ in (i) and $\|a_2\|_{L^r(\mathbb{R}^2)} < \epsilon$ in (ii). Then

$$\|(\boldsymbol{\sigma} \cdot \mathbf{A})\phi\| \leq \|a_1\phi\| + \|a_2\phi\| \leq k_\epsilon\|\phi\| + \epsilon\|\phi\|_{L^t(\mathbb{R}^n)}$$

where $t = 2n/(n-2)$ when $n \geq 3$ and $t = \frac{2r}{r-2} \in (2,\infty)$ when $n = 2$. The lemma follows from the embeddings (4.3.5) and (4.3.6). □

Lemma 4.3.2 enables us to define the operator $\mathbb{D}_A$ as the operator sum of $\boldsymbol{\sigma} \cdot \mathbf{p}$ and $\boldsymbol{\sigma} \cdot \mathbf{A}$. Its domain $\mathcal{D}(\mathbb{D}_A)$ is given by

$$\mathcal{D}(\mathbb{D}_A) = \{\phi : \phi, \ \boldsymbol{\sigma} \cdot (\mathbf{p} + \mathbf{A})\phi \in \mathcal{H}_n\}$$
$$= \mathcal{D}(\mathbb{D}_0) = D_n^1 \cap \mathcal{H}_n. \tag{4.3.11}$$

Theorem 4.3.3. *Suppose that*

$$|\mathbf{A}| \in \begin{cases} L^n(\mathbb{R}^n), & \text{when } n \geq 3, \\ L^2(\mathbb{R}^2) \cap L^r(\mathbb{R}^2), & \text{for some } r > 2, \text{ when } n = 2. \end{cases} \tag{4.3.12}$$

Then, $\mathbb{D}_A$ is the operator sum $\boldsymbol{\sigma} \cdot \mathbf{p} + \boldsymbol{\sigma} \cdot \mathbf{A}$ on the domain (4.3.11) and nul $\mathbb{D}_{tA} = 0$ except for at most a countable number of values of t in $[0, \infty)$.

Proof. The fact about $\mathbb{D}_A$ being the operator sum follows from Lemma 4.3.2. Suppose that

$$\mathbb{D}_A\phi = 0, \quad \phi \in \mathcal{D}(\mathbb{D}_A) = D_n^1 \cap \mathcal{H}_n. \tag{4.3.13}$$

Then, $\psi = \sqrt{|\mathbf{p}|}\phi \in D_n^{1/2} \cap \mathcal{H}_n$, and

$$\|\psi\|^2 = (|\mathbf{p}|\phi, \phi) \leq \frac{1}{2}(\|\mathbf{p}\phi\|^2 + \|\phi\|^2).$$

Hence,

$$(E - K)\psi = 0 \tag{4.3.14}$$

where

$$E = \frac{1}{\sqrt{|\mathbf{p}|}} (\boldsymbol{\sigma} \cdot \mathbf{p}) \frac{1}{\sqrt{|\mathbf{p}|}} : \mathcal{H}_n \longrightarrow \mathcal{H}_n \tag{4.3.15}$$

and

$$K = -\frac{1}{\sqrt{|\mathbf{p}|}} (\boldsymbol{\sigma} \cdot \mathbf{A}) \frac{1}{\sqrt{|\mathbf{p}|}} : \mathcal{H}_n \longrightarrow \mathcal{H}_n. \tag{4.3.16}$$

Furthermore, $E^2 = I$, the identity on $\mathcal{H}_n$, $E^* = E$ and, by Lemma 4.3.1, K is compact. Thus, from (4.3.14)

$$(I - EK)\psi = 0. \tag{4.3.17}$$

Since $\psi \in D_n^{1/2} \cap \mathcal{H}_n$, it follows that

$$\text{nul } \mathbb{D}_A \leq \text{nul } (I - EK) \tag{4.3.18}$$

and, hence,

$$\text{nul } \mathbb{D}_{tA} \leq \text{nul } (I - tEK).$$

The theorem therefore follows since the compact operator EK can have at most a countable number of non-zero eigenvalues. $\square$

Theorem 4.3.4. *Let (4.3.12) be satisfied. Then the set*

$$\{\mathbf{A} \; : \; \mathrm{nul}\, \mathbb{D}_{\mathbf{A}} = 0 \;\; and \;\; |\mathbf{A}| \in L^n(\mathbb{R}^n)\}$$

contains an open dense subset of $L^n(\mathbb{R}^n, \mathbb{C}^n)$.

Proof. We shall prove that with $F_A = I - EK_A$, $K_A = -\dfrac{1}{\sqrt{|\mathbf{P}|}} (\sigma \cdot \mathbf{A}) \dfrac{1}{\sqrt{|\mathbf{P}|}}$, the set

$$\{\mathbf{A} \; : \; \mathrm{nul}\, \mathrm{F}_A = 0 \;\; and \;\; |\mathbf{A}| \in L^n(\mathbb{R}^n)\} \tag{4.3.19}$$

is open in $L^n(\mathbb{R}^n, \mathbb{C}^n)$. The theorem will follow from (4.3.18) since the density of (4.3.19) is implied by Theorem 4.3.3.

Given $\epsilon > 0$, suppose $\|\mathbf{A} - \mathbf{A}_0\|_{L^n(\mathbb{R}^n)} < \epsilon$. Then

$$\|K_A - K_{A_0}\| \leq \gamma_n^2 \epsilon$$

by Lemma 4.3.1. Let nul $\mathrm{F}_{A_0} = 0$. Since EK_{A_0} is compact, this means that 1 is not an eigenvalue and there exists $\delta > 0$ such that

$$\inf_{\|f\|=1} \| (I - EK_{A_0})f \| \geq \delta.$$

Hence, if $\epsilon < \frac{\delta}{2\gamma_n^2}$

$$\inf_{\|f\|=1} \| (I - EK_A)f \| \geq \frac{\delta}{2}$$

and hence nul $\mathrm{F}_A = 0$. The theorem therefore follows. $\qquad\square$

Let $\mathcal{L} = EK$. Since $\mathcal{L}^*\mathcal{L}$ is a non-negative, self-adjoint and compact operator on $\mathcal{H}_n$, it follows from [Edmunds and Evans (1987)], Theorem II.5.6 that the number $\sharp$ of its eigenvalues $\lambda_j(\mathcal{L}^*\mathcal{L})$ that are greater than or equal to 1 satisfy

$$\sharp\{\lambda_j(\mathcal{L}^*\mathcal{L}) \geq 1\} = \max_{R \in \mathcal{K}} (\dim R),$$

where $\mathcal{K}$ is the set of all closed linear subspaces R of $\mathcal{H}_n$ which are such that

$$\|x\|^2 \leq (\mathcal{L}^*\mathcal{L}x, x), \quad \text{for all} \;\; x \in R.$$

If $\psi \in \mathcal{N}(I - \mathcal{L})$, then $\|\psi\|^2 = \|\mathcal{L}\psi\|^2 = (\mathcal{L}^*\mathcal{L}\psi, \psi)$ and so

$$\sharp\{\lambda_j(\mathcal{L}^*\mathcal{L}) \geq 1\} \geq \mathrm{nul}\,(I - \mathcal{L}).$$

Hence,

$$\sharp\{\lambda_j(|\mathcal{L}|) \geq 1\} \geq \mathrm{nul}\,(I - \mathcal{L}), \tag{4.3.20}$$

where $|\mathcal{L}|$ is the absolute value of $\mathcal{L}$.

Lemma 4.3.5. *Let (4.3.12) be satisfied. Then*

$$\mathrm{nul}\,(I - \mathcal{L}) \leq \sharp\{\lambda_j(|\mathbf{p}| - |\mathbf{A}|) \leq 0\}. \tag{4.3.21}$$

Proof.   We have

$$\mathcal{L}^*\mathcal{L} = K^*K = \left(\frac{1}{\sqrt{|\mathbf{p}|}} \, (\boldsymbol{\sigma} \cdot \mathbf{A}) \, \frac{1}{\sqrt{|\mathbf{p}|}} \right)^2.$$

The symmetric matrix $\boldsymbol{\sigma} \cdot \mathbf{A}$ has eigenvalues $\pm|\mathbf{A}|$ and is diagonalised by a unitary matrix P, thus $\boldsymbol{\sigma} \cdot \mathbf{A} = P^*\text{diag}(|\mathbf{A}|, -|\mathbf{A}|)$ P. It follows that

$$\frac{1}{\sqrt{|\mathbf{p}|}}(\boldsymbol{\sigma} \cdot \mathbf{A})\frac{1}{\sqrt{|\mathbf{p}|}} = U^* \left(\frac{1}{\sqrt{|\mathbf{p}|}} \, \text{diag} \, (|\mathbf{A}|, -|\mathbf{A}|) \, \frac{1}{\sqrt{|\mathbf{p}|}} \right) U$$

where $U = \sqrt{|\mathbf{p}|} \, P\frac{1}{\sqrt{|\mathbf{p}|}}$. Hence, $U^*U = I$ and the extension of U to $\mathcal{H}_n$ is unitary. The singular values of $\mathcal{L}$ therefore coincide with the absolute values of the eigenvalues of $\frac{1}{\sqrt{|\mathbf{p}|}} \, \text{diag} \, (|\mathbf{A}|, -|\mathbf{A}|) \, \frac{1}{\sqrt{|\mathbf{p}|}}$, and, hence, the eigenvalues of the compact, self-adjoint operator

$$L = \frac{1}{\sqrt{|\mathbf{p}|}} \, |\mathbf{A}| \, \frac{1}{\sqrt{|\mathbf{p}|}}$$

in $L^2(\mathbb{R}^n)$. This and (4.3.20) give

$$\text{nul} \, (I - \mathcal{L}) \leq \sharp\{\lambda_j(L) \geq 1\}.$$

Let R denote the linear span of the eigenvectors of L corresponding to eigenvalues $\lambda_j(L) \geq 1$. Then, for all $x \in R$,

$$([\frac{1}{\sqrt{|\mathbf{p}|}}|\mathbf{A}|\frac{1}{\sqrt{|\mathbf{p}|}}]x, x) = (Lx, x) \geq \|x\|^2$$

and, hence, for all $y \in R_1 := (1/\sqrt{|\mathbf{p}|}) \, R$,

$$(|\mathbf{A}|y, y) \geq \|\sqrt{|\mathbf{p}|}y\|^2$$

and

$$([|\mathbf{p}| - |\mathbf{A}|]y, y) \leq 0.$$

Since $1/\sqrt{|\mathbf{p}|}$ is unitary, R_1 and R have the same dimension, and the lemma is proved. $\square$

Lemma 4.3.5 and Daubechies' inequality (1.8.8) give

Corollary 4.3.6. *Let (4.3.12) be satisfied. Then*

$$\text{nul} \, \mathbb{D}_\mathbf{A} \leq c_n \int_{\mathbb{R}^n} |\mathbf{A}(\mathbf{x})|^n d\mathbf{x}, \tag{4.3.22}$$

where c_n is the best constant in Daubechies' inequality (1.8.8).

Remark 4.3.7. In [Frölich *et al.* (1986)], Theorem A1 of Appendix A, it is proved that for $|\mathbf{B}| \in L^{3/2}(\mathbb{R}^3)$,

$$\mathbf{A}(\mathbf{x}) := \frac{1}{4\pi} \int_{\mathbb{R}^3} \frac{(\mathbf{x} - \mathbf{y})}{|\mathbf{x} - \mathbf{y}|^3} \times \mathbf{B}(\mathbf{y}) d\mathbf{y} \tag{4.3.23}$$

defines the unique magnetic potential which is such that $|\mathbf{A}| \in L^3(\mathbb{R}^3)$, curl $\mathbf{A} = \mathbf{B}$, div $\mathbf{A} = 0$ in $\mathcal{D}'$, the space of distributions on $\mathbb{R}^3$, and

$$\|\mathbf{A}\|_{L^3(\mathbb{R}^3)} \le C\|\mathbf{B}\|_{L^{3/2}(\mathbb{R}^3)} \tag{4.3.24}$$

for some constant C. Hence, from (4.3.23),

$$\text{nul } \mathbb{D}_\mathbf{A} = \text{nul } \mathbb{P}_\mathbf{A} \le \text{const} \int_{\mathbb{R}^3} |\mathbf{B}|^{3/2} d\mathbf{x}. \tag{4.3.25}$$

However, this is significantly worse than (4.3.22), since $\mathbf{B}$ may be the curl of a highly oscillatory $\mathbf{A}$ for which nul $\mathbb{D}_\mathbf{A}$ is much smaller than $\|\mathbf{B}\|_{L^{3/2}(\mathbb{R}^3)}$. A natural question is: given a magnetic field $\mathbf{B}$, can one determine a gauge $\mathbf{A}$ which minimises $\|\mathbf{A}\|_{L^3(\mathbb{R}^3)}$? We require, for all smooth functions f,

$$\frac{d}{d\lambda} \int_{\mathbb{R}^3} |\mathbf{A} + \lambda \nabla f|^3 \, d\mathbf{x} = 0 \quad \text{at} \quad \lambda = 0.$$

This yields

$$\int_{\mathbb{R}^3} \text{div}(|\mathbf{A}|\mathbf{A}) \, f \, d\mathbf{x} = 0,$$

and, hence, $\text{div}(|\mathbf{A}|\mathbf{A}) = 0$.

Remark 4.3.8. Results similar in spirit to those in Theorem 4.3.4 and Corollary 4.3.6 were also obtained by Elton in [Elton (2002)]. He investigates the multiplicity of the zero modes of $\mathbb{D}_\mathbf{A}$ within the class $\mathcal{A}$ of magnetic potentials which are continuous on $\mathbb{R}^3$ and decay like $o(|\mathbf{x}|^{-1})$ at ∞. $\mathcal{A}$ is a Banach space with respect to the norm

$$\|\mathbf{A}\|_\mathcal{A} := \|(1 + |\mathbf{x}|)\mathbf{A}\|_{L^\infty(\mathbb{R}^3, \mathbb{R}^3)}$$

and contains $C_0^\infty(\mathbb{R}^3, \mathbb{R}^3)$ as a dense subset. Denoting by $\mathcal{Z}_m$ the subset of $\mathcal{A}$ consisting of those $\mathbf{A}$ for which $\mathbb{D}_\mathbf{A}$ has a zero mode of multiplicity m, i.e. nul $\mathbb{D}_\mathbf{A} = m$, the following is Theorem 1 in [Elton (2002)]:

Theorem 4.3.9. (*i*) $\mathcal{A} = \cup_{m \ge 0} \mathcal{Z}_m$.
(*ii*) $\mathcal{Z}_0$ is a dense subset of $\mathcal{A}$.
(*iii*) For any $m \in \mathbb{N}_0$ and non-empty set $\Omega \subseteq \mathbb{R}^3$, $C_0^\infty(\Omega, \mathbb{R}^3) \cap \mathcal{Z}_m \neq \emptyset$.

In [Saitō and Umeda (2009)], Saitō and Umeda determine connections between zero modes of $\mathbb{D}_\mathbf{A}$ and eigenvectors corresponding to the threshold eigenvalues ± 1 of the magnetic free Dirac operator considered in the supersymmetric form

$$\mathbb{D}_0(\mathbf{A}) := \boldsymbol{\alpha} \cdot (\mathbf{p} + \mathbf{A}) + \beta = \begin{pmatrix} 1 & \mathbb{D}_\mathbf{A} \\ \mathbb{D}_\mathbf{A} & -1 \end{pmatrix}. \tag{4.3.26}$$

It follows from their Theorems 2.1 and 2.2 that $\mathbb{D}_\mathbf{A}$ has a zero mode if and only if 1 and -1 are eigenvalues of $\mathbb{D}_0(\mathbf{A})$. If ϕ is a zero mode of $\mathbb{D}_\mathbf{A}$ then the eigenvectors of $\mathbb{D}_0(\mathbf{A})$ at $\pm$ are given respectively by

$$\begin{pmatrix} \phi \\ 0 \end{pmatrix}, \quad \begin{pmatrix} 0 \\ \phi \end{pmatrix}.$$

This is also a consequence of Theorem 7.1 in [Thaller (1992)]; see the discussions following Definition 1.1 and Corollary 2.1 in [Saitō and Umeda (2009)]. These results also combine with Theorem 1.2 in [Saitō and Umeda (2008a)] to give [Saitō and Umeda (2008a)], Theorem 4.1, that if, for some positive constant C,

$$|\mathbf{A}| \leq C \langle \mathbf{x} \rangle^{-\rho} := C \{1 + |\mathbf{x}|^2\}^{-\rho/2}, \quad (\rho > 1),$$

1 is an eigenvalue of $\mathbb{D}_0(\mathbf{A})$ and f^+ is a corresponding eigenvector, then there exists a zero mode ϕ of $\mathbb{D}_\mathbf{A}$ such that for any $\boldsymbol{\omega} \in \mathbb{S}^2$,

$$\lim_{r \to \infty} r^2 f^+(r\boldsymbol{\omega}) = \begin{pmatrix} u(\boldsymbol{\omega}) \\ 0 \end{pmatrix},$$

where

$$u(\boldsymbol{\omega}) = \frac{i}{4\pi} \int_{\mathbb{R}^3} \{(\boldsymbol{\omega} \cdot \mathbf{A}(\mathbf{y})) I_2 + i\boldsymbol{\sigma} \cdot (\boldsymbol{\omega} \times \mathbf{A}(\mathbf{y}))\} \, \phi(\mathbf{y}) d\mathbf{y}$$

and the convergence is uniform with respect to $\boldsymbol{\omega}$. If -1 is an eigenvalue with eigenvector f_-, then

$$\lim_{r \to \infty} r^2 f^-(r\boldsymbol{\omega}) = \begin{pmatrix} 0 \\ u(\boldsymbol{\omega}) \end{pmatrix}.$$

The expression for $f(r\boldsymbol{\omega})$ makes sense for each $\boldsymbol{\omega}$ since an eigenfunction of $\mathbb{D}_\mathbf{A}$ corresponding to an eigenvalue at 1 or -1 is known to be continuous under the assumption on $\mathbf{A}$, from [Saitō and Umeda (2009)], Theorem 2.1.

4.4 Decay rates of weak solutions of $\mathbb{D}_Q \psi = 0$

Here

$$\mathbb{D}_Q := \boldsymbol{\alpha} \cdot \mathbf{p} + Q, \quad \mathbf{p} = -i\boldsymbol{\nabla}, \tag{4.4.1}$$

where Q is a 4×4 matrix-valued function whose components $q_{j,k}, j, k = 1, 2, 3, 4$ satisfy

$$|q_{j,k}(\mathbf{x})| \leq C|\mathbf{x}|^{-1}, \quad |\mathbf{x}| \geq 1. \tag{4.4.2}$$

Hence, solutions of $\mathbb{D}_A \psi = 0$ will be covered, and note that the condition on Q is only assumed outside the unit ball.

Let ψ be a weak solution of $\mathbb{D}_Q \psi = 0$ that lies in $L^2(\mathbb{R}^3 \setminus \overline{B}_1, \mathbb{C}^4)$, where B_1 is the open unit ball in $\mathbb{R}^3$ with centre at the origin and $\overline{B}_1$ is its closure. Hence, for all $\phi \in C_0^\infty(\mathbb{R}^3 \setminus \overline{B}_1, \mathbb{C}^4)$,

$$I := \int \langle \mathbb{D}_Q \psi(\mathbf{x}), \phi(\mathbf{x}) \rangle d\mathbf{x} = 0, \quad \psi \in L^2(\mathbb{R}^3 \setminus \overline{B}_1, \mathbb{C}^4). \qquad (4.4.3)$$

Our objective in this section is to shed some light on the behaviour of ψ at infinity. This will be measured in terms of belonging to some L^p space at infinity, determined by means of embedding theorems for Dirac–Sobolev spaces described in Section 4.4.2 below.

4.4.1 *Reduction by inversion*

We follow the treatment in [Balinsky *et al.* (2008)] and begin by carrying out an inversion with respect to B_1. This is an involution $\mathrm{Inv} : \mathbf{x} \mapsto \mathbf{y}$, $\mathbf{y} = \mathbf{x}/|\mathbf{x}|^2$. For any function defined on $\mathbb{R}^3 \setminus B_1$, the map $M : \phi \mapsto \tilde{\phi} := \phi \, o \, \mathrm{Inv}^{-1}$ is such that $\tilde{\phi}(\mathbf{y}) = \phi(\mathbf{x})$ and yields a function on $B_1 : \phi \in C_0^\infty(\mathbb{R}^3 \setminus \overline{B}_1, \mathbb{C}^4)$ means that $\tilde{\phi} \in C_0^\infty(B_1 \setminus \{0\}, \mathbb{C}^4)$. The inversion gives

$$M\{(\boldsymbol{\alpha} \cdot \mathbf{p})\psi\}(\mathbf{y}) = |\mathbf{y}|^2 (\boldsymbol{\beta} \cdot \mathbf{p})\tilde{\psi}(\mathbf{y}), \qquad (4.4.4)$$

where $\boldsymbol{\beta} = (\beta_1, \beta_2, \beta_3)$ and

$$\beta_k(\mathbf{y}) = \sum_{j=1}^3 \alpha_j \left(\delta_{jk} - \frac{2 y_k y_j}{|\mathbf{y}|^2} \right),$$

where δ_{jk} is the Kronecker delta function. It is readily verified that the matrices $\beta_k(\mathbf{y})$ are Hermitian and satisfy

$$\beta_k(\mathbf{y})\beta_j(\mathbf{y}) + \beta_j(\mathbf{y})\beta_k(\mathbf{y}) = 2\delta_{jk} I_4.$$

Also, there exists a unitary matrix $X(\mathbf{y})$ such that $X \in C^\infty(\mathbb{R}^3 \setminus \{0\})$ and, for all $\mathbf{y} \neq 0$,

$$X(\mathbf{y})^{-1} \beta_k(\mathbf{y}) X(\mathbf{y}) = -\alpha_k, \quad k = 1, 2, 3. \qquad (4.4.5)$$

Setting $\boldsymbol{\omega} := \mathbf{y}/|\mathbf{y}|$, it is easy to verify that these conditions are satisfied by

$$X(\mathbf{y}) = \begin{pmatrix} X_2(\mathbf{y}) & O_2 \\ O_2 & X_2(\mathbf{y}) \end{pmatrix}, \qquad (4.4.6)$$

where

$$X_2(\mathbf{y}) = \begin{pmatrix} i\omega_3 & \omega_2 + i\omega_1 \\ -\omega_2 + i\omega_1 & -i\omega_3 \end{pmatrix},$$

and O_2 is the zero 2×2 matrix. Let $\tilde{\psi}(\mathbf{y}) = -X(\mathbf{y})\Psi(\mathbf{y})$. Then, from (4.4.4), we have

$$M\{(\boldsymbol{\alpha} \cdot \mathbf{p})\psi\}(\mathbf{y}) = |\mathbf{y}|^2 X(\mathbf{y})\{(\boldsymbol{\alpha} \cdot \mathbf{p})\Psi(\mathbf{y}) + Y(\mathbf{y})\Psi(\mathbf{y})\}, \qquad (4.4.7)$$

where

$$Y(\mathbf{y}) = \sum_{j=1}^{3} \alpha_k X(\mathbf{y})^{-1} \left(-i\frac{\partial}{\partial y_k} X(\mathbf{y}) \right). \qquad (4.4.8)$$

Also, a calculation of the Jacobian of the inversion gives $d\mathbf{x} = |\mathbf{y}|^{-6} d\mathbf{y}$. On applying the inversion to (4.4.3) and putting $\tilde{\phi}(\mathbf{y}) = -X(\mathbf{y})\Phi(\mathbf{y})$, we derive

$$I =$$

$$\int_{B_1} \langle |\mathbf{y}|^2 X(\mathbf{y})\{(\boldsymbol{\alpha} \cdot \mathbf{p})\Psi(\mathbf{y}) + Y(\mathbf{y})\Psi(\mathbf{y}) - \tilde{Q}(\mathbf{y})X(\mathbf{y})\Psi(\mathbf{y})\}, X(\mathbf{y})\Phi(\mathbf{y}) \rangle |\mathbf{y}|^{-6} d\mathbf{y}$$

$$= 0$$

for all $\Phi \in C_0^\infty(B_1 \setminus \{0\}, \mathbb{C}^4)$, which can be written as

$$I = \int_{B_1} \langle (\boldsymbol{\alpha} \cdot \mathbf{p})\Psi(\mathbf{y}) + Z(\mathbf{y})\Psi(\mathbf{y}), |\mathbf{y}|^{-4}\Phi(\mathbf{y}) \rangle d\mathbf{y} = 0, \qquad (4.4.9)$$

where

$$Z(\mathbf{y}) = Y(\mathbf{y}) - |\mathbf{y}|^{-2} X(\mathbf{y})^{-1} \tilde{Q}(\mathbf{y})X(\mathbf{y}). \qquad (4.4.10)$$

Equivalently, the factor $|\mathbf{y}|^{-4}$ can by removed from (4.4.9) to obtain

$$I = \int_{B_1} \langle (\boldsymbol{\alpha} \cdot \mathbf{p})\Psi(\mathbf{y}) + Z(\mathbf{y})\Psi(\mathbf{y}), \Phi(\mathbf{y}) \rangle d\mathbf{y} = 0 \qquad (4.4.11)$$

for all $\Phi \in C_0^\infty(B_1 \setminus \{0\}, \mathbb{C}^4)$.

Let $\zeta \in C^\infty(\mathbb{R}^+)$ satisfy

$$\zeta(t) = \begin{cases} 0, & \text{for } 0 < t < 1 \\ 1, & \text{for } t > 2, \end{cases}$$

and for $\mathbf{y} \in \mathbb{R}^3$, set $\zeta_n(\mathbf{y}) = \zeta(n|\mathbf{y}|)$. Then

$$\nabla \zeta_n(\mathbf{y}) = n\zeta'(n|\mathbf{y}|)\frac{\mathbf{y}}{|\mathbf{y}|}$$

and so

$$|\nabla \zeta_n(\mathbf{y})| = O\left(n\chi_{[1/n, 2/n]}(|\mathbf{y}|) \right),$$

where χ_I denotes the characteristic function of the interval I. It then follows from (4.4.11), and now for all $\Phi \in C_0^\infty(B_1, \mathbb{C}^4)$, that

$$I = \int_{B_1} \langle (\boldsymbol{\alpha} \cdot \mathbf{p})\Psi(\mathbf{y}) + Z(\mathbf{y})\Psi(\mathbf{y}), \zeta_n(\mathbf{y})\Phi(\mathbf{y}) \rangle d\mathbf{y}$$

$$= \int_{B_1} \langle \zeta_n(\mathbf{y})\{(\boldsymbol{\alpha} \cdot \mathbf{p})\Psi(\mathbf{y}) + Z(\mathbf{y})\Psi(\mathbf{y})\}, \Phi(\mathbf{y}) \rangle d\mathbf{y} = 0. \qquad (4.4.12)$$

In I,
$$\zeta_n(\mathbf{y})(\boldsymbol{\alpha}\cdot\mathbf{p})\Psi(\mathbf{y}) = (\boldsymbol{\alpha}\cdot\mathbf{p})(\zeta_n\Psi)(\mathbf{y}) - [(\boldsymbol{\alpha}\cdot\mathbf{p})\zeta_n]\Psi(\mathbf{y})$$
and
$$V_n(\mathbf{y}) := [(\boldsymbol{\alpha}\cdot\mathbf{p})\zeta_n] = O(n\chi_{[1/n,2/n]}).$$
Therefore, as $n \to \infty$,
$$\left|\int_{B_1}\langle V_n(\mathbf{y})\Psi(\mathbf{y}),\Phi(\mathbf{y})\rangle d\mathbf{y}\right| = O\left(n\int_{B_1}|\Psi(\mathbf{y})|\chi_{[1/n,2/n]}(\mathbf{y})d\mathbf{y}\right)$$
$$= O(n^{-1/2}\|\Psi\|_{L^2(B_1)}\|) \to 0.$$
Also,
$$\int_{B_1}\langle[(\boldsymbol{\alpha}\cdot\mathbf{p})+Z(\mathbf{y})]\zeta_n(\mathbf{y})\Psi(\mathbf{y}),\Phi(\mathbf{y})\rangle d\mathbf{y} = \int_{B_1}\langle\zeta_n(\mathbf{y})\Psi(\mathbf{y}),[(\boldsymbol{\alpha}\cdot\mathbf{p})+Z(\mathbf{y})]\Phi(\mathbf{y})\rangle d\mathbf{y}$$
$$\to \int_{B_1}\langle\Psi(\mathbf{y}),[(\boldsymbol{\alpha}\cdot\mathbf{p})+Z(\mathbf{y})]\Phi(\mathbf{y})\rangle d\mathbf{y}.$$
We have therefore proved that
$$\int_{B_1}\langle[(\boldsymbol{\alpha}\cdot\mathbf{p})+Z(\mathbf{y})]\Psi(\mathbf{y}),\Phi(\mathbf{y})\rangle d\mathbf{y} = 0 \tag{4.4.13}$$
for all $\Phi \in C_0^\infty(B_1,\mathbb{C}^4)$. In other words,
$$[(\boldsymbol{\alpha}\cdot\mathbf{p})+Z(\mathbf{y})]\Psi(\mathbf{y}) = 0 \tag{4.4.14}$$
in the weak sense. From (4.4.10), it follows that
$$\|Z(\mathbf{y})\|_{\mathbb{C}^4} \leq C|\mathbf{y}|^{-1}, \tag{4.4.15}$$
for some positive constant C.

4.4.2   *Embeddings of Dirac–Sobolev spaces*

Definition 4.4.1. Let Ω be an open subset of $\mathbb{R}^3$ and $1 \leq p < \infty$. The *Dirac–Sobolev* space $H_D^{1,p}(\Omega,\mathbb{C}^4)$ is defined as
$$H_D^{1,p}(\Omega,\mathbb{C}^4) := \{f : f,(\boldsymbol{\alpha}\cdot\mathbf{p})f \in L^p(\Omega,\mathbb{C}^4)\} \tag{4.4.16}$$
with norm
$$\|f\|_{D,1,p,\Omega} := \left\{\int_\Omega(|f(\mathbf{x})|_p^p + |(\boldsymbol{\alpha}\cdot\mathbf{p})f(\mathbf{x})|_p^p)d\mathbf{x}\right\}^{1/p}, \tag{4.4.17}$$
where $|\cdot|_p$ is the ℓ^p norm in $\mathbb{C}^4$, so that,
$$|f(\mathbf{x})|_p^p := \sum_{j=1}^4|f_j(\mathbf{x})|^p, \quad |(\boldsymbol{\alpha}\cdot\mathbf{p})f(\mathbf{x})|_p^p := \left|\sum_{j=1}^3 -i\alpha_j\partial_j f(\mathbf{x})\right|_p^p.$$
We denote the completion of $C_0^\infty(\Omega,\mathbb{C}^4)$ in $H_D^{1,p}(\Omega,\mathbb{C}^4)$ by $H_{D,0}^{1,p}(\Omega,\mathbb{C}^4)$ and write $\|\cdot\|_{D,1,p}$ for the $H_D^{1,p}(\mathbb{R}^3,\mathbb{C}^4)$ norm. As in the case of Sobolev spaces (see Section 1.6), $C^\infty(\Omega,\mathbb{C}^4)\cap H_D^{1,p}(\Omega,\mathbb{C}^4)$ is dense in $H_D^{1,p}(\Omega,\mathbb{C}^4)$ and when $\Omega = \mathbb{R}^3$, $H_D^{1,p}(\mathbb{R}^3,\mathbb{C}^4) = H_{D,0}^{1,p}(\mathbb{R}^3,\mathbb{C}^4)$.

The standard Sobolev space $H^{1,p}(\Omega, \mathbb{C}^4)$ is the space of $\mathbb{C}^4$-valued functions $f \in L^p(\Omega, \mathbb{C}^4)$ whose components have weak derivatives in $L^p(\Omega, \mathbb{C}^4)$, endowed with the norm

$$\|f\|_{1,p,\Omega} := \left\{ \int_\Omega (|f(\mathbf{x})|_p^p + \sum_{j=1}^3 |p_j f(\mathbf{x})|_p^p) d\mathbf{x} \right\}^{1/p}, \qquad (4.4.18)$$

where $p_j = -i\partial_j$. The closure of $C_0^\infty(\Omega, \mathbb{C}^4)$ in $H^{1,p}(\Omega, \mathbb{C}^4)$ is denoted by $H_0^{1,p}(\Omega, \mathbb{C}^4)$. Recall that we have adopted the standard notation $H^1(\Omega, \mathbb{C}^4)$ and $H_0^1(\Omega, \mathbb{C}^4)$ for $H^{1,2}(\Omega, \mathbb{C}^4)$ and $H_0^{1,2}(\Omega, \mathbb{C}^4)$, respectively.

In the case $p = 2$, on using (2.1.2), we have that for $f \in C_0^\infty(\Omega, \mathbb{C}^4)$,

$$\int_\Omega |(\boldsymbol{\alpha} \cdot \mathbf{p}) f(\mathbf{x})|_2^2 d\mathbf{x} = \int_\Omega \left\{ \sum_{j<k} (\langle \alpha_k \alpha_j f, p_j p_k f \rangle + \langle \alpha_j \alpha_k f, p_k p_j f \rangle) + \sum_{j=1}^3 |p_j f|_2^2 \right\} d\mathbf{x}$$

$$= \int_\Omega \sum_{j=1}^3 |p_j f(\mathbf{x})|_2^2 d\mathbf{x}.$$

It follows that $H_{D,0}^{1,2}(\Omega, \mathbb{C}^4) = H_0^{1,2}(\Omega, \mathbb{C}^4)$.

For general values of $p \in [1, \infty)$, we have

$$|(\boldsymbol{\alpha} \cdot \mathbf{p}) f|_p^p = | \sum_{j=1}^3 \alpha_j p_j f |_p^p$$

$$\leq \left(\sum_{j=1}^3 |\alpha_j p_j f|_p \right)^p$$

$$= \left(\sum_{j=1}^3 |p_j f|_p \right)^p$$

$$\leq 3^{p-1} \left(\sum_{j=1}^3 \left[\sum_{k=1}^4 |p_j f_k|^p \right] \right) = 3^{p-1} \sum_{j=1}^3 |p_j f|_p^p$$

and so

$$\|f\|_{D,1,p,\Omega} \leq 3^{1-1/p} \|f\|_{1,p,\Omega}.$$

Therefore, we have the continuous natural embeddings

$$H^{1,p}(\Omega, \mathbb{C}^4) \hookrightarrow H_D^{1,p}(\Omega, \mathbb{C}^4), \quad H_0^{1,p}(\Omega, \mathbb{C}^4) \hookrightarrow H_{D,0}^{1,p}(\Omega, \mathbb{C}^4). \qquad (4.4.19)$$

We have already seen above that when $p = 2$ the Dirac–Sobolev space $H_{D,0}^1(\Omega, \mathbb{C}^4)$ and Sobolev space $H_0^1(\Omega, \mathbb{C}^4)$ coincide. In the general case the following result was established in [Ichinose and Saitō]:

Theorem 4.4.2.

(i) When $1 < p < \infty$, $H_{D,0}^{1,p}(\Omega, \mathbb{C}^4) = H_0^{1,p}(\Omega, \mathbb{C}^4)$.

(ii) When $p = 1$, $H_0^{1,1}(\Omega, \mathbb{C}^4) \hookrightarrow_{\neq} H_{D,0}^{1,1}(\Omega, \mathbb{C}^4)$.

It follows from Theorem 4.4.2(i) and the Sobolev embedding theorem that, for $1 < p < 3$,

$$H_{D,0}^{1,p}(\Omega, \mathbb{C}^4) \hookrightarrow L^{p^*}(\Omega, \mathbb{C}^4), \quad p^* = 3p/(3-p). \tag{4.4.20}$$

In particular, with $\Omega = \mathbb{R}^3$, there is a positive constant C such that for all $f \in H_{D,0}^{1,p}(\mathbb{R}^3, \mathbb{C}^4)$,

$$\|f\|_{p^*} \leq C\|f\|_{D,1,p}$$
$$= C\{\|f\|_p^p + \|(\boldsymbol{\alpha} \cdot \mathbf{p})f\|_p^p\}^{1/p}. \tag{4.4.21}$$

The scale transformation $\mathbf{x} = \lambda \mathbf{y}$, $f(\mathbf{x}) =: g(\mathbf{y})$, yields

$$\|g\|_{p^*} \leq C\{\lambda\|g\|_p + \|(\boldsymbol{\alpha} \cdot \mathbf{p})g\|_p\}.$$

Since λ is arbitrary, we have the following analogue of the Sobolev inequality (1.7.1)

$$\|g\|_{p^*} \leq C\|(\boldsymbol{\alpha} \cdot \mathbf{p})g\|_p, \quad \text{for all } g \in H_{D,0}^{1,p}(\mathbb{R}^3, \mathbb{C}^4). \tag{4.4.22}$$

The result of Theorem 4.4.2 is analysed in [Balinsky *et al.*] where an interesting connection is uncovered between part (ii) of the theorem and zero modes of Weyl–Dirac operators with appropriate magnetic potentials. To be specific, let ψ be the solution (4.2.8) of the Loss–Yau equation (4.2.6) with $\lambda(\mathbf{x}) = 3(1+r^2)^{-1}$. Thus

$$\psi(\mathbf{x}) = \frac{1}{(1+r^2)^{3/2}}(1 + i\mathbf{x} \cdot \sigma)\begin{pmatrix} 1 \\ 0 \end{pmatrix}, \quad r = |\mathbf{x}|$$

and

$$(\boldsymbol{\sigma} \cdot \mathbf{p})\psi(\mathbf{x}) = \frac{3}{(1+r^2)}\psi(\mathbf{x}), \quad \mathbf{p} = -i\boldsymbol{\nabla}.$$

Let $\chi_n \in C_0^\infty(\mathbb{R}_+)$ be such that

$$\chi_n(r) = \begin{cases} 1, & 0 \leq r \leq n, \\ 0, & r \geq n+2, \end{cases}$$

and $|\chi_n'(r)| \leq 1$. Then $\psi_n := \chi_n \psi \in C_0^\infty(\mathbb{R}^3, \mathbb{C}^2)$ and for some constant positive C, independent of n,

$$\|(\boldsymbol{\sigma} \cdot \mathbf{p})\psi_n\|_{L^1(\mathbb{R}^3,\mathbb{C}^2)} = \|\chi_n(\boldsymbol{\sigma} \cdot \mathbf{p})\psi + \chi_n'(\boldsymbol{\sigma} \cdot \frac{\mathbf{x}}{r})\psi\|_{L^1(\mathbb{R}^3,\mathbb{C}^2)}$$
$$\leq C.$$

If the spaces $H_0^{1,1}(\mathbb{R}^3, \mathbb{C}^4)$ and $H_{D,0}^{1,1}(\mathbb{R}^3, \mathbb{C}^4)$ coincide, then the Sobolev embedding theorem would imply that $H_{D,0}^{1,1}(\mathbb{R}^3, \mathbb{C}^4)$ is continuously embedded in $L^{3/2}(\mathbb{R}^3, \mathbb{C}^4)$ and this is equivalent to

$$\|f\|_{L^{3/2}(\mathbb{R}^3,\mathbb{C}^2)} \leq C\|(\boldsymbol{\sigma} \cdot \mathbf{p})f\|_{L^1(\mathbb{R}^3,\mathbb{C}^2)} \tag{4.4.23}$$

for all $f \in C_0^\infty(\mathbb{R}^3, \mathbb{C}^2)$. In particular, this would mean that $\|\psi_n\|_{L^{3/2}(\mathbb{R}^3, \mathbb{C}^2)}$ is bounded independently of n, contrary to

$$\|\psi_n\|_{L^{3/2}(\mathbb{R}^3, \mathbb{C}^2)} \geq \int_{r \leq n} |\psi(\mathbf{x})| d\mathbf{x} \asymp \ln n.$$

Clearly, the properties of the above 2-spinor ψ which lead to the inequality (4.4.23) being contradicted, and hence to the recovery of Theorem 4.4.2(ii), are that $(\boldsymbol{\sigma} \cdot \mathbf{p})\psi \in L^1(\mathbb{R}^3, \mathbb{C}^2)$ and $\psi(\mathbf{x}) \asymp r^{-2}$ at infinity. In fact, it was shown in [Saitō and Umeda (2008b)] that these properties are satisfied by the zero modes of any Weyl–Dirac operator $\mathbb{D}_\mathbf{A}$ whose magnetic potential $\mathbf{A} = (A_1, A_2, A_3)$ is such that

$$A_j \text{ measurable}, \quad |A_j(\mathbf{x})| = O([1 + |\mathbf{x}|^2]^{-\rho/2}), \quad \rho > 1$$

for $j = 1, 2, 3$.

It is natural to ask if (4.4.23) can be modified to give an analogue of the Sobolev inequality in the case $p = 1$ for the standard Sobolev spaces. The answer is given by the following theorem from [Balinsky *et al.*]. In it $L^{q,\infty}$ denotes the *weak-L^q* space and

$$\|f\|_{q,\infty}^q := \sup_{t>0} t^q \mu\{\mathbf{x} \in \mathbb{R}^3 : |f(\mathbf{x})| > t\}, \tag{4.4.24}$$

where μ denotes Lebesgue measure.

Theorem 4.4.3. *For all $f \in C_0^\infty(\mathbb{R}^3, \mathbb{C}^2)$,*

$$\|f\|_{3/2,\infty} \leq C\|(\boldsymbol{\sigma} \cdot \mathbf{p})f\|_{L^1(\mathbb{R}^3, \mathbb{C}^2)} \tag{4.4.25}$$

for some positive constant C independent of f.

Proof. Let $g = (\boldsymbol{\sigma} \cdot \mathbf{p})f$. Then $g \in C_0^\infty(\mathbb{R}^3, \mathbb{C}^2)$ and

$$-\Delta f = (\boldsymbol{\sigma} \cdot \mathbf{p})^2 f = (\boldsymbol{\sigma} \cdot \mathbf{p})g.$$

Thus by (1.5.17),

$$
\begin{aligned}
f(\mathbf{x}) &= -i\mathbb{F}^{-1}\left(\frac{1}{|\cdot|^2}\mathbb{F}[(\boldsymbol{\sigma} \cdot \boldsymbol{\nabla})g]\right)(\mathbf{x}) \\
&= -i\frac{1}{4\pi}\int_{\mathbb{R}^3}\frac{1}{|\mathbf{x} - \mathbf{y}|}(\boldsymbol{\sigma} \cdot \boldsymbol{\nabla}_\mathbf{y})g(\mathbf{y})d\mathbf{y} \\
&= \frac{i}{4\pi}\sum_{j=1}^3 \sigma_j \int_{\mathbb{R}^3}\frac{\partial}{\partial y_j}\left(\frac{1}{|\mathbf{x} - \mathbf{y}|}\right)g(\mathbf{y})d\mathbf{y} \\
&= -\frac{i}{4\pi}\int_{\mathbb{R}^3}\left\{\frac{\boldsymbol{\sigma} \cdot (\mathbf{x} - \mathbf{y})}{|\mathbf{x} - \mathbf{y}|^3}\right\}g(\mathbf{y})d\mathbf{y} =: I_\sigma(g)(\mathbf{x}), \tag{4.4.26}
\end{aligned}
$$

say. Then

$$|f(\mathbf{x})| \leq \frac{1}{4\pi}\int_{\mathbb{R}^3}\frac{1}{|\mathbf{x} - \mathbf{y}|^2}|g(\mathbf{y})|d\mathbf{y} =: \frac{1}{4\pi}I(|g|)(\mathbf{x}),$$

where $I(|g|)$ is the Riesz potential of $|g|$; see [Edmunds and Evans (2004)], Section 3.5 for the terminology and properties we need. Since $|\cdot|^{-2} \in L^{3/2,\infty}(\mathbb{R}^3)$, it follows that I is of *joint weak type* $(1, 3/2; 3, \infty)$, which implies in particular that it is of *weak type* $(1, 3/2)$ and this means that there exists a constant C such that for all $u \in L^1(\mathbb{R}^3)$,

$$\|I(u)\|_{L^{3/2,\infty}(\mathbb{R}^3)} \le C\|u\|_{L^1(\mathbb{R}^3)}. \tag{4.4.27}$$

The inequality (4.4.25) follows. $\square$

It is also easily seen, through the use of a function ψ with the above properties, that the analogue of Hardy's inequality for L^1, namely

$$\int_{\mathbb{R}^3} \frac{|f(\mathbf{x})|}{|\mathbf{x}|} d\mathbf{x} \le C \int_{\mathbb{R}^3} |(\boldsymbol{\sigma} \cdot \mathbf{p})f(\mathbf{x})| d\mathbf{x}, \quad f \in C_0^\infty(\mathbb{R}^3, \mathbb{C}^2),$$

does not hold. What is true is the following result from [Balinsky *et al.*].

Theorem 4.4.4. *For all* $f \in C_0^\infty(\mathbb{R}^3, \mathbb{C}^2)$.

$$\left\| \frac{f}{|\cdot|} \right\|_{L^{1,\infty}(\mathbb{R}^3,\mathbb{C}^2)} \le \|(\boldsymbol{\sigma} \cdot \mathbf{p})f\|_{L^1(\mathbb{R}^3,\mathbb{C}^2)}. \tag{4.4.28}$$

Proof.   The inequality is an immediate consequence of Theorem 4.4.3 and the following weak Hölder type inequality on $\mathbb{R}^3$ established in [Balinsky *et al.*]. Let $u \in L^{p,\infty}, v \in L^{q,\infty}$, where $1 < p < \infty, p^{-1} + q^{-1} = 1$. Then $uv \in L^{1,\infty}$ and

$$\|uv\|_{L^{1,\infty}} \le \left((q/p)^{1/q} + (p/q)^{1/p} \right) \|u\|_{L^{p,\infty}} \|v\|_{L^{q,\infty}}. \tag{4.4.29}$$

To see this, set, for arbitrary $\varepsilon > 0$,

$$A = \{\mathbf{x} : \varepsilon |u(\mathbf{x})| > t^{1/p}\}$$

$$B = \left\{\mathbf{x} : \frac{1}{\varepsilon} |v(\mathbf{x})| > t^{1/q}\right\}$$

$$E = \{\mathbf{x} : |u(\mathbf{x})v(\mathbf{x})| > t\}.$$

Since

$$|u(\mathbf{x})v(\mathbf{x})| \le p^{-1} (\varepsilon |u(\mathbf{x})|)^p + q^{-1} \left(\frac{1}{\varepsilon} |v(\mathbf{x})|\right)^q,$$

it follows that $E \subseteq A \cup B$ and, hence,

$$t\mu(E) \le t\mu\{\mathbf{x} : \varepsilon |u(\mathbf{x})| > t^{1/p}\} + t\mu\{\mathbf{x} : \frac{1}{\varepsilon} |v(\mathbf{x})| > t^{1/q}\}. \tag{4.4.30}$$

On substituting $s := t^{1/p}/\varepsilon$, $r := \varepsilon t^{1/q}$, we have

$$t\mu(E) \le \varepsilon^p s^p \mu\{\mathbf{x} : |u(\mathbf{x})| > s\} + \varepsilon^{-q} r^q \mu\{\mathbf{x} : |v(\mathbf{x})| > r\}$$

$$\le \varepsilon^p \|u\|_{L^{p,\infty}}^p + \varepsilon^{-q} \|v\|_{L^{q,\infty}}^q.$$

The right-hand side has its minimum when $\varepsilon = (q\|v\|_{L^{q,\infty}}^q/p\|u\|_{L^{p,\infty}}^p)^{1/pq}$, which yields (4.4.29). The choice $u = f \in L^{3/2,\infty}(\mathbb{R}^3, \mathbb{C}^2), v = 1/|\cdot| \in L^{3,\infty}(\mathbb{R}^3, \mathbb{C}^2)$ and the application of Theorem 4.4.3 completes the proof of the theorem. $\square$

4.4.3 Decay of zero modes

Theorem 4.4.5. *Let ψ be such that $\psi \in L^2(\mathbb{B}_1^c, \mathbb{C}^4)$, $B_1^c := \mathbb{R}^3 \setminus B_1$, and*

$$(\boldsymbol{\alpha} \cdot \mathbf{p})\psi(\mathbf{x}) = -Q(\mathbf{x})\psi(\mathbf{x}), \tag{4.4.31}$$

where

$$\|Q(\mathbf{x})\|_{\mathbb{C}^4} = O(|\mathbf{x}|^{-1}), \quad |\mathbf{x}| \geq 1,$$

and $\| \cdot \|_{\mathbb{C}^4}$ is any norm on the space of 4×4 matrices. Then

$$\int_{B^c} |\psi(\mathbf{x})|^6 |\mathbf{x}|^6 d\mathbf{x} < \infty. \tag{4.4.32}$$

Proof. If $\theta \in C^1(\mathbb{R}^3)$ is 1 in a neighbourhood of infinity and supported in $\mathbb{R}^3 \setminus \overline{B_1}$, then $\theta\psi$ has similar properties to those of ψ. Hence, we may assume, without loss of generality, that ψ is supported in $\mathbb{R}^3 \setminus \overline{B_1}$. Moreover,

$$\int_{B_1^c} |\mathbf{x}|^2 |(\boldsymbol{\alpha} \cdot \mathbf{p})\psi(\mathbf{x})|^2 d\mathbf{x} < \infty. \tag{4.4.33}$$

In the notation of Section 4.4.1, set

$$\Psi(\mathbf{y}) := -X^{-1}(\mathbf{y})\tilde{\psi}(\mathbf{y}), \quad \tilde{\psi}(\mathbf{y}) = (M\psi)(\mathbf{y}).$$

Since $X(\mathbf{y}) \asymp 1$, we have from $\psi \in L^2(B_1^c, \mathbb{C}^4)$ that

$$\int_{B_1} |\Psi(\mathbf{y})|^2 \frac{d\mathbf{y}}{|\mathbf{y}|^6} < \infty. \tag{4.4.34}$$

Also, from (4.4.7),

$$(M(\boldsymbol{\alpha} \cdot \mathbf{p})\psi)(\mathbf{y}) = |\mathbf{y}|^2 \{X(\mathbf{y})[(\boldsymbol{\alpha} \cdot \mathbf{p})\Psi(\mathbf{y}) + Y(\mathbf{y})\Psi(\mathbf{y})]\}, \tag{4.4.35}$$

where Y is given by (4.4.8) and is readily shown to satisfy

$$\|Y(\mathbf{y})\|_{\mathbb{C}^4} \asymp 1/|\mathbf{y}|.$$

Let $\Psi(\mathbf{y}) = |\mathbf{y}|^2 \Phi(\mathbf{y})$. Then

$$(\boldsymbol{\alpha} \cdot \mathbf{p})\Psi(\mathbf{y}) = |\mathbf{y}|^2 \left\{ (\boldsymbol{\alpha} \cdot \mathbf{p})\Phi(\mathbf{y}) + O\left(\frac{|\Phi(\mathbf{y})|}{|\mathbf{y}|}\right) \right\}$$

and from (4.4.35),

$$|(M(\boldsymbol{\alpha} \cdot \mathbf{p})\psi)(\mathbf{y})| \asymp |\mathbf{y}|^4 \left\{ (\boldsymbol{\alpha} \cdot \mathbf{p})\Phi(\mathbf{y}) + O\left(\frac{|\Phi(\mathbf{y})|}{|\mathbf{y}|}\right) \right\}. \tag{4.4.36}$$

Hence, from (4.4.31),

$$\int_{B_1} |\mathbf{y}|^{-2} \left| |\mathbf{y}|^4 \left\{ (\boldsymbol{\alpha} \cdot \mathbf{p})\Phi(\mathbf{y}) + O\left(\frac{|\Phi(\mathbf{y})|}{|\mathbf{y}|}\right) \right\} \right|^2 \frac{d\mathbf{y}}{|\mathbf{y}|^6} < \infty. \tag{4.4.37}$$

Since $\int_{B_1} |\Phi(\mathbf{y})|^2 \frac{d\mathbf{y}}{|\mathbf{y}|^2} < \infty$ by (4.4.34), it follows from (4.4.37) that

$$\int_{B_1} |(\boldsymbol{\alpha} \cdot \mathbf{p})\Phi(\mathbf{y})|^2 d\mathbf{y} < \infty,$$

and so by (4.4.20), that $\Phi \in L^6(B_1, \mathbb{C}^4)$. Consequently

$$\int_{B_1} |\Psi(\mathbf{y})|^6 |\mathbf{y}|^{-12} d\mathbf{y} < \infty,$$

which yields the theorem. $\qquad \square$

Under the assumption

$$\|Q(\mathbf{x})\|_{\mathbb{C}^4} \le C\langle\mathbf{x}\rangle^{-\rho}, \quad \langle\mathbf{x}\rangle = \sqrt{1+|\mathbf{x}|^2}, \quad \rho > 1, \tag{4.4.38}$$

it is proved in [Saitō and Umeda (2008b)] that every zero mode of $\mathbb{D}_Q := (\boldsymbol{\alpha}\cdot\mathbf{p})+Q$, is continuous and decays at infinity like $\langle\mathbf{x}\rangle^{-2}$, this rate being optimal. Also, $\mathbb{D}_Q$ is shown not to have any resonances if $\rho > 3/2$, where a resonance is defined as a spinor f which is such that $\mathbb{D}_Q f = 0$ and $\langle\cdot\rangle^{-s} f \in L^2(\mathbb{R}^3, \mathbb{C}^4)$ for some $s > 0$, but $f \notin L^2(\mathbb{R}^3, \mathbb{C}^4)$.

4.5   Sobolev and CLR inequalities for Pauli operators

In Section 4.1 we noted that if the Pauli operator $\mathbb{P}_\mathbf{A}$ had a zero mode, then there could not be an analogue of the CLR inequality for the number of negative eigenvalues of $\mathbb{P}_\mathbf{A} + V$ in terms of an L^p norm of V. Our ultimate aim in this section is to show that there is a CLR-type inequality if there is no zero mode and to give a constant which involves a distance from $\mathbf{A}$ to the set in Theorem 4.3.4. First, we need the *diamagnetic inequality*.

Let $\mathbf{A} \in L^2_{\mathrm{loc}}(\mathbb{R}^n, \mathbb{R}^n)$, i.e., the components $A_j \in L^2_{\mathrm{loc}}(\mathbb{R}^n)$. We refer to the expression $\boldsymbol{\nabla}_\mathbf{A} := \boldsymbol{\nabla} + i\mathbf{A}$ as the *magnetic gradient* (with respect to $\mathbf{A}$) and to $D_\mathbf{A} := \boldsymbol{\nabla}^2_\mathbf{A}$ as the *magnetic Laplacian*. Note that for $f \in L^2_{\mathrm{loc}}(\mathbb{R}^n), \boldsymbol{\nabla}_\mathbf{A} f$ is a distribution. We define $H^1_\mathbf{A}(\mathbb{R}^n)$ to be the completion of $C^\infty_0(\mathbb{R}^n)$ with respect to the norm $\|\cdot\|_{H^1_\mathbf{A}} := (\cdot, \cdot)_{H^1_\mathbf{A}}$ determined by the inner product

$$(f, g)_{H^1_\mathbf{A}} := (f, g) + \sum_{j=1}^{n}([\partial_j + iA_j]f, [\partial_j + iA_j]g), \tag{4.5.1}$$

where $(\cdot, \cdot)$ denotes the $L^2(\mathbb{R}^n)$ inner product as usual. Note that $f \in H^1_\mathbf{A}(\mathbb{R}^n)$ does not imply in general that $f \in H^1(\mathbb{R}^n)$, or that its complex conjugate $\overline{f} \in H^1_\mathbf{A}(\mathbb{R}^n)$.

The next theorem gives the (*diamagnetic*) inequality, which has an important role in problems involving magnetic fields. It is equivalent to Kato's distributional inequality (see [Kato (1972)], Lemma A),

$$\Delta|f| \ge \mathrm{Re}\,[(\mathrm{sign}\,\overline{f})\Delta_\mathbf{A} f], \tag{4.5.2}$$

where $\mathrm{sign}\,\overline{f}(\mathbf{x}) = \overline{f}(\mathbf{x})/|f(\mathbf{x})|$, when $f(\mathbf{x}) \ne 0$ and 0 otherwise.

Theorem 4.5.1. *Let* $\mathbf{A} \in L^2_{\mathrm{loc}}(\mathbb{R}^n, \mathbb{R}^n)$ *and* $f \in H^1_\mathbf{A}(\mathbb{R}^n)$. *Then* $|f| \in H^1(\mathbb{R}^n)$ *and*

$$|\boldsymbol{\nabla}|f(\mathbf{x})|| \le |(\boldsymbol{\nabla}_\mathbf{A} f)(\mathbf{x})|, \quad \text{a.e. } \mathbf{x} \in \mathbb{R}^n. \tag{4.5.3}$$

Hence, $f \mapsto |f|$ *maps* $H^1_\mathbf{A}(\mathbb{R}^n)$ *continuously into* $H^1(\mathbb{R}^n)$ *with norm* ≤ 1.

Proof.   We first assume that $f \in C^\infty_0(\mathbb{R}^n)$ and set

$$f_\varepsilon := \sqrt{|f|^2 + \varepsilon}.$$

Then, from $f_\varepsilon^2 = |f|^2 + \varepsilon$, we have

$$2f_\varepsilon \partial_j f_\varepsilon = 2\mathrm{Re}[\overline{f}\partial_j f] = 2\mathrm{Re}[\overline{f}(\partial_j + iA_j)f]$$

and, hence,

$$\nabla f_\varepsilon = \mathrm{Re}[\frac{\overline{f}}{f_\varepsilon}\nabla_{\mathbf{A}}f] =: \mathrm{Re}[F_\varepsilon(f)], \qquad (4.5.4)$$

say. Next, we show that (4.5.4) holds for $f \in H_{\mathbf{A}}^1(\mathbb{R}^n)$. Let $f^{(m)} \in C_0^\infty(\mathbb{R}^n)$ be such that $\lim_{m\to\infty} f^{(m)} = f$ in $H_{\mathbf{A}}^1(\mathbb{R}^n)$ and $f^{(m)} \to f$ pointwise a.e. Then

$$F_\varepsilon(f) - F_\varepsilon(f^{(m)}) = \frac{\overline{f^{(m)}}}{f_\varepsilon^{(m)}}\left(\nabla_{\mathbf{A}}f - \nabla_{\mathbf{A}}f^{(m)}\right)$$

$$+ \left(\frac{\overline{f}}{f_\varepsilon} - \frac{\overline{f^{(m)}}}{f_\varepsilon^{(m)}}\right)\nabla_{\mathbf{A}}f =: I(m),$$

say. Since $|f^{(m)}/f_\varepsilon^{(m)}| \leq 1$, the first term on the right-hand side tends to zero as $m \to \infty$ and so does the second by the dominated convergence theorem, on observing that $\frac{\overline{f}}{f_\varepsilon} - \frac{\overline{f^{(m)}}}{f_\varepsilon^{(m)}} \to 0$ a.e. pointwise. Consequently $I(m) \to 0$ in $L^2(\mathbb{R}^n)$, and, hence, in $L_{loc}^1(\mathbb{R}^3)$. It follows that for all $\phi \in C_0^\infty(\mathbb{R}^n)$,

$$\int_{\mathbb{R}^n} f_\varepsilon \nabla\phi\, d\mathbf{x} = -\int_{\mathbb{R}^n} \phi\, \mathrm{Re}[F_\varepsilon(f)]d\mathbf{x} \qquad (4.5.5)$$

and, hence, (4.5.4) holds a.e. for all $f \in H_{\mathbf{A}}^1(\mathbb{R}^n)$; note that the assumption $\mathbf{A} \in L_{loc}^2(\mathbb{R}^n, \mathbb{R}^n)$ implies that ∇f, and hence ∇f_ε, lies in $L_{loc}^1(\mathbb{R}^n)$. As $\varepsilon \to 0$, $f_\varepsilon \to |f|$ uniformly, while $\overline{f}/f_\varepsilon$ is bounded and converges to sign $\overline{f}$. Thus the dominated convergence theorem applied to (4.5.5) gives

$$\int_{\mathbb{R}^n} |f|\nabla\phi\, d\mathbf{x} = -\int_{\mathbb{R}^n} \phi\, \mathrm{Re}[F_0(f)]d\mathbf{x},$$

where $F_0(\mathbf{x}) = \mathrm{sign}\,\overline{f}(\mathbf{x})\nabla_{\mathbf{A}}f(\mathbf{x})$, and

$$\nabla\,|f|(\mathbf{x}) = \mathrm{Re}[\mathrm{sign}\,\overline{f}(\mathbf{x})\,\nabla_{\mathbf{A}}f(\mathbf{x})]$$

for a.e. $\mathbf{x} \in \mathbb{R}^n$. The yields (4.5.3) and completes the proof. $\qquad\square$

In [Lieb and Loss (1997)], Section 7.20, $H_{\mathbf{A}}^1(\mathbb{R}^n)$ is defined initially as the space of functions $f : \mathbb{R}^n \to \mathbb{C}$ such that

$$f, (\partial_j + A_j)f \in L^2(\mathbb{R}^n), \quad \text{for } j = 1, 2, \cdots, n,$$

with the inner product (4.5.1). They then prove that this is a Hilbert space and that $C_0^\infty(\mathbb{R}^n)$ is a dense subspace. Therefore, the two definitions are equivalent.

In the rest of this section we assume $n = 3$. The formal Pauli operator

$$\mathbb{P}_{\mathbf{A}} = \{\boldsymbol{\sigma} \cdot (-i\nabla_{\mathbf{A}})\}^2$$

can be written as

$$\mathbb{P}_\mathbf{A} = \mathbb{S}_A + \boldsymbol{\sigma} \cdot \mathbf{B}, \tag{4.5.6}$$

where $\mathbb{S}_\mathbf{A}$ is the formal magnetic Schrödinger operator

$$\mathbb{S}_\mathbf{A} = -D_\mathbf{A} I_2 \tag{4.5.7}$$

and $\boldsymbol{\sigma} \cdot \mathbf{B}$ is the Zeeman term. For $A_j \in L^2_{\mathrm{loc}}(\mathbb{R}^3), j = 1, 2, 3$, the self-adjoint realisations $\mathbb{P}_\mathbf{A}$ and $\mathbb{S}_\mathbf{A}$ are defined as the Friedrichs extensions of (4.5.6) and (4.5.7), respectively, on $C_0^\infty(\mathbb{R}^3, \mathbb{C}^2)$. They are shown in the following lemma to have the same form domain under the assumption

$$|\mathbf{B}| \in L^{3/2}(\mathbb{R}^3). \tag{4.5.8}$$

Recall from Remark 4.3.7 that there is then a unique magnetic potential $\mathbf{A}$, which is such that $|\mathbf{A}| \in L^3(\mathbb{R}^3), \mathrm{curl}\ \mathbf{A} = \mathbf{B}, \mathrm{div}\ \mathbf{A} = 0$ and (4.3.24) is satisfied. Since a gauge transformation does not affect nul $\mathbb{P}_\mathbf{A}$, we take this $\mathbf{A}$ to be our magnetic potential.

Lemma 4.5.2. *If (4.5.8) is satisfied, then for all $\varphi \in C_0^\infty(\mathbb{R}^3, \mathbb{C}^2)$ and $\varepsilon > 0$, there exists a constant C_ε such that*

$$|((\boldsymbol{\sigma} \cdot \mathbf{B})\varphi, \varphi)| \le \varepsilon \|\mathbb{S}_\mathbf{A}\varphi\|^2 + C_\varepsilon \|\varphi\|^2. \tag{4.5.9}$$

Therefore, $\mathbb{Q}(\mathbb{P}_\mathbf{A}) = \mathbb{Q}(\mathbb{S}_\mathbf{A}) = H^1_\mathbf{A}(\mathbb{R}^3, \mathbb{C}^2)$.

Proof.   Given $\varepsilon > 0$, we may write $|\mathbf{B}| = B_1 + B_2$, where $\|B_1\|_{3/2} < \varepsilon$ and $\|B_2\|_\infty \le C_\varepsilon$. Then

$$
\begin{aligned}
|((\boldsymbol{\sigma} \cdot \mathbf{B})\varphi, \varphi)| &\le (B_1\varphi, \varphi) + (B_2\varphi, \varphi) \\
&\le \|B_1\|_{3/2}\|\varphi\|_6^2 + C_\varepsilon \|\varphi\|^2 \\
&\le \varepsilon\gamma^2 \|\boldsymbol{\nabla}|\varphi|\|^2 + C_\varepsilon \|\varphi\|^2,
\end{aligned}
$$

where γ is the norm of the embedding $H^1(\mathbb{R}^3) \hookrightarrow L^6(\mathbb{R}^3)$, and so

$$|((\boldsymbol{\sigma} \cdot \mathbf{B})\varphi, \varphi)| \le \varepsilon\gamma^2 \|\mathbb{S}_\mathbf{A}\varphi\|^2 + C_\varepsilon \|\varphi\|^2$$

on using the diamagnetic inequality. $\square$

The method in the proof of Lemma 4.3.2 also gives, for $|\mathbf{A}| \in L^3(\mathbb{R}^3)$ and $\varepsilon > 0$ arbitrary,

$$\||\mathbf{A}|\varphi\|^2 \le \varepsilon \|\varphi\|^2_{D^1_3} + C_\varepsilon \|\varphi\|^2$$

and this implies that $H^1_\mathbf{A}(\mathbb{R}^3, \mathbb{C}^2)$ coincides with $H^1(\mathbb{R}^3, \mathbb{C}^2)$ when $|\mathbf{A}| \in L^3(\mathbb{R}^3)$. Moreover, $\mathbb{P}_\mathbf{A} = \mathbb{D}^2_\mathbf{A}$, where $\mathbb{D}_\mathbf{A}$ is the Weyl–Dirac operator with domain $\mathcal{D}(\mathbb{D}) = D^1 \cap \mathcal{H}$ (see (4.3.11)): note that we have omitted the subscripts 3 from the subspaces $D^1_3, \mathcal{H}_3$.

The operator $\mathbb{S}_{\mathbf{A}}$ has no zero modes. For if $\mathbb{S}_{\mathbf{A}} f = 0$, $f \in L^2(\mathbb{R}^3)$, then, by the diamagnetic inequality,

$$0 = (\mathbb{S}_{\mathbf{A}} f, f) = \|\nabla_{\mathbf{A}} f\|^2 \geq \|\nabla|f|\|^2,$$

which implies that $f = 0$. This, in turn, implies that the operator $\mathbb{P} := \mathbb{P}_{\mathbf{A}} + |\mathbf{B}| I_2$, defined as the self-adjoint operator associated with the form $(\mathbb{P} f, f)$ on $\mathcal{Q}(\mathbb{S}_{\mathbf{A}})$, has no zero modes. For $\mathbb{P} f = 0$ and $\mathbb{P} \geq \mathbb{S}_{\mathbf{A}} + |\mathbf{B}| I_2 - \boldsymbol{\sigma} \cdot \mathbf{B} \geq \mathbb{S}_{\mathbf{A}}$ imply that $\mathbb{S}_{\mathbf{A}} f = 0$ and, hence, $f = 0$. Therefore, $\mathbb{S}_{\mathbf{A}}$ and $\mathbb{P}$ have dense domains and ranges in $\mathcal{H} = L^2(\mathbb{R}^3, \mathbb{C}^2)$. Moreover, $\mathcal{D}(\mathbb{P}^{1/2}) = \mathcal{D}(\mathbb{S}_{\mathbf{A}}^{1/2}) = \mathcal{Q}(\mathbb{S}_{\mathbf{A}})$.

We now follow a similar path to that in Section 4.3 and define spaces $D_{\mathbf{A}}^1, D_{\mathbf{B}}^1$ as the completions of $C_0^\infty(\mathbb{R}^3, \mathbb{C}^2)$ with respect to the respective norms

$$\|f\|_{D_{\mathbf{A}}^1} := \|\mathbb{S}_{\mathbf{A}}^{1/2} f\| = \|\nabla_{\mathbf{A}} f\| \tag{4.5.10}$$

$$\|f\|_{D_{\mathbf{B}}^1} = \|\mathbb{P}^{1/2} f\|. \tag{4.5.11}$$

Since $\mathbb{P} \geq \mathbb{S}_{\mathbf{A}}$, we have the natural embedding $D_{\mathbf{B}}^1 \hookrightarrow D_{\mathbf{A}}^1$, with norm ≤ 1. Also, by the diamagnetic inequality, $f \mapsto |f|$ maps $D_{\mathbf{A}}^1$ continuously into D^1, which is continuously embedded in $L^6(\mathbb{R}^3, \mathbb{C}^2)$ by the Sobolev embedding theorem. In fact the spaces $D_{\mathbf{A}}^1$ and $D_{\mathbf{B}}^1$ are isomorphic if (4.5.8) is satisfied, and we shall prove in the next lemma that $D_{\mathbf{A}}^1$ is continuously embedded in $L^6(\mathbb{R}^3, \mathbb{C}^2)$.

Lemma 4.5.3. *Let* $|\mathbf{A}| \in L^3(\mathbb{R}^3)$ *and* $|\mathbf{B}| \in L^{3/2}(\mathbb{R}^3)$. *Then*

(1) for all $f \in D_{\mathbf{B}}^1$,

$$|([\mathbf{D} \cdot \mathbf{A} + \mathbf{A} \cdot \mathbf{D}] f, f)| \leq 2\gamma \|\mathbf{A}\|_3 \|f\|_{D^1}^2 \tag{4.5.12}$$

$$\leq 2\gamma \|\mathbf{A}\|_3 \|f\|_{D_{\mathbf{B}}^1}^2, \tag{4.5.13}$$

where $\mathbf{D} = -i\nabla$, $\|\mathbf{A}\|_3 \equiv \||\mathbf{A}|\|_3$ *and* γ *is the norm of the Sobolev embedding* $D^1 \hookrightarrow L^6(\mathbb{R}^3)$.

(2)

$$D_{\mathbf{B}}^1 \hookrightarrow D_{\mathbf{A}}^1 \hookrightarrow D^1 \hookrightarrow L^6(\mathbb{R}^3, \mathbb{C}^2). \tag{4.5.14}$$

Proof. (1) Let $\phi \in C_0^\infty(\mathbb{R}^3, \mathbb{C}^2)$. Then

$$|([\mathbf{D} \cdot \mathbf{A} + \mathbf{A} \cdot \mathbf{D}] \phi, \phi)| = \left| 2\mathrm{Re} \sum_{j=1}^{3} (A_j \phi, D_j \phi) \right|$$

$$\leq 2\|\mathbf{A}\|_3 \|\phi\|_6 \|\nabla \phi\|$$

$$\leq 2\gamma \|\mathbf{A}\|_3 \|\nabla \phi\|^2.$$

Thus (4.5.12) follows by continuity, and this implies (4.5.13) once (4.5.14) is established.

(2) Let $\phi \in C_0^\infty(\mathbb{R}^3, \mathbb{C}^2)$ and $k > 1$. Then

$$k(\mathbb{S}_{\mathbf{A}} \phi, \phi) = (k-1)(-\Delta\varphi, \varphi) + (\{-\Delta + k[\mathbf{D} \cdot \mathbf{A} + \mathbf{A} \cdot \mathbf{D}] + k^2|\mathbf{A}|^2\}\phi, \phi)$$
$$+ ([k|\mathbf{A}|^2 - k^2|\mathbf{A}|^2]\phi, \phi)$$
$$= (k-1)(-\Delta\phi, \phi) + (\mathbb{S}_{k\mathbf{A}} \phi, \phi) - (k^2 - k)(|\mathbf{A}|^2 \phi, \phi)$$
$$\geq (k-1)(-\Delta\phi, \phi) - (k^2 - k)(|\mathbf{A}|^2 \phi, \phi),$$

whence

$$(k-1)\|\boldsymbol{\nabla}\phi\|^2 \leq k\|\phi\|_{D_{\mathbf{A}}^1}^2 + k(k-1)\gamma^2\|\mathbf{A}\|_3 \, \||\boldsymbol{\nabla}|\phi|\|^2$$
$$\leq \{k + k(k-1)\gamma^2\|\mathbf{A}\|_3^2\}\|\phi\|_{D_{\mathbf{A}}^1}^2,$$

by the diamagnetic inequality. Thus (4.5.14) is established, and so is (4.5.13).   □

Set

$$p[\phi] := (\mathbb{P}\phi, \phi), \quad b[\phi] := (|\mathbf{B}|\phi, \phi) \tag{4.5.15}$$

on $C_0^\infty(\mathbb{R}^3, \mathbb{C}^2)$. The self-adjoint operators associated with $p_{\mathbf{A}} := p - b$ and p are $\mathbb{P}_{\mathbf{A}}$ and $\mathbb{P}$, respectively, and as we have already seen, they have the same form domain $\mathbb{Q} = \mathcal{D}(\mathbb{P}^{1/2}) = H^1(\mathbb{R}^3, \mathbb{C}^2)$. When endowed with the graph norm

$$\|\phi\|_{\mathbb{Q}} := (\|\mathbb{P}^{1/2}\phi\|^2 + \|\phi\|^2)^{1/2},$$

$\mathbb{Q}$ is a Hilbert space that is continuously embedded in $\mathcal{H}$; it is the space $H_+(\mathbb{P}^{1/2})$ in the terminology of Section 1.2. Also $C_0^\infty(\mathbb{R}^3, \mathbb{C}^2)$ is a form core, i.e., a dense subspace of $\mathbb{Q}$. It follows that

$$\mathbb{Q} = D_{\mathbf{B}}^1 \cap \mathcal{H} \tag{4.5.16}$$

with the graph norm; the embedding $D_{\mathbf{B}}^1 \hookrightarrow L^6(\mathbb{R}^3, \mathbb{C}^2)$ guarantees the completeness, since convergent sequences in $D_{\mathbf{B}}^1$ therefore converge pointwise a.e. to their limits.

From $0 \leq b[\phi] \leq p[\phi]$, it follows that there exists a positive, bounded, self-adjoint operator $\mathcal{B}$ on $D_{\mathbf{B}}^1$ such that

$$b[\phi] = (\mathcal{B}\phi, \phi)_{D_{\mathbf{B}}^1}, \quad \phi \in D_{\mathbf{B}}^1.$$

For $\phi \in \mathcal{R}(\mathbb{P}^{1/2})$, the range of $\mathbb{P}^{1/2}$,

$$\|\mathbb{P}^{-1/2}\phi\|_{D_{\mathbf{B}}^1} = \|\phi\| \tag{4.5.17}$$

and, hence, since $\mathcal{D}(\mathbb{P}^{1/2})$ and $\mathcal{R}(\mathbb{P}^{1/2})$ are dense subspaces of $D_{\mathbf{B}}^1$ and $\mathcal{H}$, respectively, $\mathbb{P}^{-1/2}$ extends to a unitary map

$$U : \mathcal{H} \to D_{\mathbf{B}}^1, \quad U = \mathbb{P}^{-1/2} \text{ on } \mathcal{R}(\mathbb{P}^{1/2}). \tag{4.5.18}$$

By Hölder's inequality, for $f \in D_{\mathbf{B}}^1$,

$$\||\mathbf{B}|^{1/2}f\|^2 \leq \||\mathbf{B}|\|_{3/2}\|f\|_6^2 \leq \text{const.}\|f\|_{D_{\mathbf{B}}^1}^2. \tag{4.5.19}$$

Hence, the map $f \mapsto |\mathbf{B}|^{1/2}f : D_{\mathbf{B}}^1 \to \mathcal{H}$ is continuous and

$$\mathcal{S} := |\mathbf{B}|^{1/2}U : \mathcal{H} \to \mathcal{H} \tag{4.5.20}$$

is continuous.

Theorem 4.5.4.

$$\text{nul } \mathbb{P}_{\mathbf{A}} = \dim \, \{u : \mathcal{B}u = u, \, u \in D_{\mathbf{B}}^1 \cap \mathcal{H}\}$$
$$\leq \text{nul } (1 - \mathcal{S}\mathcal{S}^*). \tag{4.5.21}$$

Proof. Let $u, \phi \in \mathcal{D}(\mathbb{P}^{1/2})$. Then

$$p_{\mathbf{A}}[u, \phi] = p[u, \phi] - b[u, \phi]$$
$$= (u - \mathcal{B}u, \phi)_{D_{\mathbf{B}}^1}.$$

Hence, $u \in \mathcal{N}(\mathbb{P}_{\mathbf{A}}) \subset \mathcal{D}(\mathbb{P}^{1/2})$ if and only if $\mathcal{B}u = u$ with $u \in \mathcal{H}$. Moreover, for any $f, g \in \mathcal{H}$,

$$(\mathcal{S}f, \mathcal{S}g) = (\mathcal{B}Uf, Ug)_{D_{\mathbf{B}}^1},$$

whence

$$([\mathcal{S}^*\mathcal{S} - 1]f, g) = ([\mathcal{B} - 1]Uf, Ug)_{D_{\mathbf{B}}^1}.$$

The theorem follows since nul $[\mathcal{S}^*\mathcal{S} - 1] = $ nul $[\mathcal{S}\mathcal{S}^* - 1]$ in view of (1.2.6). $\square$

For $f \in \mathcal{R}(\mathbb{P}^{1/2}), g \in \mathcal{D}(\mathbb{P}^{1/2})$,

$$(f, \mathcal{S}^*g) = (\mathcal{S}f, g) = (|\mathbf{B}|^{1/2}Uf, g)$$
$$= (Uf, |\mathbf{B}|^{1/2}g)$$
$$= (\mathbb{P}^{-1/2}f, |\mathbf{B}|^{1/2}g);$$

note that $|\mathbf{B}|^{1/2}g \in \mathcal{H}$ by (4.5.19) and since $|\mathbf{B}| \in L^{3/2}(\mathbb{R}^3)$. Hence, $|\mathbf{B}|^{1/2}g \in \mathcal{D}(\mathbb{P}^{-1/2})$ and $\mathbb{P}^{-1/2}|\mathbf{B}|^{1/2}g = \mathcal{S}^*g$. It follows that $\mathcal{S}^* = \mathbb{P}^{-1/2}|\mathbf{B}|^{1/2}$ on $\mathcal{D}(\mathbb{P}^{1/2})$ and

$$\mathcal{S}\mathcal{S}^* = |\mathbf{B}|^{1/2}U^2|\mathbf{B}|^{1/2} \quad \text{on} \quad \mathcal{D}(\mathbb{P}^{-1/2}). \tag{4.5.22}$$

This extends by continuity to a bounded operator on $\mathcal{H}$.

Lemma 4.5.5. $\mathcal{S}$ *is compact on* $\mathcal{H}$ *and*

$$\|\mathcal{S}\|^2 \leq \gamma^2 \||\mathbf{B}|\|_{3/2}, \tag{4.5.23}$$

where γ *is the norm of* $D^1 \hookrightarrow L^6(\mathbb{R}^3, \mathbb{C}^2)$.

Proof. It is enough to prove that $|\mathbf{B}|^{1/2} : D_{\mathbf{B}}^1 \to \mathcal{H}$ is compact since $\mathcal{S} = |\mathbf{B}|^{1/2}U$ and $U : \mathcal{H} \to D_{\mathbf{B}}^1$ is unitary. Let $\{\phi_n\}$ be a sequence that converges weakly to zero in $D_{\mathbf{B}}^1$, and, hence, in D^1 by (4.5.14). It is therefore bounded, $\|\phi_n\|_{D_{\mathbf{B}}^1} \leq k$, say. Given $\varepsilon > 0$, set $|\mathbf{B}| = B_1 + B_2$, where $B_1 \in C_0^\infty(\Omega_\varepsilon)$ for some ball $\Omega_\varepsilon, B_1 \leq k_\varepsilon$, say, and $\|B_2\|_{L^{3/2}(\mathbb{R}^3)} < \varepsilon$. Then

$$\||\mathbf{B}|^{1/2}\phi_n\|^2 \leq k_\varepsilon \|\phi_n\|_{L^2(\Omega_\varepsilon, \mathbb{C}^2)}^2 + \gamma^2\|B_2\|_{L^{3/2}(\mathbb{R}^3)}\|\phi_n\|_{D_{\mathbf{B}}^1}^2$$
$$\leq k_\varepsilon \|\phi_n\|_{L^2(\Omega_\varepsilon, \mathbb{C}^2)}^2 + \gamma^2 k^2 \varepsilon.$$

The first term on the right-hand side tends to zero by the Rellich–Kondrachov theorem. Since ε is arbitrary, the lemma is proved. $\square$

Theorem 4.5.6. *Suppose that* $\mathbf{B}$ *is such that* nul $(1 - \mathcal{S}\mathcal{S}^*) = 0$, *and set*

$$\delta(\mathbf{B}) := \inf_{\|f\|=1, Uf \in \mathcal{H} \cap D_{\mathbf{B}}^1} \|[1 - \mathcal{S}^*\mathcal{S}]f\|^2. \tag{4.5.24}$$

Then $\delta(\mathbf{B}) > 0$ *and*

$$\mathbb{P}_{\mathbf{A}} \geq \delta(\mathbf{B})\mathcal{S}_{\mathbf{A}}. \tag{4.5.25}$$

Proof. Since nul $(1 - 8^*8) = $ nul $(1 - 88^*)$, the compact operator 8^*8 has no eigenvalue at 1 and consequently $\delta(\mathbf{B}) > 0$. For any f such that $f \in \mathcal{H}$ and $Uf \in \mathcal{H} \cap D_{\mathbf{B}}^1$,

$$\delta(\mathbf{B})\|f\|^2 \leq \|(1 - 8^*8)f\|^2$$
$$= \|f\|^2 - 2(8^*8f, f) + \|8^*8f\|^2.$$

Let $f = \mathbb{P}^{1/2}\phi$. Then $Uf = \phi, 8f = |\mathbf{B}|^{1/2}\phi$ and so

$$\delta(\mathbf{B})\|\mathbb{P}^{1/2}\phi\|^2 \leq \|\mathbb{P}^{1/2}\phi\|^2 - 2\||\mathbf{B}|^{1/2}\phi\|^2 + \|8^*|\mathbf{B}|^{1/2}\phi\|^2$$
$$= \|\mathbb{P}_{\mathbf{A}}^{1/2}\phi\|^2 - \||\mathbf{B}|^{1/2}\phi\|^2 + \|8^*|\mathbf{B}|^{1/2}\phi\|^2, \tag{4.5.26}$$

since $\mathbb{P} = \mathbb{P}_{\mathbf{A}} + |\mathbf{B}|I_2$ in the form sense. Also, for any $h \in \mathcal{R}(\mathbb{P}^{1/2})$,

$$\|8h\| = \||\mathbf{B}|^{1/2}\mathbb{P}^{-1/2}h\| \leq \|h\|$$

since $\mathbb{P} \geq |\mathbf{B}|I_2$, and this implies that $\|8\| \leq 1$ in view of $\mathcal{R}(\mathbb{P}^{1/2})$ being dense in $\mathcal{H}$. Thus $\|8^*\| = \|8\| \leq 1$ and from (4.5.26)

$$\delta(\mathbf{B})\|\mathbb{P}^{1/2}\phi\|^2 \leq \|\mathbb{P}_{\mathbf{A}}^{1/2}\phi\|^2,$$

whence $\mathbb{P}_{\mathbf{A}} \geq \delta(\mathbf{B})\mathbb{P} \geq \delta(\mathbf{B})\mathbb{S}_{\mathbf{A}}$. This completes the proof. $\square$

Theorem 4.5.6 has the following immediate consequences from the corresponding results for $\mathbb{S}_{\mathbf{A}}$.

Corollary 4.5.7. *Suppose that $|\mathbf{A}| \in L^3(\mathbb{R}^3)$ and $|\mathbf{B}| \in L^{3/2}(\mathbb{R}^3)$.*
(i) For all $\phi \in \mathcal{D}(\mathbb{P}_{\mathbf{A}}^{1/2}) = \mathbb{Q}$,

$$\|\mathbb{P}_{\mathbf{A}}^{1/2}\phi\|^2 \geq \frac{\delta(\mathbf{B})}{\gamma}\|\phi\|_6^2, \tag{4.5.27}$$

where γ is the norm of the embedding $D^1 \hookrightarrow L^6(\mathbb{R}^3, \mathbb{C}^2)$.
(ii) For all $\phi \in \mathbb{Q}$,

$$\|\mathbb{P}_{\mathbf{A}}^{1/2}\phi\|^2 \geq \frac{\delta(\mathbf{B})}{4}\left\|\frac{\phi}{|\cdot|}\right\|^2. \tag{4.5.28}$$

(iii) For $0 \leq V \in L^{3/2}(\mathbb{R}^3)$, the number $N(\mathbb{P}_{\mathbf{A}} - V)$ of negative eigenvalues $-\lambda_n$ of $\mathbb{P}_{\mathbf{A}} - V$ satisfies

$$N(\mathbb{P}_{\mathbf{A}} - V) \leq c[\delta(\mathbf{B})]^{-3/2} \int_{\mathbb{R}^3} V^{3/2} dx, \tag{4.5.29}$$

where c is the best constant in the CLR inequality for $-\Delta_{\mathbf{A}}$ in (1.8.7). Furthermore,

$$\sum \lambda_n^\nu \leq c[\delta(\mathbf{B})]^{-3/2} \int_{\mathbb{R}^3} V^{\nu+3/2} dx, \tag{4.5.30}$$

for any $\nu > 0$.

If any of the inequalities (4.5.27)–(4.5.29) are satisfied, then $\mathbb{P}_{\mathbf{A}}$ has no zero mode. Whether or not nul $\mathbb{P}_{\mathbf{A}} = 0$ implies that $\delta(\mathbf{B}) > 0$ is not clear. Note that the infimum in (4.5.25) is taken over the subspace of $\mathcal{H}$ in which $\mathbb{P}_{\mathbf{A}}$ and $1 - 88^*$ have the same nullity.

If $\gamma^2\|\mathbf{B}\|_{3/2} < 1$, where γ is the norm of the embedding $D^1 \hookrightarrow L^6(\mathbb{R}^3, \mathbb{C}^2)$, then from (4.5.23), we have

$$N(\mathbb{P}_{\mathbf{A}} - V) \leq c[1 - \delta^2\|\mathbf{B}\|_{L^{3/2}(\mathbb{R}^3)}]^{-3/2}\|V\|_{L^{3/2}(\mathbb{R}^3)}^{3/2}. \tag{4.5.31}$$

4.6 One-electron relativistic molecules

4.6.1 *Quasi-relativistic model*

In [Daubechies and Lieb (1983)] Daubechies and Lieb investigated the stability of one-electron molecules with K static nuclei, modelled on the Hamiltonian

$$\mathbb{H}_{1,K}(\mathbf{Z},\mathbf{R}) := (-\Delta+1)^{1/2} - \sum_{k=1}^{K} \frac{\alpha Z_k}{|\mathbf{x}-R_k|} + \sum_{j,k=1,j<k}^{K} \frac{\alpha Z_j Z_k}{|R_j-R_k|}, \qquad (4.6.1)$$

where α is the fine-structure constant, $\mathbf{x}$ is the electron position, Z_k, R_k, are the charge and position of the kth nucleus, respectively, and $\mathbf{Z} = (Z_1, \cdots, Z_K), \mathbf{R} = (R_1, \cdots, R_K)$. We separate the inter-nuclear interaction term and write

$$\mathbb{H}_{1,K}(\mathbf{Z},\mathbf{R}) = \mathbb{H}_{1,K}^0(\mathbf{Z},\mathbf{R}) + \sum_{j,k=1,j\neq k}^{K} \frac{\alpha Z_j Z_k}{|R_j-R_k|} \qquad (4.6.2)$$

where

$$\mathbb{H}_{1,K}^0(\mathbf{Z},\mathbf{R}) = (-\Delta+1)^{1/2} - \sum_{k=1}^{K} \frac{\alpha Z_k}{|\mathbf{x}-R_k|}. \qquad (4.6.3)$$

It is observed in [Daubechies and Lieb (1983)] that if the R_j are distinct, Theorem 2.2.6 can be extended to give that $\mathbb{H}_{1,K}^0(\mathbf{Z},\mathbf{R})$ is bounded below if and only if $\alpha Z_k \leq 2/\pi$ for all k, and if $\sum_{k=1}^{K} \alpha Z_k > 2/\pi$ then the ground state energy $E^0(\mathbf{Z},\mathbf{R})$ of $\mathbb{H}_{1,K}^0(\mathbf{Z},\mathbf{R})$ tends to $-\infty$ as the inter-nuclear distances tend to zero, i.e.,

$$\lim_{\delta\to 0} E^0(\mathbf{Z},\delta\mathbf{R}) = -\infty. \qquad (4.6.4)$$

The main theorem addresses the natural question of whether the nuclear repulsion is strong enough to cancel out (4.6.4). Before proceeding, the following lemma, of separate interest, is needed. It is a refinement of (1.7.8) in the case $\alpha = 1/2$ for the norm of $|\mathbf{x}|^{-1/2}|\mathbf{p}|^{-1/2}$ as an operator from $L^2(\mathbb{R}^3)$ into itself, i.e.,

$$\||\mathbf{x}|^{-1/2}|\mathbf{p}|^{-1/2}\| = (\pi/2)^{1/2}. \qquad (4.6.5)$$

Lemma 4.6.1. Let $\psi \in L^2(\mathbb{R}^3)$ be supported in the ball $B(0,R)$ centre the origin and radius R, and let $\mathcal{K}$ denote the bounded operator $2\pi^{-1}|\mathbf{x}|^{-1/2}|\mathbf{p}|^{-1}|\mathbf{x}|^{-1/2}$. Then

$$(\psi,\mathcal{K}\psi) \leq \|\psi\|^2 - \pi^{-3}R^{-2}\left(\int_{B(0,R)} |\mathbf{x}|^{-1/2}|\psi(\mathbf{x})|d\mathbf{x}\right)^2. \qquad (4.6.6)$$

Proof. We give a sketch only and refer to [Daubechies and Lieb (1983)], Lemma 2.1, for the omitted details.

It is shown that $(\psi,\mathcal{K}\psi) \leq (\psi^*,\mathcal{K}\psi^*)$, where ψ^* denotes the symmetric decreasing rearrangement of ψ. Hence, it is sufficient to prove the lemma for decreasing radially symmetric ψ. Also, by scaling, we may take $R = 1$. Thus $\psi(\mathbf{x}) = \psi(r)$ is decreasing and supported on $0 \leq r \leq 1$. For $r \geq 1$, define $\tilde{\psi}(r) = r^{-3}\psi(r^{-1})$ and

$$f(r) := \begin{cases} \psi(r), & r \leq 1, \\ \tilde{\psi}(r), & r > 1. \end{cases}$$

Then $f \in L^2(\mathbb{R}^3)$, $\|f\|^2 = 2\|\psi\|^2$ and since $\|\mathcal{K}\| = 1$ by (4.6.5), we have

$$(f, \mathcal{K}f) \leq \|f\|^2 = 2\|\psi\|^2. \tag{4.6.7}$$

Also it can be shown that

$$(f, \mathcal{K}f) = 2(\psi, \mathcal{K}\psi) + 2(\psi, L\psi), \tag{4.6.8}$$

where L is an integral operator with kernel

$$L(x, y) = \pi^{-3}|\mathbf{x}|^{-1/2}(1 + x^2y^2 - 2xy)^{-1}|y|^{-1/2}$$

and

$$(\psi, L\psi) = (8/\pi) \int_0^1 \int_0^1 r^{1/2}\psi(r)s^{1/2}\psi(s) \ln\left[\frac{1+rs}{1-rs}\right] dr ds$$

$$\geq (1/2\pi^3)\left[\int_{B(0,R)} |\mathbf{x}|^{-1/2}\psi(\mathbf{x})d\mathbf{x}\right]^2 \min_{t \in [0,1]} g(t),$$

where $g(t) = t^{-1}\ln(\frac{1+t}{1-t})$. The minimum is attained at $t = 0$ and $g(0) = 2$, whence the lemma from (4.6.7) and (4.6.8). □

Theorem 4.6.2. *Suppose that $\alpha Z_k \leq 2/\pi$ for $k = 1, \cdots, K$ and that $R_j \neq R_k$ for $j \neq k$. Then*

$$\mathbb{H}_{1,K}^0(\mathbf{Z}, \mathbf{R}) \geq -3\pi\alpha^2 \sum_{j,k=1,j<k}^K \frac{Z_jZ_k}{|R_j - R_k|}.$$

Hence $\mathbb{H}_{1,K}(\mathbf{Z}, \mathbf{R}) \geq 0$ if $\alpha \leq 1/3\pi$.

Proof.   As noted in the first paragraph of Section 4.1, it is proved in [Daubechies and Lieb (1983)] that it is sufficient to prove the result for the case when all the charges are equal and take as their common value the critical value $2/\alpha\pi$. Since $(-\Delta + 1)^{1/2} \geq (-\Delta)^{1/2} \equiv |\mathbf{p}|$, we may therefore set out to prove that for $\|\psi\| = 1$,

$$(\psi, |\mathbf{p}|\psi) - (2/\pi) \sum_{j=1}^K (\psi, |\mathbf{x} - R_j|^{-1}\psi) \geq -(12/\pi) \sum_{j,k=1,j<k}^K |R_j - R_k|^{-1} \tag{4.6.9}$$

for all $\psi \in \mathbb{Q}(|\mathbf{p}|) = H^{1/2}(\mathbb{R}^3)$. We express the problem in the form of proving that for a range $A \geq A_0$ of constants A

$$(\psi, |\mathbf{p}|\psi) \geq (2/\pi) \sum_{j=1}^K \left(\psi, \left[|\mathbf{x} - R_j|^{-1} - A\sum_{k \neq j}|R_k - R_j|^{-1}\right]\psi\right), \tag{4.6.10}$$

and our goal is to show that $A_0 = 3$ will do, which will establish (4.6.9) and hence complete the proof. We include the elegant proof of this from [Daubechies and Lieb (1983)] to enable a comparison to be made easily with the analogous result for the Brown–Ravenhall model from [Balinsky and Evans (1999)] which will be reproduced in the next section.

Let $a_+ = \max(a, 0)$, the positive part of a. Then

$$\sum_{j=1}^{K} a_j \leq \sum_{j=1}^{K} (a_j)_+ \leq \left(\sum_{j=1}^{K} (a_j)_+^{1/2} \right)^2$$

and hence (4.6.10) will be satisfied if

$$(\psi, |\mathbf{p}|\psi) \geq (\psi, [\sum_{j=1}^{K} W_j]^2 \psi), \tag{4.6.11}$$

where

$$W_j(\mathbf{x}) = \left[\frac{2}{\pi} \left(|\mathbf{x} - R_j|^{-1} - A \sum_{k \neq j} |R_k - R_j|^{-1} \right) \right]_+^{1/2}.$$

For any A, W_j has support in a ball $B(R_j, t_j)$ centre R_j and radius t_j, where

$$t_j^{-1} = A \sum_{k \neq j} |R_k - R_j|^{-1}. \tag{4.6.12}$$

If $A > 2$, the balls $B(R_j, t_j)$ are disjoint since $t_j < A^{-1}|R_k - R_j|$ for any $k \neq j$ and, hence,

$$|R_k - R_j| - (t_k + t_j) \geq (1 - 2/A)|R_k - R_j|. \tag{4.6.13}$$

We assume from now on that $A > 2$.

Let f_j denote the characteristic function of $B(R_j, t_j)$. Since $W_j = W_j f_j$ and $W_j(\mathbf{x}) \leq (2/\pi)^{1/2}|\mathbf{x} - R_j|^{-1/2}$, we have

$$\sum_j W_j(\mathbf{x}) \leq \left(\frac{2}{\pi} \right)^{1/2} \sum_j |\mathbf{x} - R_j|^{-1/2} f_j(\mathbf{x}) =: W(\mathbf{x}). \tag{4.6.14}$$

Since $|\mathbf{x} - R_j|^{-1/2} f_j(\mathbf{x}) \leq |\mathbf{x} - R_j|^{-1/2}$ and $|\mathbf{x}|^{-1/2}|\mathbf{p}|^{-1/2}$ is bounded, it follows that $W|\mathbf{p}|^{-1/2}$ is a bounded operator on $L^2(\mathbb{R}^3)$. In fact, we next show that it has norm ≤ 1 if A is large enough. Let $\psi \in C_0^\infty(\mathbb{R}^3)$ be such that $W\psi \in L^2(\mathbb{R}^3)$. Then

$$\||\mathbf{p}|^{-1/2}W\psi\|^2 = (W\psi, |\mathbf{p}|^{-1}W\psi)$$

$$= (2/\pi) \sum_{j=1}^{K} (|\mathbf{x} - R_j|^{-1/2} f_j \psi, |\mathbf{p}|^{-1}|\mathbf{x} - R_j|^{-1/2} f_j \psi)$$

$$+ (2/\pi) \sum_{j,k=1, j \neq k}^{K} (|\mathbf{x} - R_j|^{-1/2} f_j \psi, |\mathbf{p}|^{-1}[|\mathbf{x} - R_k|^{-1/2} f_k \psi])$$

$$=: I_1 + I_2, \tag{4.6.15}$$

say. By Lemma 4.6.1,

$$I_1 = (2/\pi) \sum_{j=1}^{K} (f_j \psi, |\mathbf{x} - R_j|^{-1/2}|\mathbf{p}|^{-1}|\mathbf{x} - R_j|^{-1/2} f_j \psi)$$

$$\leq \sum_{j=1}^{K} \|f_j \psi\|^2 - \pi^{-3} \sum_{j=1}^{K} t_j^{-2} m_j^2 \tag{4.6.16}$$

where

$$m_j = \int |\mathbf{x} - R_j|^{-1/2} f_j(\mathbf{x})|\psi(\mathbf{x})|d\mathbf{x}.$$

To deal with I_2 we first recall that $(\frac{1}{|\mathbf{p}|}\psi)(\mathbf{x}) := (\mathbb{F}^{-1}\frac{1}{|\cdot|}\mathbb{F}\psi)(\mathbf{x})$, so that by (1.5.15)

$$\left(\frac{1}{|\mathbf{p}|}\psi\right)(\mathbf{x}) = \mathbb{F}^{-1}[\frac{1}{|\cdot|}\hat{\psi}](\mathbf{x})$$

$$= \frac{1}{2\pi^2} \int_{\mathbb{R}^3} \frac{1}{|\mathbf{x} - \mathbf{y}|^2} \psi(\mathbf{y})d\mathbf{y}. \tag{4.6.17}$$

This gives

$$(2/\pi)(|\mathbf{x} - R_j|^{-1/2} f_j\psi, |\mathbf{p}|^{-1}[|\mathbf{x} - R_k|^{-1/2} f_k\psi])$$
$$= (1/\pi^3) \int |\mathbf{x} - R_j|^{-1/2} f_j(\mathbf{x})\psi(\mathbf{x}) \left[\int |\mathbf{x} - \mathbf{y}|^{-2}|\mathbf{y} - R_k|^{-1/2} f_k(\mathbf{y})\psi(\mathbf{y})d\mathbf{y}\right] d\mathbf{x}$$
$$\leq \pi^{-3}[1 - 2/A]^{-2} m_j m_k |R_j - R_k|^{-2}, \tag{4.6.18}$$

since (4.6.13) implies that

$$|\mathbf{x} - \mathbf{y}| \geq R_j + R_k - (t_j + t_k) \geq (1 - 2/A)|R_j - R_k|.$$

We therefore have from (4.6.16) and (4.6.18)

$$\||\mathbf{p}|^{-1/2} W\psi\|^2 \leq \|\psi\|^2 - \pi^{-3} \sum_{j=1}^{K} t_j^{-2} m_j^2$$

$$+ \pi^{-3}[1 - 2/A]^{-2} \sum_{j,k=1,j\neq k}^{K} m_j m_k |R_j - R_k|^{-2}.$$

On using the inequality $m_j m_k \leq (m_j^2 + m_k^2)/2$ this gives

$$\||\mathbf{p}|^{-1/2} W\psi\|^2 \leq \|\psi\|^2 - \sum_{j=1}^{K} m_j^2 b_j,$$

where

$$b_j = \pi^{-3} t_j^{-2} - \pi^{-3}[1 - 2/A]^{-2} \sum_{k\neq j} |R_k - R_j|^{-2}$$

$$\geq \pi^{-3}[A^2 - (1 - 2/A)^{-2}] \sum_{k\neq j} |R_k - R_j|^{-2},$$

by (4.6.12). If $A \geq 3$, then $A^2 \geq (1 - 2/A)^{-2}$ and $b_j \geq 0$. We have, therefore, proved that if $A \geq 3$, $\||\mathbf{p}|^{-1/2} W\psi\|^2 \leq \|\psi\|^2$ for all $\psi \in C_0^\infty(\mathbb{R}^3)$ and consequently $\|W|\mathbf{p}|^{-1/2}\| = \||\mathbf{p}|^{-1/2} W\| \leq 1$. It follows that

$$|\mathbf{p}| - W^2 = |\mathbf{p}|^{1/2}[1 - (|\mathbf{p}|^{-1/2} W)(W|\mathbf{p}|^{-1/2})]|\mathbf{p}|^{1/2} \geq 0.$$

Since $W^2 \geq (\sum_{j=1}^{K} W_j)^2$, (4.6.11) is established and the theorem is proved. $\square$

4.6.2 The Brown–Ravenhall model

The counterpart to the problem in Section 4.6.1 for the Brown–Ravenhall operator concerns the Hamiltonian

$$\mathbb{B}_{1,K}(\mathbf{Z},\mathbf{R}) = \Lambda_+ \left(\mathbb{D}_0 - \sum_{k=1}^{K} \frac{\alpha Z_k}{|\mathbf{x} - R_k|} + \sum_{k,l=1,k<l}^{K} \frac{\alpha Z_k Z_l}{|R_k - R_l|} \right) \Lambda_+, \qquad (4.6.19)$$

where, in our usual notation, $\mathbb{D}_0$ is the free Dirac operator and Λ_+ is the projection onto the positive spectral subspace of $\mathbb{D}_0$. The Hilbert space in which $\mathbb{B}_{1,K}(\mathbf{Z},\mathbf{R})$ acts is $\mathcal{H}_+ := \Lambda_+ L^2(\mathbb{R}^3, \mathbb{C}^4)$. As in Section 4.6.1, we separate out the inter-nuclear interaction term and write

$$\mathbb{B}_{1,K}(\mathbf{Z},\mathbf{R}) = \mathbb{B}^0_{1,K}(\mathbf{Z},\mathbf{R}) + \sum_{k,l=1,k<l}^{K} \frac{\alpha Z_k Z_l}{|R_k - R_l|}, \qquad (4.6.20)$$

where

$$\mathbb{B}^0_{1,K}(\mathbf{Z},\mathbf{R}) = \Lambda_+ \left(\mathbb{D}_0 - \sum_{k=1}^{K} \frac{\alpha Z_k}{|\mathbf{x} - R_k|} \right) \Lambda_+. \qquad (4.6.21)$$

The operator defined by (4.6.19) has form domain $H^{1/2}(\mathbb{R}^3, \mathbb{C}^4)$ and is the restriction to $\mathcal{H}_+$ of $\mathbb{H}_{1,K}(\mathbf{Z},\mathbf{R})I_4$, where $\mathbb{H}_{1,K}(\mathbf{Z},\mathbf{R})$ is given by (4.6.1). Hence if $Z_k \leq 2/\pi\alpha$ for all k and $\alpha \leq 1/3\pi$, Theorem 4.6.2 implies that $\mathbb{B}_{1,K}(\mathbf{Z},\mathbf{R})$ is non-negative. The problem tackled in [Balinsky and Evans (1999)] was to extend the range of nuclear charges to $Z_k \leq \gamma_c/\alpha$, where $\gamma_c = 2(\pi/2 + 2/\pi)^{-1}$ is the critical value of γ in Theorem 2.3.7 for the Brown–Ravenhall operator $\mathfrak{b}$ (and hence $\mathbb{B}$) to be bounded below, and to determine an appropriate bound on α. The operator $\mathbb{B}^0_{1,K}(\mathbf{Z},\mathbf{R})$ is non-negative if the $Z_k \leq \gamma_c/\alpha$ and the argument in Theorem 3.3.1 remains valid to prove that the essential spectrum of $\mathbb{B}^0_{1,K}(\mathbf{Z},\mathbf{R})$ is $[1,\infty)$. Hence, $\mathbb{B}_{1,K}(\mathbf{Z},\mathbf{R})$ has essential spectrum

$$\left[1 + \sum_{k,l=1,k<l}^{K} \frac{\alpha Z_k Z_l}{|R_k - R_l|}, \infty \right).$$

The analogue of Theorem 4.6.2 is

Theorem 4.6.3. *Suppose that* $\alpha Z_k \leq \gamma_c$, $k = 1, 2, \ldots, K$, *and* $\alpha \leq \gamma_c/(4 + 2\sqrt{1 + \pi/2})$, *where* $\gamma_c = 2/\left(\frac{\pi}{2} + \frac{2}{\pi}\right)$, *and* $R_k \neq R_l$ *for* $k \neq l$. *Then* $\mathbb{B}_{1,K}(\mathbf{Z},\mathbf{R})$ *is stable in the sense that there exists a constant* C *such that*

$$\mathbb{B}_{1,K}(\mathbf{Z},\mathbf{R}) \geq CK.$$

The range $\alpha \leq \gamma_c/(4 + 2\sqrt{1 + \pi/2}) \approx 0.125721$ *includes the physical value of* α.

Just as in Section 4.6.1, Lemma 4.6.1 had a pivotal role in the proof of Theorem 4.6.2, the proof of Theorem 4.6.3 rests heavily on an analogous lemma, but the argument involving rearrangement inequalities, which was used to establish Lemma 4.6.1, is not available for the 2-spinors now involved, and there is an additional complication in that the operator $(\mathbf{p} \cdot \sigma)/|\mathbf{p}|^2$ which appears, does not have a positive kernel, unlike $1/|\mathbf{p}|$. To overcome these obstacles, new ideas were developed in [Balinsky and Evans (1999)], which we now present. Some preparatory work is first necessary.

We recall from (2.3.8) that a spinor ψ in the positive spectral subspace of $\mathbb{D}_0$ satisfies

$$\widehat{\psi}(\mathbf{p}) = \frac{1}{n(p)} \begin{pmatrix} [e(p)+1]\, u(\mathbf{p}) \\ (\mathbf{p} \cdot \sigma)\, u(\mathbf{p}) \end{pmatrix}, \tag{4.6.22}$$

where $u \in L^2(\mathbb{R}^3, \mathbb{C}^2)$ is a Pauli spinor, $e(p) = \sqrt{p^2 + 1}$ and $n(p) = [2e(p)(e(p) + 1)]^{1/2}$. Conversely any Dirac spinor of the form (4.6.22) is in the image of $\mathcal{H}_+$ under the Fourier transform. Also

$$(\psi, \mathbb{B}^0_{1,1}(Z)\psi) = (u, \mathfrak{b}_{1,1}(Z)u),$$

where $\mathfrak{b}_{1,1}(Z)$ is given by (2.3.13) and (2.3.14), with $\gamma = \alpha Z$. In Lemma 3.3.4, $\mathfrak{b}_{1,1}(Z)$ is proved to have the same domain in $L^2(\mathbb{R}^3, \mathbb{C}^2)$ as $\mathfrak{b}^0_{1,1}(Z)$, which is defined in the form sense by

$$\mathfrak{b}^0_{1,1}(Z) = |\mathbf{p}| - \frac{1}{2}\gamma \Big(W + PWP\Big), \qquad \gamma = \alpha Z, \tag{4.6.23}$$

where

$$(Pu)(\mathbf{p}) = \frac{\mathbf{p} \cdot \sigma}{|\mathbf{p}|} u(\mathbf{p}), \tag{4.6.24}$$

$$(Wu)(\mathbf{p}) = \frac{1}{2\pi^2} \int\limits_{R^3} u(\mathbf{p}') \frac{d\mathbf{p}'}{|\mathbf{p} - \mathbf{p}'|^2}. \tag{4.6.25}$$

Also, $\mathfrak{b}_{1,1}(Z) - \mathfrak{b}^0_{1,1}(Z)$ can be extended to a bounded operator on $L^2(\mathbb{R}^3, \mathbb{C}^2)$.

While the Brown–Ravenhall operator is simply expressed in momentum space, it will be necessary for us to work in $\mathbf{x}$-space with the operator

$$\widetilde{\mathfrak{b}}^0_{1,1}(Z) = \mathbb{F}^{-1}\mathfrak{b}^0_{1,1}(Z)\mathbb{F}$$

$$= |\mathbf{p}| - \frac{1}{2}\gamma \Big(\frac{1}{|\mathbf{x}|} + \frac{\mathbf{p} \cdot \sigma}{|\mathbf{p}|} \frac{1}{|\mathbf{x}|} \frac{\mathbf{p} \cdot \sigma}{|\mathbf{p}|}\Big). \tag{4.6.26}$$

Here we have used (1.5.16) to give

$$(\mathbb{F}^{-1}W\mathbb{F}\, u)(\mathbf{x}) = \frac{1}{|\mathbf{x}|} u(\mathbf{x})$$

and $\mathbf{p}$, $1/|\mathbf{x}|$ stand for $-i\nabla$ and multiplication by $1/|\mathbf{x}|$, respectively. Hereafter we omit the tilde in $\widetilde{\mathfrak{b}}^0_{1,1}$.

The operator $\frac{\mathbf{p}\cdot\boldsymbol{\sigma}}{|\mathbf{p}|}$ is a unitary (and self-adjoint) involution; let $P_\pm$ denote the projections onto its eigenspaces corresponding to its eigenvalues at ± 1. Then, we have $P_\pm = \frac{1}{2}\left(1 \pm \frac{\mathbf{p}\cdot\boldsymbol{\sigma}}{|\mathbf{p}|}\right)$ and

$$P_\pm |\mathbf{p}| = |\mathbf{p}| P_\pm, \quad P_\pm \frac{1}{|\mathbf{p}|^{1/2}} = \frac{1}{|\mathbf{p}|^{1/2}} P_\pm, \tag{4.6.27}$$

$$P_\pm^2 = P_\pm, \quad P_\pm^* = P_\pm. \tag{4.6.28}$$

Lemma 4.6.4. *Let* $\mathcal{K} := \gamma_c \frac{1}{|\mathbf{x}|^{1/2}} P_+ \frac{1}{|\mathbf{p}|} P_+ \frac{1}{|\mathbf{x}|^{1/2}}$. *Then* $\mathcal{K} \geq 0$ *and*

$$\sup_{\|\psi\|=1} \left(\psi, \mathcal{K}\psi\right) = 1.$$

Proof. The operator $\mathfrak{b}_{1,1}^0(Z)$ is non-negative if and only if $\gamma \leq \gamma_c$ and, for $\phi \in H^{1/2}(\mathbb{R}^3, \mathbb{C}^2) = \mathbb{Q}(|\mathbf{p}|I_2)$,

$$\left(\phi, P_+ \mathfrak{b}_{1,1}^0(Z) P_+ \phi\right) = \left(P_+ \phi, |\mathbf{p}| P_+ \phi\right) - \gamma\left(\phi, P_+ \frac{1}{|\mathbf{x}|} P_+ \phi\right).$$

Hence, with $\psi = |\mathbf{p}|^{1/2}\phi$, $\gamma \leq \gamma_c$ is equivalent to

$$\|P_+\psi\|^2 = \left(P_+ \frac{1}{|\mathbf{p}|^{1/2}}\psi, |\mathbf{p}| P_+ \frac{1}{|\mathbf{p}|^{1/2}}\psi\right)$$

$$\geq \gamma\left(\frac{1}{|\mathbf{p}|^{1/2}}\psi, P_+ \frac{1}{|\mathbf{x}|} P_+ \frac{1}{|\mathbf{p}|^{1/2}}\psi\right)$$

$$= \gamma\left\|\frac{1}{|\mathbf{x}|^{1/2}} P_+ \frac{1}{|\mathbf{p}|^{1/2}}\psi\right\|^2. \tag{4.6.29}$$

Let $A := \frac{1}{|\mathbf{x}|^{1/2}} P_+ \frac{1}{|\mathbf{p}|^{1/2}}$. Then $A^* := \frac{1}{|\mathbf{p}|^{1/2}} P_+ \frac{1}{|\mathbf{x}|^{1/2}}$ and $\mathcal{K} = \gamma_c A A^*$. It follows from (4.6.29), and since $A = A P_+$, that

$$\sup_{\|\psi\|=1} \left(\psi, \mathcal{K}\psi\right) = \gamma_c \|A^*\|^2 = 1.$$

$\square$

From

$$[(\cdot \times \mathbf{p})u(\cdot)]^\wedge(\mathbf{p}) = \left(-i\boldsymbol{\nabla}_\mathbf{p} \times \mathbf{p}\right)\hat{u}(\mathbf{p})$$

and

$$[(\mathbf{x} \times \cdot)u(\cdot)]^\vee(\mathbf{x}) = \left(-i\boldsymbol{\nabla}_\mathbf{x} \times \mathbf{x}\right)\check{u}(\mathbf{x})$$

where $\check{u} = \mathbb{F}^{-1}$ it follows that in both $\mathbf{x}$ and $\mathbf{p}$ spaces, the total angular momentum operator $\mathbf{J} \equiv (J_1, J_2, J_3) = \mathbf{x} \times \mathbf{p} + \frac{1}{2}\boldsymbol{\sigma} = \mathbf{L} + \frac{1}{2}\boldsymbol{\sigma}$ depends only on the angular coordinates, and commutes with the Brown–Ravenhall operators $\mathfrak{b}_{1,1}(Z)$ and $\mathfrak{b}_{1,1}^0(Z)$ (see Section 2.1.1). We also recall from Section 2.1.1 that the $\Omega_{l,m,s}$ are simultaneous eigenvectors of J^2, J_3 and L^2 with corresponding eigenvectors $(l+s)(l+s+1), m$ and $l(l+1)$, respectively. From

$$(\mathbf{J})^2 = (\mathbf{L} + \frac{1}{2}\boldsymbol{\sigma})^2 = L^2 + \frac{3}{4} + \boldsymbol{\sigma} \cdot \mathbf{L}$$

we have $\boldsymbol{\sigma} \cdot \mathbf{L} + 1 = J^2 - L^2 + \frac{1}{4}$. It follows from (2.1.37) and (2.1.38) that the $\Omega_{l,m,s}$ and $\Omega_{l+2s,m,-s}$ are, respectively, eigenvectors of $\boldsymbol{\sigma} \cdot \mathbf{L} + 1$ corresponding to the eigenvalue k, where

$$k = \begin{cases} l+1, & \text{if } s = 1/2, \\ -l, & \text{if } s = -1/2. \end{cases}$$

We have the isomorphism

$$\mathcal{H} = L^2(\mathbb{R}^3, \mathbb{C}^2) \longrightarrow L^2((0,\infty); r^2 dr) \otimes L^2(\mathbb{S}^3, \mathbb{C}^2);$$

any $f \in \mathcal{H}$ can be written, with $r = |\mathbf{x}|$, $\omega = \mathbf{x}/|\mathbf{x}|$,

$$f(\mathbf{x}) = \sum_{(l,m,s) \in \mathfrak{I}} c_{l,m,s}(r) \Omega_{l,m,s}(\omega) \tag{4.6.30}$$

and

$$\|f\|^2 = \sum_{(l,m,s) \in \mathfrak{I}} \int_0^\infty |c_{l,m,s}(r)|^2 r^2 dr \tag{4.6.31}$$

where $\mathfrak{I}$ is the index set defined in (2.1.27).

The Fourier transform takes angular momentum channels into angular momentum channels, and so (4.6.30) and (4.6.31) continue to hold when $\mathbf{x}$ is replaced by $\mathbf{p}$. We also need the following identities:

$$\int_{\mathbb{S}^2} \int_{\mathbb{S}^2} \frac{1}{|\mathbf{x} - \mathbf{y}|^2} \Omega^*_{l,m,s}(\omega_x) \, \Omega_{l',m',s'}(\omega_y) d\omega_x d\omega_y =$$

$$= \frac{2\pi}{|\mathbf{x}||\mathbf{y}|} Q_l \left(\frac{|\mathbf{x}|^2 + |\mathbf{y}|^2}{2|\mathbf{x}||\mathbf{y}|} \right) \delta_{ll'} \delta_{mm'} \delta_{ss'}, \tag{4.6.32}$$

$$\left(\frac{\mathbf{x} \cdot \boldsymbol{\sigma}}{|\mathbf{x}|} \right) \Omega_{l,m,s}(\omega) = \Omega_{l+2s,m,-s}(\omega), \quad \omega = \frac{\mathbf{x}}{|\mathbf{x}|}, \tag{4.6.33}$$

$$\frac{1}{|\mathbf{p}|} \left[\frac{g(r)}{r} \Omega_{l,m,s}(\omega) \right] = \left\{ \frac{1}{\pi r} \int_0^\infty g(s) Q_l \left(\frac{r^2 + s^2}{2rs} \right) ds \right\} \Omega_{l,m,s}(\omega), \tag{4.6.34}$$

$$(\boldsymbol{\sigma} \cdot \mathbf{p}) [ig(r) \Omega_{l,m,s}(\omega)] = \Omega_{l+2s,m,-s}(\omega) \left[\frac{dg}{dr} + \frac{1-k}{r} g(r) \right]. \tag{4.6.35}$$

The first is established in Lemma 2.2.2 and the second in (2.1.28). The identity (4.6.34) follows on using (4.6.17), (2.2.5) and the orthogonality properties of the spherical harmonics. To prove (4.6.35) we use the argument in [Greiner (1990)], p. 171. To begin, we have by (2.1.17) and (2.1.28),

$$(\boldsymbol{\sigma} \cdot \mathbf{p})[ig\Omega_{l,m,s}] = [(\boldsymbol{\sigma} \cdot \mathbf{p})(ig)]\Omega_{l,m,s} + ig(\boldsymbol{\sigma} \cdot \mathbf{p})\Omega_{l,m,s}$$

$$= \frac{dg}{dr}(\boldsymbol{\sigma} \cdot \mathbf{e}_r)\Omega_{l,m,s} + ig(\boldsymbol{\sigma} \cdot \mathbf{p})\Omega_{l,m,s}$$

$$= \frac{dg}{dr}\Omega_{l+2s,m,-s} + ig(\boldsymbol{\sigma} \cdot \mathbf{p})\Omega_{l,m,s}. \qquad (4.6.36)$$

The use of (2.1.28) again and then the identity (2.1.35) gives

$$(\boldsymbol{\sigma} \cdot \mathbf{p})\Omega_{l,m,s} = (\boldsymbol{\sigma} \cdot \mathbf{p})(\boldsymbol{\sigma} \cdot \mathbf{e}_r)\Omega_{l+2s,m,-s}$$

$$= \{(\mathbf{p} \cdot \mathbf{e}_r) + i\boldsymbol{\sigma} \cdot (\mathbf{p} \times \mathbf{e}_r)\}\Omega_{l+2s,m,-s}$$

$$= \{-i\mathbf{x} \cdot \boldsymbol{\nabla} - 3i - i\boldsymbol{\sigma} \cdot (\mathbf{x} \times \mathbf{p})\}\frac{1}{r}\Omega_{l+2s,m,-s}$$

$$= -i\left\{\frac{2}{r} + \frac{1}{r}(\boldsymbol{\sigma} \cdot \mathbf{L})\right\}\Omega_{l+2s,m,-s}$$

$$= -i\left\{\frac{1}{r} - \frac{k}{r}\right\}.$$

On substituting this in (4.6.36) we obtain (4.6.35).

It follows from (4.6.34) and (4.6.35) that $1/|\mathbf{p}|$ preserves the angular momentum channels, but $(\mathbf{p} \cdot \boldsymbol{\sigma})/|\mathbf{p}|$ does not. However, $(\mathbf{p} \cdot \boldsymbol{\sigma})/|\mathbf{p}|$ and P_+ have the following invariant subspaces. We can write the decomposition represented by (4.6.30) as

$$\mathcal{H} \equiv L^2(\mathbb{R}^3, \mathbb{C}^2) = \bigoplus_{l \geq 0,\ m=-l-1/2,\ldots,l+1/2} \mathcal{H}_{l,m},$$

where $\mathcal{H}_{l,m}$ is the space of functions of the form

$$c_{l,m,1/2}(r)\Omega_{l,m,1/2}(\omega) + c_{l+1,m,-1/2}(r)\Omega_{l+1,m,-1/2}(\omega). \qquad (4.6.37)$$

Then, by (4.6.34) and (4.6.35)

$$P_+ : \mathcal{H}_{l,m} \to \mathcal{H}_{l,m}. \qquad (4.6.38)$$

We are now in a position to state the analogue of Lemma 4.6.1.

Lemma 4.6.5. *Let* $\mathcal{K} = \gamma_c \frac{1}{|\mathbf{x}|^{1/2}} P_+ \frac{1}{|\mathbf{p}|} P_+ \frac{1}{|\mathbf{x}|^{1/2}}$. *Then for all* $\psi \in L^2(\mathbb{R}^3, \mathbb{C}^2)$ *with support in* $B(0, R)$ *we have*

$$(\psi, \mathcal{K}\psi) \leq \|\psi\|^2 - \frac{C_0}{R^2}\left(\int_{\mathbb{R}^3} |\psi(\mathbf{x})|\frac{d\mathbf{x}}{\sqrt{|\mathbf{x}|}}\right)^2, \qquad (4.6.39)$$

where $C_0 = 1/(\pi^3 + 4\pi)$.

To prove it we need some preliminary lemmas, the first of which seems to be of particular interest

Lemma 4.6.6.

$$\frac{\mathbf{p} \cdot \boldsymbol{\sigma}}{|\mathbf{p}|^2} \leq \frac{\mathbf{x} \cdot \boldsymbol{\sigma}}{|\mathbf{x}|}\frac{1}{|\mathbf{p}|}\frac{\mathbf{x} \cdot \boldsymbol{\sigma}}{|\mathbf{x}|}. \qquad (4.6.40)$$

Proof. Let

$$\tilde{\mathcal{K}} = \frac{\gamma_c}{2} \frac{1}{|\mathbf{x}|^{1/2}} \left(\frac{1}{|\mathbf{p}|} + \frac{\mathbf{x} \cdot \boldsymbol{\sigma}}{|\mathbf{x}|} \frac{1}{|\mathbf{p}|} \frac{\mathbf{x} \cdot \boldsymbol{\sigma}}{|\mathbf{x}|} \right) \frac{1}{|\mathbf{x}|^{1/2}}. \tag{4.6.41}$$

We shall prove that for all $\psi \in L^2(\mathbb{R}^3, \mathbb{C}^2)$

$$(\psi, \mathcal{K}\psi) \le \left(\psi, \tilde{\mathcal{K}}\psi \right),$$

whence the lemma.

Let $\psi \in \mathcal{H}_{l,m}$ be of the form

$$\frac{1}{r} f(r)\Omega_{l,m,1/2}(\boldsymbol{\omega}) + \frac{1}{r} g(r)\Omega_{l+1,m,-1/2}(\boldsymbol{\omega}) \tag{4.6.42}$$

with $f, g \in C_0^1(0, \infty)$; such spinors are dense in $\mathcal{H}_{l,m}$. In

$$(\psi, \mathcal{K}\psi) = \frac{\gamma_c}{2} \left(\frac{1}{|\mathbf{x}|^{1/2}} \psi, \frac{1}{|\mathbf{p}|} \frac{1}{|\mathbf{x}|^{1/2}} \psi \right) + \frac{\gamma_c}{2} \left(\frac{1}{|\mathbf{x}|^{1/2}} \psi, \frac{\mathbf{p} \cdot \boldsymbol{\sigma}}{|\mathbf{p}|^2} \frac{1}{|\mathbf{x}|^{1/2}} \psi \right)$$

it follows from (4.6.34) and (4.6.35) that the first term is diagonal and the second off-diagonal in f and g. The diagonal term is

$$\frac{\gamma_c}{4\pi^2} \int\limits_{\mathbb{R}^3} \int\limits_{\mathbb{R}^3} \frac{\bar{f}(r)}{r^{3/2}} \frac{f(r')}{r'^{3/2}} \frac{1}{|\mathbf{x} - \mathbf{x}'|^2} \Omega^*_{l,m,1/2}(\boldsymbol{\omega})\Omega_{l,m,1/2}(\boldsymbol{\omega}')d\mathbf{x}d\mathbf{x}'$$

$$+ \frac{\gamma_c}{4\pi^2} \int\limits_{\mathbb{R}^3} \int\limits_{\mathbb{R}^3} \frac{\bar{g}(r)}{r^{3/2}} \frac{g(r')}{r'^{3/2}} \frac{1}{|\mathbf{x} - \mathbf{x}'|^2} \Omega^*_{l+1,m,-1/2}(\boldsymbol{\omega})\Omega_{l+1,m,-1/2}(\boldsymbol{\omega}')d\mathbf{x}d\mathbf{x}'$$

$$= \frac{\gamma_c}{2\pi} \int\limits_0^\infty \int\limits_0^\infty \frac{\bar{f}(r)}{\sqrt{r}} \frac{f(r')}{\sqrt{r'}} Q_l\left(\frac{r^2 + r'^2}{2rr'} \right) drdr' +$$

$$+ \frac{\gamma_c}{2\pi} \int\limits_0^\infty \int\limits_0^\infty \frac{\bar{g}(r)}{\sqrt{r}} \frac{g(r')}{\sqrt{r'}} Q_{l+1}\left(\frac{r^2 + r'^2}{2rr'} \right) drdr' \tag{4.6.43}$$

from (4.6.32). The off-diagonal term in $(\psi, \mathcal{K}\psi)$ is

$$\frac{\gamma_c}{8\pi} \int\limits_{\mathbb{R}^3} \int\limits_{\mathbb{R}^3} \frac{\bar{f}(r')}{r'^{3/2}} \Omega^*_{l,m,1/2}(\boldsymbol{\omega}') \frac{1}{|\mathbf{x} - \mathbf{x}'|} \mathbf{p} \cdot \boldsymbol{\sigma} \left\{ \frac{g(r)}{r^{3/2}} \Omega_{l+1,m,-1/2}(\boldsymbol{\omega}) \right\} d\mathbf{x}d\mathbf{x}'$$

$$+ \frac{\gamma_c}{8\pi} \int\limits_{\mathbb{R}^3} \int\limits_{\mathbb{R}^3} \frac{\bar{g}(r')}{r'^{3/2}} \Omega^*_{l+1,m,-1/2}(\boldsymbol{\omega}') \frac{1}{|\mathbf{x} - \mathbf{x}'|} \mathbf{p} \cdot \boldsymbol{\sigma} \left\{ \frac{f(r)}{r^{3/2}} \Omega_{l,m,1/2}(\boldsymbol{\omega}) \right\} d\mathbf{x}d\mathbf{x}'$$

$$=: \gamma_c\left(I_1 + I_2 \right) \tag{4.6.44}$$

say; in (4.6.44) we have used the fact that $|\mathbf{p}|^{-2}$ has kernel $(4\pi|\mathbf{x} - \mathbf{x}'|)^{-1}$; cf. (4.6.17). We shall also need

$$\int\limits_{\mathbb{S}^2} \int\limits_{\mathbb{S}^2} \frac{1}{|\mathbf{x} - \mathbf{x}'|} \Omega^*_{l,m,s}(\boldsymbol{\omega}_x)\Omega_{l,m,s}(\boldsymbol{\omega}_{x'})d\boldsymbol{\omega}_x d\boldsymbol{\omega}_{x'} = \frac{4\pi}{2l + 1} \frac{r_<^l}{r_>^{l+1}}, \tag{4.6.45}$$

where $r_< = \min(|\mathbf{x}|, |\mathbf{x}'|)$ and $r_> = \max(|\mathbf{x}|, |\mathbf{x}'|)$. In I_1, we set $G(r) = g(r)/r^{3/2}$. From (4.6.35) we have

$$(\mathbf{p} \cdot \boldsymbol{\sigma})\{G(r)\Omega_{l+1,m,-1/2}(\boldsymbol{\omega})\} = i\Omega_{l,m,1/2}(\boldsymbol{\omega})\left(\frac{dG}{dr} + \frac{l+2}{r}G(r)\right),$$

and, on substituting in (4.6.44), we get

$$I_1 = \frac{1}{2(2l+1)}\int_0^\infty\int_0^\infty \frac{\bar{f}(r')}{r'^{3/2}}i\frac{r_<^l}{r_>^{l+1}}\left(\frac{dG}{dr} + \frac{l+2}{r}G(r)\right)r^2 r'^2 \, dr dr'$$

$$= \frac{i}{2(2l+1)}\int_0^\infty\int_0^\infty \bar{f}(r')\left(\frac{r_<}{r_>}\right)^{l+1/2}\left\{\frac{dg}{dr} + \frac{l+1/2}{r}g(r)\right\} dr dr'$$

$$= \frac{i}{2}\int_0^\infty\int_0^r \bar{f}(r')g(r)\left(\frac{r'}{r}\right)^{l+1/2}\frac{dr dr'}{r}. \qquad (4.6.46)$$

Similarly

$$I_2 = \frac{-i}{2}\int_0^\infty\int_0^r \bar{g}(r)f(r')\left(\frac{r'}{r}\right)^{l+\frac{1}{2}}\frac{dr dr'}{r}. \qquad (4.6.47)$$

Hence, for $\psi \in \mathcal{H}_{l,m}$ given by (4.6.42), we have

$$(\psi, \mathcal{K}\psi) = \frac{\gamma_c}{2\pi}\int_0^\infty\int_0^\infty \frac{\bar{f}(r)}{\sqrt{r}}\frac{f(r')}{\sqrt{r'}}Q_l\left(\frac{r^2 + r'^2}{2rr'}\right) dr dr'$$

$$+ \frac{\gamma_c}{2\pi}\int_0^\infty\int_0^\infty \frac{\bar{g}(r)}{\sqrt{r}}\frac{g(r')}{\sqrt{r'}}Q_{l+1}\left(\frac{r^2 + r'^2}{2rr'}\right) dr dr'$$

$$- \gamma_c \text{Im}\left[\int_0^\infty\int_0^r \bar{f}(r')g(r)\left(\frac{r'}{r}\right)^{l+\frac{1}{2}}\frac{dr dr'}{r}\right]. \qquad (4.6.48)$$

In

$$(\psi, \tilde{\mathcal{K}}\psi) = \frac{\gamma_c}{2}\left(\frac{1}{|\mathbf{x}|^{1/2}}\psi, \frac{1}{|\mathbf{p}|}\frac{1}{|\mathbf{x}|^{1/2}}\psi\right) + \frac{\gamma_c}{2}\left(\frac{\mathbf{x}\cdot\boldsymbol{\sigma}}{|\mathbf{x}|}\frac{1}{|\mathbf{x}|^{1/2}}\psi, \frac{1}{|\mathbf{p}|}\frac{\mathbf{x}\cdot\boldsymbol{\sigma}}{|\mathbf{x}|}\frac{1}{|\mathbf{x}|^{1/2}}\psi\right)$$

$$\qquad\qquad (4.6.49)$$

the first term is the same as that in $(\psi, \mathcal{K}\psi)$, whilst in the second we have from (4.6.33)

$$\left(\boldsymbol{\sigma}\cdot\frac{\mathbf{x}}{|\mathbf{x}|}\right)\psi(x) = -\frac{1}{r}f(r)\Omega_{l+1,m,-1/2}(\boldsymbol{\omega}) - \frac{1}{r}g(r)\Omega_{l,m,1/2}(\boldsymbol{\omega}).$$

Hence, both terms in (4.6.49) are diagonal and

$$
(\psi, \tilde{\mathcal{K}}\psi) = \frac{\gamma_c}{2\pi} \int_0^\infty \int_0^\infty \frac{\bar{f}(r)}{\sqrt{r}} \frac{f(r')}{\sqrt{r'}} Q_l\left(\frac{r^2 + r'^2}{2rr'}\right) dr dr'
$$

$$
+ \frac{\gamma_c}{2\pi} \int_0^\infty \int_0^\infty \frac{\bar{g}(r)}{\sqrt{r}} \frac{g(r')}{\sqrt{r'}} Q_{l+1}\left(\frac{r^2 + r'^2}{2rr'}\right) dr dr'
$$

$$
+ \frac{\gamma_c}{2\pi} \int_0^\infty \int_0^\infty \frac{\bar{g}(r)}{\sqrt{r}} \frac{g(r')}{\sqrt{r'}} Q_l\left(\frac{r^2 + r'^2}{2rr'}\right) dr dr'
$$

$$
+ \frac{\gamma_c}{2\pi} \int_0^\infty \int_0^\infty \frac{\bar{f}(r)}{\sqrt{r}} \frac{f(r')}{\sqrt{r'}} Q_{l+1}\left(\frac{r^2 + r'^2}{2rr'}\right) dr dr'. \qquad (4.6.50)
$$

Therefore, from (4.6.48) and (4.6.50),

$$
(\psi, \tilde{\mathcal{K}}\psi) - (\psi, \mathcal{K}\psi) = \frac{\gamma_c}{2\pi} \int_0^\infty \int_0^\infty \frac{\bar{g}(r)}{\sqrt{r}} \frac{g(r')}{\sqrt{r'}} Q_l\left(\frac{r^2 + r'^2}{2rr'}\right) dr dr'
$$

$$
+ \frac{\gamma_c}{2\pi} \int_0^\infty \int_0^\infty \frac{\bar{f}(r)}{\sqrt{r}} \frac{f(r')}{\sqrt{r'}} Q_{l+1}\left(\frac{r^2 + r'^2}{2rr'}\right) dr dr'
$$

$$
+ \gamma_c \, \text{Im}\left[\int_0^\infty \int_0^r \bar{f}(r')g(r) \left(\frac{r'}{r}\right)^{l+1/2} \frac{dr dr'}{r}\right]. \qquad (4.6.51)
$$

We need to prove that the right-hand side of (4.6.51) is non-negative. For this we use the Mellin transform $\mathcal{M}$.

Let $\widehat{Q}_l(z) = \tilde{Q}_l(z)/\sqrt{z}$, where $\tilde{Q}_l(x) := Q_l(\frac{1}{2}[x + \frac{1}{x}])$. Thus, from (1.5.21) and (1.5.22)

$$
\widehat{Q}_l^\sharp(z) = \frac{1}{\sqrt{2\pi}} \int_0^\infty p^{-1-iz} \tilde{Q}_l(p) dp
$$

$$
= \frac{1}{2} \sqrt{\frac{\pi}{2}} \frac{\Gamma([l+1-iz]/2)\Gamma([l+1+iz]/2)}{\Gamma([l+2-iz]/2)\Gamma([l+2+iz]/2)}.
$$

We then have in the first term of (4.6.48),

$$
\frac{1}{\sqrt{2\pi}} \int_0^\infty \int_0^\infty \frac{\bar{f}(r)}{\sqrt{r}} \frac{f(r')}{\sqrt{r'}} Q_l\left(\frac{r^2 + r'^2}{2rr'}\right) dr dr' = \int_0^\infty \bar{f}(r)\left(f \star \widehat{Q}_l\right)(r) dr
$$

$$
= \int_{-\infty}^\infty |f^\sharp(s)|^2 \widehat{Q}_l^\sharp(s) ds,
$$

since $\mathcal{M} : L^2(\mathbb{R}_+) \to L^2(\mathbb{R})$ is unitary. Similarly for the second term of (4.6.48). In (4.6.46) and (4.6.47) we set $\Theta(x) = x^{l+1/2}\chi(x)$, where χ is the characteristic function of $[0, 1]$. Then,

$$I_1 = \frac{i}{2} \int_0^\infty \int_0^\infty \bar{f}(r')g(r)\Theta\left(\frac{r'}{r}\right)\frac{dr\,dr'}{r}$$

$$= i\sqrt{\frac{\pi}{2}} \int_0^\infty \bar{f}(r')\,(g \star \Theta)\,(r')dr'$$

$$= i\sqrt{\frac{\pi}{2}} \int_{-\infty}^\infty \overline{f^\sharp}(s)g^\sharp(s)\Theta^\sharp(s)ds,$$

where

$$\Theta^\sharp(s) = \frac{1}{\sqrt{2\pi}} \int_0^1 p^{-is+l}dp = \frac{1}{\sqrt{2\pi}}(1 + l - is)^{-1}.$$

Similarly,

$$I_2 = -i\sqrt{\frac{\pi}{2}} \int_{-\infty}^\infty f^\sharp(s)\overline{g^\sharp}(s)\,\overline{\Theta^\sharp}(s)ds.$$

Hence, we can write

$$(\psi, \mathcal{K}\psi) = \frac{\gamma_c}{\sqrt{2\pi}} \int_{-\infty}^\infty \left(\overline{f^\sharp}(s), \overline{g^\sharp}(s)\right) \begin{pmatrix} \widehat{Q}_l^\sharp(s) & i\pi\Theta^\sharp(s) \\ -i\pi\overline{\Theta^\sharp}(s) & \widehat{Q}_{l+1}^\sharp(s) \end{pmatrix} \begin{pmatrix} f^\sharp(s) \\ g^\sharp(s) \end{pmatrix} ds. \qquad (4.6.52)$$

Since $(\psi, \mathcal{K}\psi) \geq 0$, the smooth 2×2-matrix in (4.6.52) is pointwise non-negative, and hence so is the matrix

$$\begin{pmatrix} \widehat{Q}_{l+1}^\sharp(s) & -i\pi\Theta^\sharp(s) \\ i\pi\overline{\Theta^\sharp}(s) & \widehat{Q}_l^\sharp(s) \end{pmatrix}$$

since the two matrices have the same eigenvalues. Hence,

$$\gamma_c \int_{-\infty}^\infty \left(\overline{f^\sharp}(s), \overline{g^\sharp}(s)\right) \begin{pmatrix} \widehat{Q}_{l+1}^\sharp(s) & -i\pi\Theta^\sharp(s) \\ i\pi\overline{\Theta^\sharp}(s) & \widehat{Q}_l^\sharp(s) \end{pmatrix} \begin{pmatrix} f^\sharp(s) \\ g^\sharp(s) \end{pmatrix} ds \geq 0.$$

But this is precisely the right-hand side of (4.6.51). Hence, $(\psi, \mathcal{K}\psi) \leq (\psi, \tilde{\mathcal{K}}\psi)$ for all the $C_0^1(\mathbb{R}^3)$ functions ψ in $\mathcal{H}_{l,m}$ considered, and, hence, for all $\psi \in \mathcal{H}_{l,m}$ since $\mathcal{K}$ and $\tilde{\mathcal{K}}$ are bounded. Since $\mathcal{K}$ and $\tilde{\mathcal{K}}$ map $\mathcal{H}_{l,m}$ into itself, it follows that $\mathcal{K} \leq \tilde{\mathcal{K}}$ on $\mathcal{H}$ and, hence, the lemma is proved. $\qquad \square$

Lemma 4.6.7. *For all* $\psi \in L^2(\mathbb{R}^3, \mathbb{C}^2)$

$$(\psi, \tilde{\mathcal{K}}\psi) \leq \|\psi\|^2.$$

Proof. First observe from (4.6.50) that since $Q_0 \geq Q_1 \geq \ldots \geq 0$, it is sufficient to prove the result for $\psi \in \mathcal{H}_{0,m}$. In (4.6.50), with $l = 0$, we may use Hilbert's inequality in Theorem 1.7.2 since $\frac{1}{\sqrt{r}\sqrt{r'}} Q_l\left(\frac{r^2+r'^2}{2rr'}\right)$ is homogeneous of degree -1. We have

$$(\psi, \tilde{\mathcal{K}}\psi) \leq \frac{\gamma_c}{2\pi} \left\{ k_1 \int_0^\infty (|f(r)|^2 + |g(r)|^2) dr + k_2 \int_0^\infty (|f(r)|^2 + |g(r)|^2) dr \right\},$$

where, by (2.2.8),

$$k_1 = \int_0^\infty Q_0\left(\frac{1}{2}\left[r + \frac{1}{r}\right]\right) \frac{dr}{r} = \int_0^\infty \ln\left|\frac{r+1}{r-1}\right| \frac{dr}{r} = \frac{\pi^2}{2}$$

and

$$k_2 = \int_0^\infty Q_1\left(\frac{1}{2}\left[r + \frac{1}{r}\right]\right) \frac{dr}{r} = \int_0^\infty \left\{\frac{1}{2}\left(r + \frac{1}{r}\right) \ln\left|\frac{r+1}{r-1}\right| - 1\right\} \frac{dr}{r} = 2.$$

Hence,

$$(\psi, \tilde{\mathcal{K}}\psi) \leq \int_0^\infty (|f(r)|^2 + |g(r)|^2) dr = \|\psi\|^2$$

and the lemma is proved. $\qquad\square$

Lemma 4.6.8. *Let*

$$\psi(\mathbf{x}) = \sum_{(l,m,s)\in\mathcal{I}} \frac{1}{r} f_{l,m,s}(r) \Omega_{l,m,s}(\boldsymbol{\omega})$$

$$\phi(\mathbf{x}) = \frac{1}{r} \sqrt{\sum_{(l,m,s)\in\mathcal{I}} |f_{l,m,s}(r)|^2} \; \Omega_{0,1/2,1/2}(\boldsymbol{\omega}), \quad \Omega_{0,1/2,1/2}(\boldsymbol{\omega}) = \frac{1}{2\sqrt{\pi}}\begin{pmatrix}1\\0\end{pmatrix}.$$

Then

$$\|\psi\| = \|\phi\|, \tag{4.6.53}$$

$$(\psi, \tilde{\mathcal{K}}\psi) \leq (\phi, \tilde{\mathcal{K}}\phi), \tag{4.6.54}$$

$$\int_{\mathbb{R}^3} \frac{1}{\sqrt{|\mathbf{x}|}} |\psi(\mathbf{x})| d\mathbf{x} \leq \int_{\mathbb{R}^3} \frac{1}{\sqrt{|\mathbf{x}|}} |\phi(\mathbf{x})| d\mathbf{x}. \tag{4.6.55}$$

Proof. The equality (4.6.53) is obvious. With

$$\psi_{l,m}(\mathbf{x}) = \frac{1}{r}f_{l,m,1/2}(r)\Omega_{l,m,1/2}(\boldsymbol{\omega}) + \frac{1}{r}f_{l+1,m,-1/2}(r)\Omega_{l+1,m,-1/2}(\boldsymbol{\omega})$$

we have from (4.6.50)

$$(\psi_{l,m}, \tilde{\mathcal{K}}\psi_{l,m}) \le \frac{\gamma_c}{2\pi}\int_0^\infty\int_0^\infty \left\{ \frac{1}{\sqrt{r}\sqrt{r'}}\left(|f_{l,m,1/2}(r)||f_{l,m,1/2}(r')|\right.\right.$$

$$+ |f_{l+1,m,-1/2}(r)||f_{l+1,m,-1/2}(r')|)$$

$$\left.\left.\times\left(Q_0\left(\frac{r^2+r'^2}{2rr'}\right) + Q_1\left(\frac{r^2+r'^2}{2rr'}\right)\right)\right)\right\}drdr'.$$

Hence,

$$(\psi, \tilde{\mathcal{K}}\psi) = \sum_{l,m}(\psi_{l,m}, \tilde{\mathcal{K}}\psi_{l,m})$$

$$\le \frac{\gamma_c}{2\pi}\sum_{(l,m,s)\in\mathcal{I}}\int_0^\infty\int_0^\infty \frac{1}{\sqrt{r}\sqrt{r'}}|f_{l,m,s}(r)||f_{l,m,s}(r')|$$

$$\times\left[Q_0\left(\frac{r^2+r'^2}{2rr'}\right) + Q_1\left(\frac{r^2+r'^2}{2rr'}\right)\right]drdr'$$

$$\le \frac{\gamma_c}{2\pi}\int_0^\infty\int_0^\infty \frac{1}{\sqrt{rr'}}\left(\sum_{(l,m,s)\in\mathcal{I}}|f_{l,m,s}(r')|^2\right)^{1/2}\left(\sum_{(l,m,s)\in\mathcal{I}}|f_{l,m,s}(r)|^2\right)^{1/2}$$

$$\times\left[Q_0\left(\frac{r^2+r'^2}{2rr'}\right) + Q_1\left(\frac{r^2+r'^2}{2rr'}\right)\right]drdr'$$

$$= (\phi, \tilde{K}\phi)$$

again by (4.6.50) with $g = 0$, $f = \left(\sum_{(l,m,s)\in I}|f_{l,m,s}|^2\right)^{1/2}$, $l = 0$ and $m = 1/2$.

Finally, if $d\mu$ denotes the Haar measure on the orthogonal group O(3), we have for $g \in$ O(3)

$$\int_{\mathbb{R}^3}\frac{1}{\sqrt{|\mathbf{x}|}}|\psi(\mathbf{x})|d\mathbf{x} = \int_{\mathbb{R}^3}\frac{1}{\sqrt{|\mathbf{x}|}}|\psi(g\mathbf{x})|d\mathbf{x}$$

$$=: I(g)$$

$$= \int_{O(3)}I(g)d\mu(g)$$

$$= \int_{\mathbb{R}^3}\frac{1}{\sqrt{|\mathbf{x}|}}d\mathbf{x}\left(\int_{O(3)}|\psi(g\mathbf{x})|d\mu(g)\right)$$

$$\leq \int\limits_{\mathbb{R}^3} \frac{1}{\sqrt{|\mathbf{x}|}} \left(\int\limits_{O(3)} |\psi(g\mathbf{x})|^2 d\mu(g) \right)^{1/2} d\mathbf{x}$$

$$= \frac{1}{2\sqrt{\pi}} \int\limits_{\mathbb{R}^3} \frac{1}{\sqrt{|\mathbf{x}|}} \left(\frac{1}{r^2} \sum_{(l,m,s)\in\mathfrak{J}} |f_{l,m,s}(r)|^2 \right)^{1/2} d\mathbf{x}$$

$$= \int\limits_{\mathbb{R}^3} \frac{1}{\sqrt{|\mathbf{x}|}} |\phi(\mathbf{x})| d\mathbf{x}.$$

The proof is therefore complete.   □

Proof of Lemma 4.6.5.

In view of Lemmas 4.6.4–4.6.8 and a change of scale, it is enough to prove (4.6.39) with $\mathcal{K}$ replaced by $\tilde{\mathcal{K}}$ and ψ of the form

$$\psi(\mathbf{x}) = \frac{f(r)}{r} \Omega_{0,1/2,1/2}(\boldsymbol{\omega}),$$

with $f \geq 0$ and support $f \subset B(0,1)$, that is, from (4.6.48)

$$(\psi, \tilde{\mathcal{K}}\psi) \equiv \frac{\gamma_c}{2\pi} \int\limits_0^\infty \int\limits_0^\infty \frac{f(r)f(r')}{\sqrt{r}\sqrt{r'}} \left[Q_0\left(\frac{r^2+r'^2}{2rr'} \right) + Q_1\left(\frac{r^2+r'^2}{2rr'} \right) \right] dr dr'$$

$$\leq \int\limits_0^1 |f(r)|^2 dr \; - \; C_0 \left(2\sqrt{\pi} \int\limits_0^1 \sqrt{r} f(r) dr \right)^2. \qquad (4.6.56)$$

Define

$$\tilde{f}(r) := \begin{cases} f(r), & r \in [0,1] \\ \frac{1}{r} f\left(\frac{1}{r}\right), & r > 1. \end{cases}$$

Then

$$2 \int\limits_0^1 |f(r)|^2 dr = \int\limits_0^\infty |\tilde{f}(r)|^2 dr \qquad (4.6.57)$$

and, from Lemma 4.6.7,

$$\frac{\gamma_c}{2\pi} \int\limits_0^\infty \int\limits_0^\infty \frac{\tilde{f}(r)\tilde{f}(r')}{\sqrt{r}\sqrt{r'}} \left[Q_0\left(\frac{r^2+r'^2}{2rr'} \right) + Q_1\left(\frac{r^2+r'^2}{2rr'} \right) \right] dr dr'$$

$$\leq \int\limits_0^\infty |\tilde{f}(r)|^2 dr = 2 \int\limits_0^1 |f(r)|^2 dr. \qquad (4.6.58)$$

The left-hand side can be written as

$$2\gamma_c \Bigg\{ \frac{1}{2\pi} \int_0^1 \int_0^1 \frac{f(r)f(r')}{\sqrt{r}\sqrt{r'}} \left[Q_0\left(\frac{r^2+r'^2}{2rr'}\right) + Q_1\left(\frac{r^2+r'^2}{2rr'}\right) \right] dr dr'$$

$$+ \frac{1}{2\pi} \int_0^1 \int_1^\infty \frac{f(r)f(1/r')}{\sqrt{r}\, r'\sqrt{r'}} \left[Q_0\left(\frac{r^2+r'^2}{2rr'}\right) + Q_1\left(\frac{r^2+r'^2}{2rr'}\right) \right] dr dr' \Bigg\}$$

$$= 2\gamma_c (J_1 + J_2)$$

say. We have

$$J_2 = \frac{1}{2\pi} \int_0^1 \int_0^1 \frac{f(r)f(r')}{\sqrt{r}\sqrt{r'}} \left[Q_0\left(\frac{1}{2}[rr' + \frac{1}{rr'}]\right) + Q_1\left(\frac{1}{2}[rr' + \frac{1}{rr'}]\right) \right] dr dr'$$

$$\geq \frac{1}{2\pi} \int_0^1 \int_0^1 \sqrt{r} f(r) \sqrt{r'} f(r') \left[\frac{1}{rr'} Q_0\left(\frac{1}{2}[rr' + \frac{1}{rr'}]\right) \right] dr dr'$$

$$\geq \frac{1}{\pi} \int_0^1 \int_0^1 \sqrt{r} f(r) \sqrt{r'} f(r') dr dr' \quad (4.6.59)$$

since $\inf_{0 \leq u \leq 1} \left[\frac{1}{u} \ln\left(\frac{1+u}{1-u}\right) \right] = 2$. Hence, from (4.6.58) and (4.6.59),

$$2 \int_0^1 |f(r)|^2 dr \geq 2(\psi, \tilde{K}\psi) + 2\gamma_c \left(\frac{1}{\pi} \int_0^1 \int_0^1 \sqrt{r} f(r) \sqrt{r'} f(r') dr dr' \right),$$

whence

$$(\psi, \tilde{\mathcal{K}}\psi) \leq \|\psi\|^2 - \frac{\gamma_c}{\pi} \left(\int_0^1 \sqrt{r} f(r) dr \right)^2$$

$$= \|\psi\|^2 - C_0 \left(\int_{\mathbb{R}^3} |\psi(\mathbf{x})| \frac{d\mathbf{x}}{\sqrt{|\mathbf{x}|}} \right)^2$$

with $C_0 = \gamma_c/4\pi^2 = (\pi^3 + 4\pi)^{-1}$. $\qquad\square$

Proof of Theorem 4.6.3.

Now that we have available Lemma 4.6.5, we are able to proceed along similar lines to those in the proof of Theorem 4.6.2. The argument to justify the sufficiency of taking $Z_k = Z_c$ for all k, continues to hold.

We shall prove that when $\alpha Z_k = \alpha Z = \gamma_c$, $k = 1, \ldots, K$, there exist constants A and C such that

$$(\psi, \mathbb{B}^0_{1,K}(\mathbf{Z}, \mathbf{R})\psi) \geq -A\gamma_c \sum_{j \neq k} \frac{1}{|R_j - R_k|} \|\psi\|^2 + CK\|\psi\|^2 \quad (4.6.60)$$

for all $\psi \in \Lambda_+ H^1(\mathbb{R}^3, \mathbb{C}^4)$.

We have

$$V(\mathbf{x}) := \sum_{k=1}^{K} \left\{ \frac{\alpha Z_k}{|\mathbf{x} - R_k|} - A\gamma_c \sum_{j \neq k} \frac{1}{|R_j - R_k|} \right\}$$

$$\leq \left(\sum_{k=1}^{K} V_k^2(\mathbf{x}) \right)$$

$$\leq \left(\sum_{k=1}^{K} V_k(\mathbf{x}) \right)^2$$

$$=: U(\mathbf{x}), \qquad (4.6.61)$$

where

$$V_k^2(\mathbf{x}) := \gamma_c \left[\frac{1}{|\mathbf{x} - R_k|} - A\sum_{j \neq k} \frac{1}{|R_j - R_k|} \right]_+,$$

and the subscript $+$ denotes the positive part. Hence, (4.6.60) is satisfied if

$$(\psi,\ \mathbb{D}_0\psi) \geq (\psi,\ U\psi) + CK\|\psi\|^2. \qquad (4.6.62)$$

We now translate the problem to one for 2-spinors, using (4.6.22). In view of the remarks concerning $\mathfrak{b}_{1,1}^0(Z)$ in (4.6.23) (applied K times with the origin shifted to R_k, $k = 1, \ldots, K$, to accommodate the K nuclei), (4.6.62) will follow if we can prove that for all $\phi \in H^1(\mathbb{R}^3, \mathbb{C}^2)$

$$\left(\phi,\ [|\mathbf{p}| - \frac{1}{2}(U + \frac{\mathbf{p}\cdot\boldsymbol{\sigma}}{|\mathbf{p}|} U \frac{\mathbf{p}\cdot\boldsymbol{\sigma}}{|\mathbf{p}|})]\phi \right) \geq 0. \qquad (4.6.63)$$

Clearly, $\frac{\mathbf{p}\cdot\boldsymbol{\sigma}}{|\mathbf{p}|}$ commutes with $U + \frac{\mathbf{p}\cdot\boldsymbol{\sigma}}{|\mathbf{p}|} U \frac{\mathbf{p}\cdot\boldsymbol{\sigma}}{|\mathbf{p}|} =: \tilde{U}$ and if L is the operator $\phi(x_1, x_2, x_3) \to \sigma_3\phi(x_1, x_2, -x_3)$ we have $L(\frac{\mathbf{p}\cdot\boldsymbol{\sigma}}{|\mathbf{p}|}) = -(\frac{\mathbf{p}\cdot\boldsymbol{\sigma}}{|\mathbf{p}|})L$, and $L\tilde{U} = \tilde{\tilde{U}}L$, where $\tilde{\tilde{U}}$ coincides with $\tilde{U}$ with a minus sign inserted before the third components of the R_k. Thus we need only prove (4.6.63) for $\phi \in P_+ H^{1/2}(\mathbb{R}^3, \mathbb{C}^2)$, where P_+ is the projection onto the eigenspace at 1 for $\frac{\mathbf{p}\cdot\boldsymbol{\sigma}}{|\mathbf{p}|}$. In other words, it is sufficient to prove that for all $\phi \in H^{1/2}(\mathbb{R}^3, \mathbb{C}^2)$

$$(\phi, |\mathbf{p}|\phi) \geq (\phi,\ P_+ U P_+ \phi),$$

or, equivalently, for all $u \in \mathcal{H}$,

$$\|u\|^2 \geq \left(u,\ \frac{1}{\sqrt{|\mathbf{p}|}} P_+ U P_+ \frac{1}{\sqrt{|\mathbf{p}|}} u \right). \qquad (4.6.64)$$

The functions V_k in (4.6.61) are supported in balls $B(R_k, t_k)$, where

$$t_k^{-1} = A \sum_{j \neq k} \frac{1}{|R_j - R_k|}. \qquad (4.6.65)$$

Hence,

$$|R_j - R_k| - (t_j + t_k) \geq \left(1 - \frac{2}{A}\right)|R_j - R_k| \qquad (4.6.66)$$

so that the balls are disjoint if $A > 2$.

Let χ_k denote the characteristic function of $B(R_k, t_k)$. Then

$$V_k^2(\mathbf{x}) = V_k^2(\mathbf{x})\chi_k(\mathbf{x}) \leq \frac{\gamma_c}{|\mathbf{x} - R_k|}$$

and

$$\sqrt{U(\mathbf{x})} \leq \gamma_c^{1/2} \sum_{k=1}^{K} \{|\mathbf{x} - R_k|^{-1/2}\chi_k(\mathbf{x})\} =: \sqrt{U_1(\mathbf{x})}.$$

Clearly (4.6.64) is satisfied if, for all $u \in \mathcal{H}$

$$\|u\|^2 \geq \left(u, \frac{1}{\sqrt{|\mathbf{p}|}}P_+U_1P_+\frac{1}{\sqrt{|\mathbf{p}|}}u\right).$$

With $T := \frac{1}{\sqrt{|\mathbf{p}|}}P_+\sqrt{U_1}$, this becomes

$$(u, TT^*u) \leq \|u\|^2,$$

and so $\|T^*\| \leq 1$, which in turn is implied by

$$(u, T^*Tu) \leq \|u\|^2, \quad u \in \mathcal{H},$$

or

$$\left(u, \sqrt{U_1}P_+\frac{1}{|\mathbf{p}|}P_+\sqrt{U_1}\,u\right) \leq \|u\|^2, \quad u \in \mathcal{H}. \qquad (4.6.67)$$

Thus, we need to prove that for all $u \in \mathcal{H}$

$$\|u\|^2 \geq \gamma_c \sum_{j,k=1}^{K} \left(u, \chi_j(\mathbf{x})\frac{1}{|\mathbf{x} - R_j|^{1/2}}P_+\frac{1}{|\mathbf{p}|}P_+\chi_k(\mathbf{x})\frac{1}{|\mathbf{x} - R_k|^{1/2}}u\right)$$

$$= \gamma_c \sum_{j=1}^{K} \left(u, \chi_j(\mathbf{x})\frac{1}{|\mathbf{x} - R_j|^{1/2}}P_+\frac{1}{|\mathbf{p}|}P_+\chi_j(\mathbf{x})\frac{1}{|\mathbf{x} - R_j|^{1/2}}u\right)$$

$$+ \gamma_c \sum_{j\neq k} \left(u, \chi_j(\mathbf{x})\frac{1}{|\mathbf{x} - R_j|^{1/2}}P_+\frac{1}{|\mathbf{p}|}P_+\chi_k(\mathbf{x})\frac{1}{|\mathbf{x} - R_k|^{1/2}}u\right)$$

$$= I_1 + I_2$$

say. By Lemma 4.6.5, with $C_0 = (\pi^3 + 4\pi)^{-1}$,

$$I_1 \leq \sum_{j=1}^{K} \left\{\|\chi_j u\|^2 - \frac{C_0}{t_j^2}\left(\int\frac{1}{|\mathbf{x} - R_j|^{1/2}}|(\chi_j u)(\mathbf{x})|dx\right)^2\right\}.$$

In I_2

$$P_+\frac{1}{|\mathbf{p}|}P_+ = \frac{1}{2}\left(\frac{1}{|\mathbf{p}|} + \frac{\mathbf{p}\cdot\sigma}{|\mathbf{p}|^2}\right);$$

$\frac{1}{|\mathbf{p}|}$, $\frac{\mathbf{p}\cdot\boldsymbol{\sigma}}{|\mathbf{p}|^2}$ have kernels $\frac{1}{2\pi^2|\mathbf{x}-\mathbf{y}|^2}$, $\frac{i}{4\pi}\frac{(\mathbf{x}-\mathbf{y})\cdot\boldsymbol{\sigma}}{|\mathbf{x}-\mathbf{y}|^{3/2}}$, respectively. Thus

$$I_2 \le \frac{1}{4\pi}\left(\frac{1}{\pi}+\frac{1}{2}\right)\gamma_c \sum_{j\ne k}\int_{\mathbb{R}^3}\int_{\mathbb{R}^3}\frac{1}{|\mathbf{x}-R_j|^{1/2}}|(\chi_j u)(\mathbf{x})| \times$$

$$\times \frac{1}{|\mathbf{x}-\mathbf{y}|^2}\frac{1}{|\mathbf{y}-R_k|^{1/2}}|(\chi_k u)(\mathbf{y})|d\mathbf{x}d\mathbf{y}$$

$$\le \frac{1}{4\pi}\left(\frac{1}{\pi}+\frac{1}{2}\right)\gamma_c\left[1-\frac{2}{A}\right]^{-2}\sum_{j\ne k}|R_j-R_k|^{-2}m_j m_k$$

by (4.6.66), where

$$m_j := \int_{\mathbb{R}^3}\frac{\chi_j(\mathbf{x})|u(\mathbf{x})|}{|\mathbf{x}-R_j|^{1/2}}d\mathbf{x}.$$

Thus

$$I_1 + I_2 \le \|u\|^2 - C_0\sum_{j=1}^{K}\frac{1}{t_j^2}m_j^2 + \frac{1}{4\pi}\left(\frac{1}{\pi}+\frac{1}{2}\right)\gamma_c\left[1-\frac{2}{A}\right]^{-2}\times$$

$$\times \sum_{j\ne k}|R_j-R_k|^{-2}(1/2)(m_j^2+m_k^2)$$

$$\le \|u\|^2$$

if

$$\frac{C_0}{t_j^2} - \frac{1}{4\pi}\left(\frac{1}{\pi}+\frac{1}{2}\right)\gamma_c\left[1-\frac{2}{A}\right]^{-2}\sum_{j\ne k}|R_j-R_k|^{-2}\ge 0$$

for all j. On substituting (4.6.65), the last inequality is satisfied if

$$A^2 \ge \left(1+\frac{\pi}{2}\right)\left(1-\frac{2}{A}\right)^{-2},$$

that is $A \ge 2+\sqrt{1+\pi/2}$. We have, therefore, established (4.6.60), and can take $A = 2+\sqrt{1+\pi/2}$. Consequently, we have

$$\mathbb{B}_{1,K}(\mathbf{Z},\mathbf{R}) \ge -2A\gamma_c\sum_{j<k}\frac{1}{|R_k-R_j|} + \alpha Z_c^2\sum_{j<k}\frac{1}{|R_k-R_j|} + CK \ge CK$$

if $2\alpha[2+\sqrt{1+\pi/2}] \le \gamma_c$, that is $\alpha \le \frac{2\pi}{(\pi^2+4)(2+\sqrt{1+\pi/2})} \approx 0.125721$. The proof is therefore complete. $\square$

4.7 Stability of matter in magnetic fields

Our focus in this section is on techniques that have proved to be effective for establishing the stability of matter subject to magnetic fields. We recall from Section 4.1 that, as was shown to be the case in the last section for relativistic molecules, when magnetic fields are present, stability requires a bound on the fine-structure constant α as well as on $Z\alpha$ or $Z\alpha^2$.

4.7.1 *Non-relativistic matter*

The Hamiltonian to be considered for N electrons and K static nuclei having charge Z in a magnetic field $\mathbf{B} = \text{curl } \mathbf{A}$ is that given in (4.1.8), namely

$$\mathbb{P}_{N,K}(\mathbf{A}) = \sum_{j=1}^{N} \mathbb{P}_{\mathbf{A}}(\mathbf{x}_j) + \alpha V_C + H_F, \tag{4.7.1}$$

where $\mathbb{P}_{\mathbf{A}} = [\boldsymbol{\sigma} \cdot (-i\boldsymbol{\nabla} + \mathbf{A})]^2$, the Pauli operator and $H_F = (8\pi\alpha^2)^{-1} \int_{\mathbb{R}^3} |\mathbf{B}(\mathbf{x})|^2 d\mathbf{x}$ is the field energy, which, as noted before, has to be included for stability for large magnetic fields. We follow the treatment in [Lieb *et al.* (1995)] in which use is made of the relativistic stability result

$$\sum_{j=1}^{N} |-i\boldsymbol{\nabla} + \mathbf{A}(\mathbf{x}_j)| + \kappa V_C \geq 0 \tag{4.7.2}$$

proved in [Lieb and Yau (1988)], to give a lower bound for the Coulomb potential term V_C in (4.7.1) and thus obtain

$$\mathbb{P}_{N,K}(\mathbf{A}) \geq \tilde{P} := \sum_{j=1}^{N} T_j + H_F, \tag{4.7.3}$$

where the T_j are copies of the one-body operator

$$T := \mathbb{P}_{\mathbf{A}} - \kappa^{-1} |\frac{1}{i} \boldsymbol{\nabla} + \mathbf{A}|. \tag{4.7.4}$$

The inequality (4.7.2) is proved in the corollary to Theorem 1 in [Lieb and Yau (1988)] under the conditions (appropriate to here)

$$0 < 2\kappa \leq 0.032, \quad Z\kappa \leq 1/\pi. \tag{4.7.5}$$

It follows from the Pauli Exclusion Principle that the ground state energy E of $\mathbb{P}_{N,K}(\mathbf{A})$ is bounded below by $\overline{E}_N + H_F$, where

$$\overline{E}_N = 2 \sum_{j=1}^{[N/2]} \lambda_j$$

and $\lambda_1 \leq \lambda_2 \leq \cdots$ are the eigenvalues of T repeated according to multiplicity. Define

$$N_\lambda(T) := \#\{j : \lambda_j \leq \lambda\}.$$

Let $\mu > 0$, and suppose there exists k such that $-\lambda_{k+1} < \mu \leq -\lambda_k, 1 \leq k \leq [N/2]$. Then

$$\int_\mu^\infty N_{-\lambda}(T) d\lambda = \sum_{j=1}^{k} j(-\lambda_j + \lambda_{j+1})$$

$$= -\sum_{j=1}^{k} \lambda_j + k\lambda_{k+1}$$

and so

$$\sum_{j=1}^{k} \lambda_j = k\lambda_{k+1} - \int_{\mu}^{\infty} N_{-\lambda}(T)d\lambda. \tag{4.7.6}$$

This gives

$$\sum_{j=1}^{[N/2]} \lambda_j = k\lambda_{k+1} + \sum_{j=k+1}^{[N/2]} \lambda_j - \int_{\mu}^{\infty} N_{-\lambda}(T)d\lambda$$

$$\geq -\mu N/2 - \int_{\mu}^{\infty} N_{-\lambda}(T)d\lambda \tag{4.7.7}$$

since $\lambda_j > -\mu$ for $j \geq k+1$. We therefore have

$$\overline{E}_N \geq -N\mu - 2\int_{\mu}^{\infty} N_{-\lambda}(T)d\lambda. \tag{4.7.8}$$

If $\mu > -\lambda_k$ for all $1 \leq k \leq [N/2]$, then $\overline{E}_N \geq -N\mu$, and, hence, (4.7.8) holds always.

Since $\mathbb{P}_{\mathbf{A}} > 0$, and $\mathbb{P}_{\mathbf{A}} = (-i\boldsymbol{\nabla} + \mathbf{A})^2 + \boldsymbol{\sigma} \cdot \mathbf{B}$, it follows that, for $\lambda \geq \mu$,

$$\mathbb{P}_{\mathbf{A}} \geq \mu\lambda^{-1}\mathbb{P}_{\mathbf{A}}$$
$$= \mu\lambda^{-1}(-i\boldsymbol{\nabla} + \mathbf{A})^2 - \mu\lambda^{-1}|\mathbf{B}|.$$

By the Cauchy–Schwarz inequality, for any $\kappa > 0, \psi \in \mathcal{D}([\mathbf{p}+\mathbf{A}]^2)$, with $\mathbf{p} := -i\boldsymbol{\nabla}$,

$$(|\mathbf{p} + A|\psi, \psi) \leq \||\mathbf{p} + A|\psi\|\|\psi\|$$
$$= ([\mathbf{p} + A]^2\psi, \psi)^{1/2}\|\psi\|$$
$$\leq (1/3\lambda\kappa)([\mathbf{p} + A]^2\psi, \psi) + (3\lambda\kappa/4)\|\psi\|^2.$$

The choice $\mu = 4/3\kappa^2$ gives

$$T \geq \lambda^{-1}\kappa^{-2}(\mathbf{p} + \mathbf{A})^2 - (4/3)\lambda^{-1}\kappa^{-2}|\mathbf{B}| - (3\lambda/4)$$

and, hence,

$$T + \lambda \geq \lambda^{-1}\kappa^{-2}\left\{(\mathbf{p} + \mathbf{A})^2 - V\right\},$$

where $V(\mathbf{x}) = (4/3)|\mathbf{B}(\mathbf{x})| - (\lambda^2\kappa^2/4)$. This implies that

$$N_{-\lambda}(T) \leq N_0(T_\lambda),$$

where $T_\lambda := (\mathbf{p} + \mathbf{A})^2 - V$ and so the CLR inequality (1.8.7) gives

$$N_{-\lambda}(T) \leq L_3 \int_{\mathbb{R}^3} \left[\frac{4|\mathbf{B}(\mathbf{x})|}{3} - \frac{\lambda^2\kappa^2}{4}\right]_+^{3/2} d\mathbf{x}, \tag{4.7.9}$$

where L_3 is Lieb's bound, $L_3 = 0.1156$.

On substituting (4.7.9) in (4.7.8) and inverting the order of integration, we get

$$\overline{E}_N \geq -N\mu - 2L_3 \int_\mu^\infty \int_{|\mathbf{B}(\mathbf{x})| \geq 3\lambda^2\kappa^2/16} \left[\frac{4}{3}|\mathbf{B}(\mathbf{x})| - \frac{\lambda^2\kappa^2}{4}\right]^{3/2} d\mathbf{x}\,d\lambda$$

$$\geq -N\mu - 2L_3 \int_{\mathbb{R}^3} d\mathbf{x} \int_0^{\sqrt{16|\mathbf{B}(\mathbf{x})|/3\kappa^2}} \left[\frac{4}{3}|\mathbf{B}(\mathbf{x})| - \frac{\lambda^2\kappa^2}{4}\right]^{3/2} d\lambda$$

$$= -N\mu - \frac{4\pi L_3}{3\kappa} \int_{\mathbb{R}^3} |\mathbf{B}(\mathbf{x})|^2 d\mathbf{x}.$$

Therefore, the ground state energy E satisfies

$$E \geq \overline{E}_N + H_F$$

$$\geq -N\mu + \left\{\frac{1}{8\pi\alpha^2} - \frac{4\pi L_3}{3\kappa}\right\} \int_{\mathbb{R}^3} |\mathbf{B}(\mathbf{x})|^2 d\mathbf{x}$$

$$\geq -N\mu$$

if $\kappa \geq (32\pi^2/3\alpha^2)L_3 = 12.2\alpha^2$ when $L_3 = 0.1156$. On inserting this in the Lieb–Yau conditions (4.7.5), we have that $E \geq -N\mu$, where $\mu = (4/3)\kappa^{-2}$, and, hence, stability, if

$$2\alpha \leq 0.071, \quad 2Z\alpha^2 \leq 0.052. \tag{4.7.10}$$

For $\alpha = 1/137$, this requires $Z \leq 490$. If (4.7.10) is assumed, the choice $\kappa = \min\{0.0315/2, (\pi Z)^{-1}\}$ yields the lower bound

$$E \geq -N\max\{5380, 13.2Z^2\}. \tag{4.7.11}$$

Note that (4.7.11) holds for all magnetic fields $\mathbf{B}$. In [Lieb *et al.* (1995)], it is also established by a direct method that does not involve the use of (4.7.2), that

$$E \geq -2.62^{2/3}\max\{Q(Z)^2, Q(11.4)^2\}N^{1/3}K^{2/3}, \tag{4.7.12}$$

where $Q(t) = t + \sqrt{2t} + 2.2$, provided that

$$2Z\alpha^2 \leq 0.082, \quad \text{and} \quad 2\alpha \leq 0.12.$$

It is shown in [Lieb (1976)] that this gives the correct dependence of the ground state energy E on Z for $Z \geq 1$.

The following Lieb–Thirring inequality is proved in [Lieb *et al.* (1995)] by the same technique

Theorem 4.7.1. *Let* $\lambda_1 \leq \lambda_2 \leq \cdots$ *be the negative eigenvalues of* $\mathbb{P}_\mathbf{A} - V, V \geq 0$. *Then, for all* $0 < \gamma < 1$,

$$\sum |\lambda_j| \leq a_\gamma \int_{\mathbb{R}^3} V(\mathbf{x})^{5/2}d\mathbf{x} + b_\gamma \left(\int_{\mathbb{R}^3} |\mathbf{B}(\mathbf{x})|^2 d\mathbf{x}\right)^{3/4} \left(\int_{\mathbb{R}^3} V(\mathbf{x})^4 d\mathbf{x}\right)^{1/4} \tag{4.7.13}$$

where $a_\gamma(2^{3/2}/5)(1-\gamma)^{-1}L_3$, $b_\gamma = 3^{1/4}2^{-9/4}\pi\gamma^{-3/8}(1-\gamma)^{-5/8}L_3$, *and* $L_3 = 0.1156$.

Proof. As in (4.7.6), it follows that

$$\sum \lambda_j = -\int_0^\infty N_{-\lambda}(\mathbb{P}_\mathbf{A} - V) d\lambda$$

$$= \left\{ -\left[\int_0^\mu + \int_\mu^\infty \right] N_{-\lambda}(\mathbb{P}_\mathbf{A} - V) d\lambda \right\}$$

$$= I_1 + I_2, \tag{4.7.14}$$

say, where μ is a parameter which is to be optimised. For $0 < \lambda \le \mu$,

$$\mathbb{P}_\mathbf{A} - V + \lambda \ge (\mathbf{p} + \mathbf{A})^2 - |\mathbf{B}| - V + \lambda$$

and, hence, by the CLR inequality (1.8.7)

$$N_{-\lambda}(\mathbb{P}_\mathbf{A} - V) \le N_0((\mathbf{p} + \mathbf{A})^2 - |\mathbf{B}| - V + \lambda)$$

$$\le L_3 \int_{\mathbb{R}^3} (|\mathbf{B}(\mathbf{x})| + V(\mathbf{x}) - \lambda)_+^{3/2} \, d\mathbf{x}. \tag{4.7.15}$$

For $\lambda \ge \mu$, we take

$$\mathbb{P}_\mathbf{A} \ge \mu \lambda^{-1} \mathbb{P}_\mathbf{A} \ge \mu \lambda^{-1} [(\mathbf{p} + \mathbf{A})^2 - |\mathbf{B}|]$$

to get

$$N_{-\lambda}(\mathbb{P}_\mathbf{A} - V) \le N_{-\lambda}(\mu \lambda^{-1}[(\mathbf{p} + \mathbf{A})^2 - |\mathbf{B}|] - V)$$

$$= N_0([(\mathbf{p} + \mathbf{A})^2 - |\mathbf{B}|] - \mu^{-1}\lambda V + \mu^{-1}\lambda^2)$$

$$\le L_3 \int_{\mathbb{R}^3} (|\mathbf{B}(\mathbf{x})| + \mu^{-1}\lambda V(\mathbf{x}) - \mu^{-1}\lambda^2)_+^{3/2} \, d\mathbf{x}. \tag{4.7.16}$$

We now claim that, for any $0 < \gamma < 1$, the integrand in (4.7.15) is bounded above by

$$\sqrt{2} \left([|\mathbf{B}(\mathbf{x})| - \gamma \lambda^2/\mu]_+^{3/2} + [V(\mathbf{x}) - (1 - \gamma)\lambda]_+^{3/2} \right).$$

To verify this, we first write

$$|\mathbf{B}| + V - \lambda = [|\mathbf{B}| - \gamma \mu^{-1}\lambda^2]_+ + [V - (1 - \gamma)\lambda] + \gamma \lambda[(-1 + \mu^{-1}\lambda],$$

which gives, since $\lambda < \mu$,

$$(|\mathbf{B}| + V - \lambda)_+ \le ([|\mathbf{B}| - \gamma \mu^{-1}\lambda^2+])_+ + ([V - (1 - \gamma)\lambda])_+.$$

The assertion follows on observing that $(1 + x)^{3/2} \le \sqrt{2}(1 + x^{3/2})$. Similarly, on writing

$$|\mathbf{B}| + \mu^{-1}\lambda V - \mu^{-1}\lambda^2 = (|\mathbf{B}| - \gamma \mu^{-1}\lambda^2) + (\mu^{-1}\lambda V - (1 - \gamma)\mu^{-1}\lambda^2),$$

we see that the integrand in (4.7.16) is bounded above by

$$\sqrt{2} \left\{ (|\mathbf{B}| - \gamma \mu^{-1}\lambda^2)_+^{3/2} + (\mu^{-1}\lambda V - (1 - \gamma)\mu^{-1}\lambda^2)_+^{3/2} \right\}.$$

On substituting these bounds in (4.7.15) and changing the order of integration, we obtain

$$\sum |\lambda_j| \leq \sqrt{2} L_3 \int_{\mathbb{R}^3} \left\{ \int_0^\infty [|\mathbf{B}(\mathbf{x})| - \gamma\mu^{-1}\lambda^2]_+^{3/2} d\lambda \right.$$

$$+ \int_0^\mu [V(\mathbf{x}) - (1-\gamma)\lambda]_+^{3/2} d\lambda + \int_\mu^\infty [\mu^{-1}\lambda V(\mathbf{x}) - (1-\gamma)\mu^{-1}\lambda^2]_+^{3/2} d\lambda \left.\right\} d\mathbf{x}$$

$$\leq \sqrt{2} L_3 \int_{\mathbb{R}^3} \left\{ \int_0^\infty [|\mathbf{B}(\mathbf{x})| - \gamma\mu^{-1}\lambda^2]_+^{3/2} d\lambda \right.$$

$$+ \int_0^\infty [V(\mathbf{x}) - (1-\gamma)\lambda]_+^{3/2} d\lambda + \int_0^\infty [\mu^{-1}\lambda V(\mathbf{x}) - (1-\gamma)\mu^{-1}\lambda^2]_+^{3/2} d\lambda \left.\right\} d\mathbf{x}$$

$$= \sqrt{2} L_3 \int_{\mathbb{R}^3} \{J_1 + J_2 + J_3\},$$

say. The following are readily verified:

$$J_1 = \left(\frac{\gamma}{\mu}\right)^{3/2} \int_0^\infty [\mu\gamma^{-1}|\mathbf{B}(\mathbf{x})| - \lambda^2]_+^{3/2} d\lambda$$

$$= \frac{3\pi\mu^{1/2}}{16\gamma^{1/2}} |\mathbf{B}(\mathbf{x})|^2;$$

$$J_2 = (1-\gamma)^{3/2} \int_0^\infty [(1-\gamma)^{-1} V(\mathbf{x}) - \lambda]_+^{3/2} d\lambda$$

$$= \frac{2}{5(1-\gamma)} V(\mathbf{x})^{5/2};$$

$$J_3 = \frac{(1-\gamma)^{3/2}}{\mu^{3/2}} \int_0^\infty [(1-\gamma)^{-1} V(\mathbf{x}) - \lambda]_+^{3/2} \lambda^{3/2} d\lambda$$

$$= \frac{3\pi}{128(1-\gamma)^{5/2}\mu^{3/2}} V(\mathbf{x})^4.$$

We therefore have

$$\sum |\lambda_j| \leq \sqrt{2} L_3 \int_{\mathbb{R}^3} \left\{ \frac{2}{5(1-\gamma)} V(\mathbf{x})^{5/2} + \frac{3\pi\mu^{1/2}}{16\gamma^{1/2}} |\mathbf{B}(\mathbf{x})|^2 \right.$$

$$+ \left. \frac{3\pi}{128\mu^{3/2}(1-\gamma)^{5/2}} V(\mathbf{x})^4 \right\} d\mathbf{x}.$$

The theorem follows on optimising this with respect to μ. $\qquad\square$

The paper [Lieb *et al.* (1995)] should be consulted for further details of the methods and results in this section and for important background information.

4.7.2 *Relativistic matter*

The Hamiltonian considered in this case is that studied in [Lieb *et al.* (1997)], namely

$$\mathbb{B}_{N,K}(\mathbf{A}) = \Lambda_+ \left(\sum_{j=1}^N \mathbb{D}_0^{(j)}(\mathbf{A}) + \alpha V_C + H_F \right) \Lambda_+, \qquad (4.7.17)$$

where

$$\mathbb{D}_0^{(j)}(\mathbf{A}) = \boldsymbol{\alpha} \cdot (-i\boldsymbol{\nabla} + \mathbf{A}(\mathbf{x}_j)) + \beta$$

and $H_F = (8\pi\alpha)^{-1} \int_{\mathbb{R}^3} |\mathbf{B}(\mathbf{x})|^2 d\mathbf{x}$, the field energy.

It is proved in [Lieb *et al.* (1997)], Theorem 2, that if Λ_+ is the projection $\Lambda_{+,N}$ onto

$$\mathcal{H}_N = \bigwedge_{j=1}^{N} \mathcal{H}_+^{(j)},$$

where $\mathcal{H}_+^{(j)}$ is the positive spectral subspace of the free Dirac operator $\mathbb{D}_0^{(j)}$ acting in the space of the jth electron, the Hamiltonian in (4.7.17) is unstable for any fixed α, if the particle numbers N, K are arbitrarily large. However, it is shown that there is stability for a suitable range of values of α and Z if $\Lambda_+ = \Lambda_{+,N}(\mathbf{A})$, the projection onto

$$\mathcal{H}_N(\mathbf{A}) = \bigwedge_{j=1}^{N} \mathcal{H}_+^{(j)}(\mathbf{A})$$

where $\mathcal{H}_+^{(j)}(\mathbf{A})$ is the positive spectral subspace of $\mathbb{D}_0^{(j)}(\mathbf{A})$. We shall sketch a proof of this last result from [Lieb *et al.* (1997)], but see the original paper for a fuller treatment and discussion.

The proof uses the following special case of an inequality due to Birman, Koplienko and Solomyak in [Birman *et al.* (1975)]; a proof of this special case is given in [Lieb *et al.* (1997)], Appendix A.

Lemma 4.7.2. *Let C, D be self-adjoint operators which are such that $(C^2 - D^2)_-^{1/2}$ is trace class. Then*

$$\text{trace } (C - D)_- \leq \text{trace } (C^2 - D^2)_-^{1/2}. \tag{4.7.18}$$

Recall that the negative part A_- of a self-adjoint operator A is defined to be $A_- := (1/2)(|A| - A)$. It is in trace class if its eigenvalues $\lambda_n, n \in \mathbb{N}$, lie in the sequence space ℓ^1. The following application of (4.7.18) is noted in [Lieb *et al.* (1997)]. A special case of the Lieb–Thirring inequality (1.9.3) is

$$\sum |\lambda_n((\mathbf{p}+\mathbf{A})^2 - V)|^{1/2} = \text{trace}[(\mathbf{p}+\mathbf{A})^2 - V]_-^{1/2} \leq L_{1/2,3} \int_{\mathbb{R}^3} V_+(\mathbf{x})^2 d\mathbf{x}, \tag{4.7.19}$$

where $L_{1/2,3} \leq 0.06003$ and the $\lambda_n((\mathbf{p}+\mathbf{A})^2 - V)$ are the negative eigenvalues of the exhibited operator. On applying Lemma 4.7.2, we obtain the relativistic inequality proved by Daubechies in [Daubechies (1983)], namely,

$$\sum |\lambda_n(|\mathbf{p} + \mathbf{A}| - V)| = \text{trace}[|\mathbf{p} + \mathbf{A}| - V]_- \leq L_{1/2,3} \int_{\mathbb{R}^3} V_+(\mathbf{x})^4 d\mathbf{x}. \tag{4.7.20}$$

The main theorem proved in [Lieb *et al.* (1997)] is

Theorem 4.7.3. *Let $Z < 2/(\pi\alpha)$ and $\alpha \le \alpha_c$, where α_c is the unique solution of the equation*

$$(8\pi L_{1/2,3}\alpha_c)^{2/3} = 1 - \alpha_c^2/\tilde{\alpha}_c^{\,2},$$

with $\tilde{\alpha}_c := [(\pi/2)Z + 2.2159.2^{1/3}Z^{2/3} + 1.0307 \cdot 2^{1/3}]^{-1}$. Then

$$\mathbb{B}_{N,K}(\mathbf{A}) = \Lambda_{+,N}(\mathbf{A})\left(\sum_{j=1}^{N} \mathbb{D}_0^{(j)}(\mathbf{A}) + \alpha V_C + H_F\right)\Lambda_{+,N}(\mathbf{A}) \ge 0.$$

When $\alpha = 1/137$, this requires $Z \le 56$.

Proof. As in Section 4.6.1, the first step is to replace the Coulomb potential V_C by N copies of a one-body operator. On the fermionic Hilbert space $\bigwedge_{j=1}^{N} H^{1/2}(\mathbb{R}^3, \mathbb{C}^2)$,

$$\sum_{j=1}^{N} |\mathbf{p}_j + \mathbf{A}_j| + \tilde{\alpha} V_C \ge 0$$

is proved in [Lieb *et al.* (1997)] to hold if $\tilde{\alpha} \le \tilde{\alpha}_c$. This yields

$$\mathbb{B}_{N,K}(\mathbf{A}) \ge \sum_{j=1}^{N} \mathbb{B}_j + H_F,$$

where the $\mathbb{B}_j$ are copies of

$$\mathbb{B} := \Lambda_+(\mathbf{A})\mathbb{D}_\mathbf{A}\Lambda_+(\mathbf{A}) - \Lambda_+(\mathbf{A})\kappa|\mathbf{p} + \mathbf{A}|\Lambda_+(\mathbf{A}), \qquad (4.7.21)$$

with $\kappa = \alpha/\tilde{\alpha}_c$, and $\Lambda_+(\mathbf{A})$ the projection onto the positive spectral subspace of the operator $\mathbb{D}_\mathbf{A} = \mathbb{D} + \boldsymbol{\alpha} \cdot \mathbf{A}$. A lower bound for $\mathbb{B}_{N,K}(\mathbf{A}) - H_F$ is therefore given by the sum of the negative eigenvalues of $\mathbb{B}$, i.e., $-\text{trace}\,(\mathbb{B}_-)$. We substitute $C = \Lambda_+(\mathbf{A})\mathbb{D}_\mathbf{A}\Lambda_+(\mathbf{A}), D = \kappa\Lambda_+(\mathbf{A})|\mathbf{p} + \mathbf{A}|\Lambda_+(\mathbf{A})$ in (4.7.18). Since $\Lambda_+(\mathbf{A})$ commutes with $\mathbb{D}_\mathbf{A}$, and $\Lambda_+(\mathbf{A})^2$ is the identity, we have that $C^2 = \Lambda_+(\mathbf{A})\mathbb{D}_\mathbf{A}^2\Lambda_+(\mathbf{A})$. Furthermore, setting $X := \kappa|\mathbf{p} + \mathbf{A}|$ it follows that

$$D^2 = \Lambda_+(\mathbf{A})X\Lambda_+(\mathbf{A})X\Lambda_+(\mathbf{A}) \le \Lambda_+(\mathbf{A})X^2\Lambda_+(\mathbf{A}),$$

since $X\Lambda_+(\mathbf{A})X \le X^2$ on account of $0 \le \Lambda_+(\mathbf{A}) \le 1$ and X being self-adjoint. Thus,

$$C^2 - D^2 \ge \Lambda_+(\mathbf{A})[\mathbb{D}_\mathbf{A}^2 - X^2]\Lambda_+(\mathbf{A}).$$

In view of the final paragraph of Section 1.4, this implies

$$\text{trace}\,(\mathbb{B}_-) \le \text{trace}\,(\Lambda_+(\mathbf{A})S\Lambda_+(\mathbf{A}))_-^{1/2}, \quad S := \mathbb{D}_\mathbf{A}^2 - \kappa^2|\mathbf{p} + \mathbf{A}|^2. \qquad (4.7.22)$$

Next observe from (2.1.4) that

$$\mathbb{D}_\mathbf{A} = \begin{pmatrix} 1 & \boldsymbol{\sigma} \cdot (\mathbf{p} + \mathbf{A}) \\ \boldsymbol{\sigma} \cdot (\mathbf{p} + \mathbf{A}) & -1 \end{pmatrix},$$

and this gives

$$\mathbb{D}_\mathbf{A}^2 = \begin{pmatrix} \mathbb{P}_\mathbf{A} + 1 & 0 \\ 0 & \mathbb{P}_\mathbf{A} + 1 \end{pmatrix},$$

where the Pauli operator satisfies

$$\mathbb{P}_\mathbf{A} \geq (\mathbf{p} + \mathbf{A})^2 - |\mathbf{B}|.$$

It follows that

$$\text{trace } (\mathbb{B}_-) \leq \text{trace } \left[(1 - \kappa^2)(\mathbf{p} + \mathbf{A})^2 - |\mathbf{B}| \right]_-^{1/2}. \tag{4.7.23}$$

The Lieb–Thirring inequality (4.7.19) therefore gives

$$\text{trace } (\mathbb{B}_-) \leq 4(1 - \kappa^2)^{-3/2} L_{1/2,3} \int_{\mathbb{R}^3} |\mathbf{B}(\mathbf{x})|^2 d\mathbf{x}. \tag{4.7.24}$$

The 4 appears because the trace is over 4-dimensional spinors. In fact, it is shown in [Lieb *et al.* (1997)] that the factor 4 can be replaced by 2. With this modification, we have that

$$\mathbb{B} \geq \left\{ \frac{1}{8\pi\alpha} - 2(1 - \frac{\alpha^2}{\tilde{\alpha}_c^2})^{-3/2} L_{1/2,3} \right\} \int_{\mathbb{R}^3} |\mathbf{B}(\mathbf{x})|^2 d\mathbf{x}$$

$$\geq 0$$

if

$$(16\pi\alpha L_{1/2,3})^{2/3} \leq 1 - \alpha^2/\tilde{\alpha}_c^2. \tag{4.7.25}$$

The left-hand side of (4.7.25) is increasing in α and the right-hand side is decreasing. The unique value of α which gives equality is α_c. The proof is therefore complete.

$\square$

Bibliography

Adam, C., Muratori, B. and Nash, C. (1999). Zero modes of the Dirac operator in three dimensions, *Phys. Rev. D (3)* **60**, 12, p. 125001.

Adam, C., Muratori, B. and Nash, C. (2000a). Degeneracy of the zero modes of the Dirac operator in three dimensions, *Phys. Lett. B* **485**, 1-3, pp. 314–318.

Adam, C., Muratori, B. and Nash, C. (2000b). Multiple zero modes of the Dirac operator in three dimensions, *Phys. Rev. D (3)* **62**, 8, p. 085026.

Adam, C., Muratori, B. and Nash, C. (2000c). Particle creation via relaxing hypermagnetic knots, *Phys. Rev. D* **62**, 10, p. 105027.

Adam, C., Muratori, B. and Nash, C. (2000d). Zero modes in finite range magnetic fields, *Modern Phys. Lett. A* **15**, 25, pp. 1577–1581.

Adam, C., Muratori, B. and Nash, C. (2003). Chern–Simons action for zero-mode supporting gauge fields in three dimensions, *Phys. Rev. D (3)* **67**, 8, p. 087703.

Arai, M. (1975). On essential self-adjointness of Dirac operators, *RIMS, Kokyuroku, Kyoto Univ.* **242**, pp. 10–21.

Arai, M. (1983). On essential self-adjointness, distinguished self-adjoint extensions and essential spectrum of Dirac operators with matrix valued potentials, *Publ. Res. Inst. Math. Sci.* **19**, 1, pp. 33–57.

Arai, M. and Yamada, O. (1982). Essential self-adjointness and invariance of the essential spectrum for Dirac operators, *Publ. Res. Inst. Math. Sci.* **18**, 3, pp. 973–985.

Aronszajn, N. (1957). A unique continuation theorem for solutions of elliptic partial differential equations or inequalities of second order, *J. Math. Pures Appl.* **36**, pp. 235–249.

Aubin, T. (1976). Problèmes isopérimetriques et espaces de Sobolev, *J. Differential Geometry* **11**, 4, pp. 573–598.

Avron, J., Herbst, I. and Simon, B. (1978). Schrödinger operators with magnetic fields. I. General interactions, *Duke Math. J.* **45**, 4, pp. 847–883.

Balinsky, A. and Evans, W. D. (1998). On the virial theorem for the relativistic operator of Brown and Ravenhall and the absence of embedded eigenvalues, *Lett. Math. Phys.* **44**, 3, pp. 233–248.

Balinsky, A. and Evans, W. D. (1999). Stability of one-electron molecules in the Brown–Ravenhall model, *Commun. Math. Phys.* **202**, 2, pp. 481–500.

Balinsky, A. and Evans, W. D. (2001). On the zero modes of Pauli operators, *J. Funct. Anal.* **179**, 1, pp. 120–135.

Balinsky, A. and Evans, W. D. (2002a). On the spectral properties of the Brown–Ravenhall operator, *J. Comput. Appl. Math.* **148**, 1, pp. 239–255.

Balinsky, A. and Evans, W. D. (2002b). On the zero modes of Weyl–Dirac operators and their multiplicity, *Bull. London Math. Soc.* **34**, 2, pp. 236–242.

Balinsky, A. and Evans, W. D. (2003). Zero modes of Pauli and Weyl–Dirac operators, *Contemp. Math.* **327**, pp. 1–9.

Balinsky, A., Evans, W. D. and Saito, Y. (2008). Dirac–Sobolev inequalities and estimates for the zero modes of massless Dirac operators, *J. Math. Phys.* **49**, 4, p. 043514.

Balinsky, A., Evans, W. D. and Umeda, T. (2010). The Dirac–Hardy and Dirac–Sobolev inequalities in L^1, *Preprint*.

Bär, C. and Strohmaier, A. (2001). Semi-bounded restrictions of Dirac type operators and the unique continuation property, *Differential Geom. Appl.* **15**, 2, pp. 175–182.

Berthier, A. and Georgescu, V. (1987). On the point spectrum of Dirac operators, *J. Funct. Anal.* **71**, 2, pp. 309–338.

Birman, M., Koplienko, L. and Solomyak, M. (1975). Estimates for the spectrum of the difference between fractional powers of two self-adjoint operators, *Soviet Math.* **19**, 3, pp. 1–6, translation of Izvestija vyssich.

Booß-Bavnbeck, B. (2000). Unique continuation property for Dirac operators, revised, *Contemp. Math.* **258**, pp. 21–32.

Bouzouina, A. (2002). Stability of the two-dimensional Brown–Ravenhall operator, *Proc. Roy. Soc. Edinburgh Sect. A* **132**, 5, pp. 1133–1144.

Bratelli, O., Kishimoto, A. and Robinson, D. (1980). Positivity and monotonicity properties of c_0-semigroups, i, *Commun. Math. Phys.* **75**, 1, pp. 67–84.

Brown, G. and Ravenhall, D. (1951). On the interaction of two electrons, *Proc. Roy. Soc. London Ser. A* **208**, pp. 552–559.

Burenkov, V. I. and Evans, W. D. (1998). On the evaluation of the norm of an integral operator associated with the stability of one-electron atoms, *Proc. Roy. Soc. Edinburgh Sect. A* **128**, 5, pp. 993–1005.

Chernoff, P. R. (1973). Essential self-adjointness of powers of generators of hyperbolic equations, *J. Funct. Anal.* **12**, pp. 401–414.

Chernoff, P. R. (1977). Schrödinger and Dirac operators with singular potentials and hyperbolic equations, *Pacific J. Math.* **72**, 2, pp. 361–382.

Conlon, J. G. (1984). The ground state energy of a classical gas, *Commun. Math. Phys.* **94**, 4, pp. 439–458.

Cwikel, M. (1977). Weak type estimates for singular values and the number of bound states of Scrödinger operators, *Ann. Math.* **106**, 1, pp. 93–100.

Cycon, H. L., Froese, R. G., Kirsh, W. and Simon, B. (1987). *Schrödinger operators with application to quantum mechanics and global geometry*, Texts and Monographs in Physics (Springer-Verlag, Berlin).

Daubechies, I. (1983). An uncertainty principle for fermions with generalized kinetic energy, *Commun. Math. Phys.* **90**, 4, pp. 511–520.

Daubechies, I. and Lieb, E. H. (1983). One-electron relativistic molecules with Coulomb interaction, *Commun. Math. Phys.* **90**, 4, pp. 497–510.

Dolbeault, J., Esteban, M. J. and Séré, E. (2000a). On the eigenvalues of operators with gaps. Application to Dirac operators, *J. Funct. Anal.* **174**, 1, pp. 208–226.

Dolbeault, J., Esteban, M. J. and Séré, E. (2000b). Variational characterization for eigenvalues of Dirac operators, *Calc. Var. Partial Differential Equations* **10**, 4, pp. 321–347.

Dyson, F. J. and Lenard, A. (1967). Stability of matter. I, *J. Math. Phys.* **8**, 3, pp. 423–434.

Dyson, F. J. and Lenard, A. (1968). Stability of matter. II, *J. Math. Phys.* **9**, 5, pp. 698–711.

Edmunds, D. E. and Evans, W. D. (1987). *Spectral Theory and Differential Operators*, Oxford Mathematical Monographs (Oxford University Press, The Clarendon Press,

Oxford University Press, New York).

Edmunds, D. E. and Evans, W. D. (2004). *Hardy operators, function spaces and embeddings*, Springer Monographs in Mathematics (Springer-Verlag, Berlin).

Elton, D. M. (2000). New examples of zero modes, *J. Phys. A* **33**, 41, pp. 7297–7303.

Elton, D. M. (2002). The local structure of zero mode producing magnetic potentials, *Comm. Math. Phys.* **229**, 1, pp. 121–139.

Erdős, L. and Solovej, J. P. (2000). On the kernel of $Spin^c$ Dirac operators on $\mathbb{S}^3$ and $\mathbb{R}^3$, *AMS/IP Stud. Adv. Math.* **16**, pp. 111–119.

Erdős, L. and Solovej, J. P. (2001). The kernel of Dirac operators on $\mathbb{S}^3$ and $\mathbb{R}^3$, *Rev. Math. Phys.* **13**, 10, pp. 1247–1280.

Esteban, M. J. and Loss, M. (2007). Self-adjointness of Dirac operators via Hardy–Dirac inequalities, *J. Math. Phys.* **48**, 11, p. 112107.

Esteban, M. J. and Loss, M. (2008). *Self-adjointness via partial Hardy-like inequalities*, Mathematical Results in Quantum Mechanics, pp. 41–47 (World Sci. Publ., Hackensack, NJ).

Evans, W. D. (1970). On the unique self-adjoint extension of the Dirac operator and the existence of the Green matrix, *Proc. London Math. Soc.* **20**, pp. 537–557.

Evans, W. D., Perry, P. and Siedentop, H. (1996). The spectrum of relativistic one-electron atoms according to Bethe and Salpeter, *Commun. Math. Phys.* **178**, 3, pp. 733–746.

Federer, H. and Fleming, W. (1960). Normal and integral currents, *Ann. of Math.* **72**, pp. 458–520.

Fefferman, C. L. and de la Llave, R. (1986). Relativistic stability of matter. I, *Rev. Mat. Iberoamericana* **2**, 1-2, pp. 119–213.

Frölich, J., Lieb, E. and Loss, M. (1986). Stability of Coulomb systems with magnetic fields I. The one-electron atom, *Commun. Math. Phys.* **104**, 2, pp. 251–270.

Gagliardo, E. (1958). Proprietà di alcune classi di funzioni di piu variabili, *Ricerche Mat.* **7**, pp. 102–137.

Gradshteyn, I. and Ryzhik, I. M. (1965). *Tables of Integrals, Series and Products*, 1st edn. (Academic Press, New York).

Greiner, W. (1990). *Relativistic Quantum Mechanics, Theoretical Physics – Text and Exercise Books*, Vol. 3, 1st edn. (Springer, Berlin, Heidelberg, New York).

Griesemer, M., Lewis, R. T. and Siedentop, H. (1999). A minimax principle for the eigenvalues in spectral gaps: Dirac operators with Coulomb potentials, *Doc. Math.* **4**, pp. 275–283.

Griesemer, M. and Siedentop, H. (1999). A minimax principle for the eigenvalues in spectral gaps, *J. London Math. Soc.* **60**, 2, pp. 490–500.

Gustafson, K. and Weidmann, J. (1969). On the essential spectrum, *J. Math. Anal. Appl.* **25**, pp. 121–127.

Hardekopf, G. and Sucher, J. (1985). Critical coupling constants for relativistic wave equations and vacuum breakdown in quantum electrodynamics, *Phys. Rev. A* **31**, 4, pp. 2020–2029.

Hardy, G., Littlewood, J. E. and Pólya, G. (1959). *Inequalities* (Cambridge University Press, Cambridge).

Helffer, B., Nourrigat, J. and Wang, X. P. (1989). Sur le spectre de l'équation de Dirac (dans $\mathbb{R}^3$ ou $\mathbb{R}^2$) avec champ magnétique, *Ann. sci. l'E.N.S.* **22**, pp. 515–533.

Herbst, I. (1977). Spectral theory of the operator $(p^2 + m^2)^{1/2} - ze^2/r$, *Commun. Math. Phys.* **53**, 3, pp. 285–294.

Hoever, G. and Siedentop, H. (1999). Stability of the Brown–Ravenhall operator, *Math. Phys. Elec. J.* **5**, pp. 1–11.

Hundertmark, D., Laptev, A. and Weidl, T. (2000). New bounds on the Lieb–Thirring constants, *Inventiones mathematicae* **140**, 2, pp. 693–704.

Ichinose, T. and Saitō, Y. (2010). Dirac–Sobolev spaces and Sobolev spaces, *Preprint arXiv:1001.0275v1 [math.AP]*.

Jakabussa-Amundsen, D. H. (2004). *Spectral Theory of the Atomic Dirac Operator in the No-Pair Formalism* (Dissertation LMU München).

Jörgens, K. (1972). *Perturbations of the Dirac operator, in Proceedings of the conference on the theory of ordinary and partial differential equations, Dundee, Scotland, 1972, p. 87-102, eds. W. N. Everitt and B. D. Sleeman, Lecture Notes in Mathematics*, Vol. 280 (Springer-Verlag, Berlin).

Kato, T. (1972). Schrödinger operators with singular potentials, *Israel J. Math.* **13**, 1-2, pp. 135–148.

Kato, T. (1976). *Perturbation Theory for Linear Operators, Grundlehren der mathematischen Wissenschaften*, Vol. 132, 2nd edn. (Springer-Verlag, Berlin, Heidelberg, New York).

Klaus, M. (1980). Dirac operators with several Coulomb singularities, *Helv. Phys. Acta.* **53**, 3, pp. 463–482.

Klaus, M. and Wüst, R. (1979). Characterization and uniqueness of distinguished self-adjoint extensions of Dirac operators, *Commun. Math. Phys.* **64**, 2, pp. 171–176.

Landgren, J. J. and Reitö, P. A. (1979). An application of the maximum principle to the study of essential self-adjointness of Dirac operators, I, *J. Math. Phys.* **20**, 11, pp. 2204–2211.

Landgren, J. J., Reitö, P. A. and Klaus, M. (1980). An application of the maximum principle to the study of essential self-adjointness of Dirac operators, II, *J. Math. Phys.* **21**, 5, pp. 1210–1217.

Li, P. and Yau, S.-T. (1983). On the Schrödinger equation and the eigenvalue problem, *Commun. Math. Phys.* **88**, 3, pp. 309–318.

Lieb, E. H. (1976). Bounds on the eigenvalues of the Laplace and Schrödinger operators, *Bull. Amer. Math. Soc.* **82**, 5, pp. 751–753.

Lieb, E. H. and Loss, M. (1986). Stability of Coulomb systems with magnetic fields ii. the many-electron atom and the one-electron molecule, *Commun. Math. Phys.* **104**, 2, pp. 271–282.

Lieb, E. H. and Loss, M. (1997). *Analysis, Graduate Studies in Mathematics*, Vol. 14 (American Mathematical Society, Providence, RI).

Lieb, E. H., Loss, M. and Siedentop, H. (1996). Stability of relativistic matter via Thomas–Fermi theory, *Helv. Phys. Acta.* **69**, 5-6, pp. 974–984.

Lieb, E. H., Loss, M. and Solovej, J. P. (1995). Stability of matter in magnetic fields, *Phys. Rev. Lett.* **75**, 6, pp. 985–989.

Lieb, E. H., Siedentop, H. and Solovej, J. P. (1997). Stability and instability of relativistic electrons in classical electromagnetic fields, *J. Stat. Phys.* **89**, 1-2, pp. 37–59.

Lieb, E. H. and Thirring, W. (1975). Bounds for the kinetic energy of fermions which proves the stability of matter, *Phys. Rev. Lett.* **35**, 11, pp. 687–689, Erratum 36 (16), 11116.

Lieb, E. H. and Thirring, W. (1976). *Inequalities for the moments of the eigenvalues of the Schrödinger Hamiltonian and their relation to Sobolev inequalities*, Studies in Mathematical Physics (Princeton University Press, Princeton), essays in Honor of V. Bargmann.

Lieb, E. H. and Yau, H.-T. (1988). The stability and instability of relativistic matter, *Commun. Math. Phys.* **118**, 2, pp. 177–213.

Loss, M. and Yau, H.-T. (1986). Stability of Coulomb systems with magnetic fields iii. zero energy bound states of the Pauli operator, *Commun. Math. Phys.* **104**, 2, pp.

283–290.

Magnus, W., Oberhettinger, F. and Soni, R. P. (1966). *Formulas and Theorems for the Special Functions of Mathematical Physics, Die Grundlehren der mathematischen Wissenshaften*, Vol. 52, 3rd edn. (Springer-Verlag, New York).

Maz'ya, V. (1960). Classes of domains and imbedding theorems for function spaces, *Dokl. Akad. Nauk. SSSR, English transl.: Sov. Math. Dokl. (1960), 1 882-885* **133**, pp. 527–530.

Meyers, N. G. and Serrin, J. (1964). *H = W*, *Proc. Nat. Acad. Sci. USA* **51**, pp. 1055–1056.

Nenciu, G. (1976). Self-adjointness and invariance of the essential spectrum for Dirac operators defined as quadratic forms, *Commun. Math. Phys.* **48**, 3, pp. 235–247.

Nenciu, G. (1977). Distinguished self-adjoint extensions for Dirac operators with potential dominated by multi-center Coulomb potentials, *Helv. Phy. Acta.* **50**, 1, pp. 1–3.

Nirenberg, L. (1959). On elliptic partial differential equations, *Ann. Sc. Norm. Super. Pisa* **13**, pp. 1–48.

Oberhettinger, F. (1974). *Tables of Mellin Transforms* (Springer-Verlag, New York).

Rastall, P. (1964). Quaternions in relativity, *Rev. Mod. Phys.* **36**, pp. 820–832.

Raynal, J. C., Roy, S. M., Singh, V., Martin, A. and Stubbe, J. (1994). The "Herbst Hamiltonian" and the mass of boson stars, *Phys. Lett. B* **320**, 1-2, pp. 105–109.

Reed, M. and Simon, B. (1978). *Methods of Modern Mathematical Physics, vol. IV: Analysis of Operators* (Academic Press, New York).

Roe, J. (1988). *Elliptic operators, topology and asymptotic methods*, Science and Technical (Longman, New York).

Rozenbljum, G. V. (1972). The distribution of the discrete spectrum for singular differential operators, *Dokl. Akad. Nauk SSSR* **202**, pp. 1012–1015.

Rozenbljum, G. V. and Solomyak, M. (1998). The Cwikel–Lieb–Rozenbljum estimator for generators of positive semigroups and semigroups dominated by positive semigroups, *St. Petersburg Math. J.* **9**, 6, pp. 1195–1211.

Saitō, Y. and Umeda, T. (2008a). The asymptotic limits of zero modes of massless Dirac operators, *Lett. Math. Phys.* **83**, 1, pp. 97–106.

Saitō, Y. and Umeda, T. (2008b). The zero modes and zero resonances of massless Dirac operators, *Hokkaido Math. J.* **37**, 2, pp. 363–388.

Saitō, Y. and Umeda, T. (2009). Eigenfunctions at the threshold energies of magnetic Dirac operators, *Preprint arXiv:0905.0961v2 [math.SP]*.

Schechter, M. (1966). On the essential spectrum of an arbitrary operator I, *J. Math. Anal. Appl.* **13**, pp. 205–215.

Schmincke, U. W. (1972a). Distinguished self-adjoint extensions of Dirac operators, *Math. Z.* **129**, pp. 335–349.

Schmincke, U. W. (1972b). Essential self-adjointness of Dirac operators with a strongly singular potential, *Math. Z.* **126**, pp. 71–81.

Sobolev, S. L. (1938). On a theorem of functional analysis, *Mat. Sb. and Amer. Math. Soc. Transl. II Ser. (1963), 34, 39-68* **46**, pp. 471–497.

Stein, E. M. (1970). *Singular Integrals and Differentiability Properties of Functions* (Princeton University Press, Princeton).

Stein, E. M. and Weiss, G. (1971). *Introduction to Fourier Analysis on Euclidean Spaces* (Princeton University Press, Princeton).

Talenti, G. (1976). Best constant in Sobolev inequality, *Ann. Mat. Pura Appl.* **110**, pp. 353–372.

Thaller, B. (1992). *The Dirac Equation*, Texts and Monographs in Physics (Springer-Verlag, Berlin, Heidelberg, New York).

Thaller, B. (1994). A criterion for essential self-adjointness, *J. Operator Theory* **31**, pp.

351–361.

Titchmarsh, E. C. (1986). *Introduction to the Theory of Fourier Integrals*, 3rd edn. (Chelsea Publishing Co., New York).

Tix, C. (1997a). Lower bound for the ground state energy of the no-pair Hamiltonian, *Phys. Lett. B* **405**, 3-4, pp. 293–296.

Tix, C. (1997b). Self-adjointness and spectral properties of a pseudo-relativistic Hamiltonian due to Brown and Ravenhall, *Mathematical Physics Preprint Archive mp-arc 97-441*.

Tix, C. (1998). Strict positivity of a relativistic Hamiltonian due to Brown and Ravenhall, *Bull. London Math. Soc.* **30**, pp. 283–290.

Vogelsang, V. (1987). Remark on essential selfadjointness of Dirac operators with Coulomb potentials, *Math. Z.* **196**, pp. 517–521.

Weder, R. A. (1974). Spectral properties of one-body relativistic spin-zero Hamiltonians, *Ann. Inst. H. Poincaré Sect. A (N.S.)* **20**, pp. 211–220.

Weder, R. A. (1975). Spectral analysis of pseudodifferential operators, *J. Funct. Anal.* **20**, 4, pp. 319–337.

Weidmann, J. (1967). The virial theorem and its application to the spectral theory of Schrödinger operators, *Bull. American Math. Soc.* **73**, pp. 452–456.

Weidmann, J. (1971). Oscillationsmethoden für Systeme gewöhnlicher Differentialgleichungen, *Math. Z.* **119**, pp. 349–373.

Weidmann, J. (1980). *Linear Operators in Hilbert Spaces* (Springer-Verlag, Berlin, Heidelberg, New York).

Weidmann, J. (1982). Absolut stetiges Spektrum bei Sturm–Liouville-Operatoren und Dirac-Systemen, *Math. Z.* **180**, pp. 423–427.

Weyl, H. (1909). Über gewöhnliche Differentialgleichungen mit Singularitäten und die zugehörigen Entwicklungen willkürlicher Functionen, *Math. Ann.* **68**, pp. 220–269.

Whittaker, E. T. and Watson, G. N. (1940). *A Course of Modern Analysis*, 4th edn. (Cambridge: The University Press).

Wüst, R. (1971). Generalisations of Rellich's theorem on perturbations of (essentially) self-adjoint operators, *Math. Z.* **119**, pp. 276–280.

Wüst, R. (1973). A convergence theorem for self-adjoint operators applicable to Dirac operators with cutoff potentials, *Math. Z.* **131**, pp. 339–349.

Wüst, R. (1975). Distinguished self-adjoint extensions of Dirac operators constructed by means of cutoff potentials, *Math. Z.* **141**, pp. 93–98.

Wüst, R. (1977). Dirac operators with strongly singular potentials, *Math. Z.* **152**, pp. 259–271.

Yaouanc, A. L., Oliver, L. and Raynal, J.-C. (1997). The Hamiltonian $(p^2 + m^2)^{1/2} - \alpha/r$ near the critical value $\alpha_c = 2/\pi$,, *J. Math. Phys.* **38**, pp. 3997–4012.

Index

List of Symbols